Optimal reliability design

Optimal reliability design provides a detailed introduction to systems reliability and reliability optimization. State-of-the-art techniques for maximizing system reliability are described, focusing on component reliability enhancement and redundancy arrangement. The authors present several case studies and show how optimization techniques are applied in practice. They also pay particular attention to finding methods that give the optimal tradeoff between reliability and cost.

The book begins with a review of key background material, and a discussion of a range of optimization models. The authors go on to cover optimization tools, such as heuristics, discrete optimization, nonlinear programming, mixed integer programming, optimal arrangement, and metaheuristic algorithms. They also describe the computational implementation of these tools. The case studies are taken from a broad spectrum of engineering applications, including microelectronics fabrication, software development, and nuclear reactor maintenance.

Many numerical examples are included, and the book contains over 180 homework exercises. It is suitable as a textbook for graduate-level courses in reliability engineering and operations research. It will also be a valuable reference for practicing engineers.

Way Kuo is Wisenbaker Chair of Engineering, Associate Vice Chancellor, and Executive Associate Dean of Engineering at Texas A&M University. He is a Fellow of the American Society for Quality, the Institute of Industrial Engineers, and the Institute of Electrical and Electronics Engineers. He is the co-author of three previous books and the Chief Editor of *IEEE Transactions on Reliability*.

V. Rajendra Prasad is a researcher at Knowledge Based Systems, Inc. He previously served as a consultant to several corporations in the manufacturing and processing industries, being a faculty member of the Statistical Quality Control and Operations Research Division of the Indian Statistical Institute.

Frank A. Tillman is Chairman and Chief Executive Officer of IBX, Inc., having previously been a Professor in the Department of Industrial Engineering at Kansas State University. He has performed many consultancy projects for industrial and governmental organizations. He is a Fellow of the Institute of Industrial Engineers.

Ching-Lai Hwang was formerly a Professor of Industrial Engineering at Kansas State University and President of IBES, Inc. He was a world-renowned scholar and author of eight previous books on reliability, optimization, and decision making. Professor Hwang died in January, 1999.

OPTIMAL RELIABILITY DESIGN

Fundamentals and applications

Way Kuo
Texas A&M University

V. Rajendra Prasad
KBSI

Frank A. Tillman
IBX, Inc.; Kansas State University

Ching-Lai Hwang
Formerly IBES, Inc.

CAMBRIDGE
UNIVERSITY PRESS

CAMBRIDGE UNIVERSITY PRESS
Cambridge, New York, Melbourne, Madrid, Cape Town, Singapore, São Paulo

Cambridge University Press
The Edinburgh Building, Cambridge CB2 2RU, UK

Published in the United States of America by Cambridge University Press, New York

www.cambridge.org
Information on this title: www.cambridge.org/9780521781275

First published 2001
This digitally printed first paperback version 2006

A catalogue record for this publication is available from the British Library

Library of Congress Cataloguing in Publication data

Optimal reliability design : fundamentals and applications / Way Kuo . . . [et al.].
   p. cm.
  ISBN 0 521 78127 2 (hardbound)
  1. Reliability (Engineering) I. Kuo, Way, 1951–
TA169.O72 2000
620′.00452–dc21 00-021911 CIP

ISBN-13 978-0-521-78127-5 hardback
ISBN-10 0-521-78127-2 hardback

ISBN-13 978-0-521-03191-2 paperback
ISBN-10 0-521-03191-5 paperback

Contents

8 Reliability–redundancy allocation 208

9 Component assignment in reliability systems 236

10 Reliability systems with multiple objectives 275

Figures

Tables

Preface

Appearing on the scene in the late 1940s and early 1950s, reliability engineering was first applied to communication and transportation systems. Much of the early work was confined to the analysis of system reliability. Traditionally, theories were developed hypothetically without taking into account the problems encountered by the designers and users. For example, although tradeoffs between cost and performance are the key issue in both conventional and modern industries, they have not been either systematically addressed by practitioners or completely reported by the analysts. Further, many existing reliability problems cannot be solved by using existing theories largely due to the irrelevance of the paradigms proposed in the past. With the heightened quality consciousness faced by industry since the 1980s, we are now more intent on finding ways to enhance product and process reliability in the design and maintenance stages. To remain competitive, the guarantee of high system reliability at a competitive cost is essential. In this book, we introduce the fundamentals and applications of systems reliability and reliability optimization. These fundamentals can serve as useful design tools. Examples are illustrated for various algorithms and extended case studies are described throughout the book.

In Chapter 1 of this book, we present a complete picture of the fundamentals of reliability systems including several important aspects of system configurations and modeling. We illustrate how high reliability, through careful design and analysis, can be built into systems within the limits of economical and physical constraints. Information discussed in this chapter is used extensively in the later chapters on reliability optimization. Applications of the modeling techniques can be found in the case studies as well. Some important principles for enhancing system reliability are:

1. to keep the system as simple as is compatible with the performance requirements;

2. to increase the reliability of the components in the system;

3. to use parallel redundancy for the less reliable components;

4. to use standby redundancy which can be switched to active components when failures occur;

5. to use repair maintenance where failed components are replaced but not automatically switched in;

6. to use preventative maintenance such that components are replaced by new ones whenever they fail, or at some fixed time interval, whichever comes first;

7. to use better arrangement for exchangeable components;

8. to use large safety factors or management programs for product improvement; and

9. to use burned-in components that have high infant-mortality as indicated by Kuo et al. [174].

Implementation of the above steps often consumes resources. Consequently, system-reliability optimization problems are in general quite difficult to solve. In this book, a thorough discussion of the problems and techniques of reliability optimization is presented. Reliability optimization problems are classified as: (1) redundancy allocation problems where the decision variables are the number of redundancies, (2) reliability allocation problems where the decision variables are the component reliabilities, (3) reliability–redundancy allocation problems where the decision variables are a combination of the number of redundancies and the component reliabilities, and (4) component assignment problems where the arrangement of components in the system can make a difference in system reliability. In view of the computational complexity, it has been shown that even a simple redundancy allocation problem in a series system with linear constraints is NP-hard. NP-hard problems are very unlikely to have computationally efficient algorithms for exact optimal solutions.

Chapter 2 presents an overview of eight types of reliability optimization problems along with methods to reduce the size of the problem under consideration. Most of the system-reliability optimization problems are nonlinear programming ones representing special structures where the decision variables are integers, real numbers between 0 and 1, or a mix of both. Although many algorithms have been proposed for nonlinear programming problems, only a few have proven effective when applied to large-scale problems. None has proven to be sufficiently superior to the others to allow it to be classified as "the algorithm" for solving general nonlinear programming problems. Several powerful heuristics developed for redundancy allocation problems, along with a majority of the well-known mathematical programming methods, have been intelligently adopted for reliability optimization. In fact, some heuristics are also based on nonlinear optimization methods. During the 1990s, metaheuristic algorithms were developed to solve a variety of reliability optimization problems. Applications of genetic algorithms, tabu search, simulated annealing, and multiple objective optimization methods are also presented here. The techniques for reliability optimization are broadly classified as follows:

1. Heuristics for redundancy allocation

2. Metaheuristic algorithms for redundancy allocation: including genetic algorithms, tabu search, and simulated annealing techniques

3. Exact algorithms for redundancy allocation

   (a) Dynamic programming

   (b) Implicit enumeration

   (c) Branch-and-bound methods

(d) Lexicographic search procedure

(e) Integer programming

(f) Mixed integer programming

(g) Nonlinear programming techniques

4. Approximation methods for reliability–redundancy allocation (based on nonlinear programming)

5. Multiple objective optimization methods

6. Optimal assignment of components in coherent system

7. Effort function optimization

Chapters 3–8 provide a description of these methods and illustrations. Based on Chapters 3–7, several reliability–redundancy allocation methods are introduced in Chapter 8. Problems presented in Chapter 8 are of the mixed integer programming type where system designers have the option to specify the design parameters. Optimal assignment of interchangeable components in reliability systems in which a global optimal solution always exists is presented in Chapter 9. Problems described in Chapter 9 are not the classical mathematical programming types; instead they are a special case of order statistics. For special systems that have k-out-of-n and consecutive k-out-of-n configurations, Chapter 9 presents various optimal design patterns.

Reliability optimization with multiple criteria and objectives is discussed in Chapter 10. Other methods for reliability optimization, such as effort function minimization methods, are described in Chapter 11.

Details of a number of case studies are presented in Chapters 12–15. Burn-in optimization under limited capacity for products incorporating state-of-the-art integrated circuits is presented in Chapter 12. An actual case including the tradeoffs between burn-in schedule, burn-in facility, and reliability enhanced through burn-in is discussed. The increasingly complex modern systems use large quantities of software. Optimization of software design with regard to software reliability is addressed in Chapter 13. Here common cause exists among different redundant software developed by different software firms. Design parameters are of the difficult mixed integer type that include the determination of the software debugging time and the number of redundant software to be selected.

Chapter 14 presents a case study of an optimal scheduled-maintenance policy in which a realization of the multi-criteria decision-making method is applied to a reliability optimization problem. Three other important optimization case studies are presented in Chapter 15, including one involving maintenance models, a case involving a pressurized water reactor (PWR) coolant system, and one focusing on the design of a gas pipe line.

In the case studies, it is apparent that optimal solutions may not be obtainable. Furthermore, we realize that there exist uncertainties in estimating parameters of both reliability and the economic and physical constraints encountered in reliability optimization problems. However, an optimal thought process, instead of an *ad hoc* approach, is necessary to reach a sound decision based on the fundamentals and algorithms presented in Chapters 3–11.

Problems are included at the end of each chapter and four appendices are included at the end of the book.

Acknowledgments

We would like to acknowledge the National Science Foundation, Army Research Office, Office of Naval Research, Air Force Office for Scientific Research, National Research Council, Fulbright Foundation, the Texas Advanced Technology Program, Bell Labs, Hewlett Packard, and IBM. This manuscript grows from the research and development projects, supported in part by the above agencies during the past 25 years, granted to Kuo, Hwang, and Tillman.

The first draft of this book has been used and thoroughly examined in graduate classes attended by John Yuan of the National Tsinghua University; Taeho Kim of Korea Telecom; and Kyungmee Oh Kim, Dong Ju Lee, Jeongbae Kim, Chang Woo Kang, and Chunghun Ha of Texas A&M University. We are very grateful for their valuable suggestions and criticisms.

We acknowledge input to this manuscript from Alice Smith of Auburn University, Guy Curry of Texas A&M University, Mitsuo Gen of the Ashikaga Institute of Technology, S. J. Chen of HCI, International, and Y. J. Lai of Phillips 66.

Mary Ann Dickson edited the manuscript; Chunghun Ha and Wendy Luo verified the references; and Jill L. Grotefendt helped with formatting the text files of this manuscript.

We have obtained permission to use material or figures from the following *IEEE Transactions on Reliability*: Way Kuo and V. R. R. Prasad, "An annotated overview of system reliability optimization." **R49**(2): 176–191, June 2000, © 2000 IEEE; D. Chi and Way Kuo, "Burn-in optimization under reliability and capacity restrictions." **R38**(2): 193–198, June 1989, © 1989 IEEE; F. A. Tillman, C. L. Hwang, and Way Kuo, "Determining component reliability and redundancy for optimum system reliability." **R26**(3): 162–165, September 1977, © 1977 IEEE.

1 Introduction to reliability systems

1.1 Background

Reliability engineering appeared on the scene in the late 1940s and early 1950s and was first applied to communication and transportation systems. Much of the early reliability work was confined to the analysis of performance aspects of systems. In the past half a century, numerous well-written books on reliability have been made available. Among the system-oriented reliability texts, refer to Grosh [116] for an introductory understanding, Barlow and Proschan [23] for a theoretical foundation, and O'Connor [256] and Kapur and Lamberson [151] for a practical engineering approach. The primary goal of the reliability engineer has always been to find the best way to increase system reliability. Accepted principles for doing this include: (1) keeping the system as simple as is compatible with the performance requirements; (2) increasing the reliability of the components in the system; (3) using parallel redundancy for the less reliable components; (4) using standby redundancy which is switched to active components when failure occurs; (5) using repair maintenance where failed components are replaced but not automatically switched in, as in (4); (6) using preventive maintenance such that components are replaced by new ones whenever they fail, or at some fixed interval, whichever comes first; (7) using better arrangement for exchangeable components; (8) using large safety factors or a product improvement management program; and (9) selecting burn-in components that have high infant-mortality. Implementation of the above steps to improve system reliability will normally consume resources. A balance between system reliability and resource consumption is essential.

1.2 General description of the problem

There are a number of measures that indicate the performance of a system. The most popular are reliability, availability, and system effectiveness. Reliability is the probability of successful operation, whereas availability is the probability that the system is operational and available when it is needed. System effectiveness is the overall capability of a system to accomplish its mission and is determined by calculating the product of reliability and availability. In this book, we will concentrate on ways to improve system reliability through the optimal allocation of redundancies and/or reliabilities to systems which may be subject to constraints.

One definition reads: "Reliability is the probability that a system will perform satisfactorily for at least a given period of time when used under stated conditions" [4]. Therefore, the probability that a system successfully performs as designed is called "system reliability," or the "probability of survival." Throughout this book, unreliability refers to the probability of failure.

System reliability is a measure of how well a system meets its design objective, and it is usually expressed in terms of the reliabilities of the subsystems or components. The following definitions are necessary to further discuss this concept. A "part" or "element" is the least subdivision of a system, or an item that cannot ordinarily be disassembled without being destroyed. A "component" is a collection of parts which represent a self-contained element of a complete operating system. "Unit" and "component" are synonymous; they are assembled to form a "stage" or "subsystem." A "system" can then be characterized as a group of stages or subsystems integrated to perform one or more specified operational functions.

In describing the reliability of a given system, it is necessary to specify: (1) the equipment failure process; (2) the system configuration which describes how the equipment is connected and the rules of operation; and (3) the state in which the system is defined to be failed. The equipment failure process describes the probability law governing those failures. The system configuration, on the other hand, defines the manner in which the system-reliability function will behave. The third consideration in developing the reliability function for a nonmaintained system is to define the conditions of system failure.

1.3   System hardware, human factors, software, and environment

Many systems, designed to accomplish their intended mission, consist of hardware components operated by humans. The hardware may malfunction or wear-out, hence degrading the system performance. The element of human factors is usually more unpredictable than the hardware component and as a result it has greater influence on the overall system performance. Furthermore, almost all modern systems contain a very significant portion of software to help with smooth control of the system operation. For example, all automobiles manufactured in 1999 use fast processing integrated circuits programmed by failure-prone software. Performance such as reliability of hardware, software, and humans varies from environment to environment. Regarding hardware, operating conditions including temperature and humidity will determine reliability whereas the user's profile will determine the reliability of software. Needless to say, human performance is heavily affected by the work environment including temperature, humidity, vibration, etc.

Conceptually, to guarantee the high performance of a modern system one needs to guarantee high performance of hardware, software, and humans in a number of most-likely exposed environments. Such a schematic is shown in Figure 1.1. In reality, one

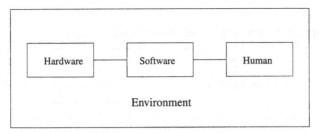

Constraints: physical and economic

Figure 1.1. A schematic to partition the elements of system performance

searches for optimal performance under various constraints, but the nature of hardware, software, and human failures makes it very difficult to generalize a solution procedure.

1.3.1 Hardware reliability

The reliability of a product is strongly influenced by decisions made during the design process. Poor design can affect product reliability and will greatly increase life-cycle cost (LCC). Therefore, it is important that design disciplines are used to minimize the failure possibility. For a piece of hardware, failure rate changes with time: in the early stage, the failure rate decreases; during the useful life, the failure rate remains constant; and after wear-out, the failure rate increases. The three types of failure rate form a bathtub curve. The different failure patterns of the bathtub curve can be described by different statistical distributions. Kuo et al. [174] have offered details on improving processes with high infant-mortalities.

Eventually, all hardware products including mechanical, electrical, and electronic components or systems follow the failure pattern described by the bathtub curve. Unless one addresses such an issue early in the design stage, achieving high reliability can be very expensive, particularly when the product is complex or involves relatively untried technology.

1.3.2 Human factors

In most situations, humans and machines are linked in one system. Accidents and malfunctions occur in most systems; and, therefore, there are procedures for reporting them. In such reporting, a large proportion of the incidents are often erroneously assigned to "human error" or "human reliability." Recently, the emphasis has been on developing techniques for predicting human reliability. Present effort is focused on developing a more academic methodology which applies to practical human–machine systems. For example, a life-cycle approach to predicting and evaluating human–machine reliability was developed under US Navy sponsorship [188]. Techniques for allocating, predicting, and evaluating achieved human reliability have been developed as part of the Navy program.

Table 1.1. The differences between hardware and software reliability

Category	Hardware reliability	Software reliability
Fundamental concept	Due to physical effects	Due to programmer errors (or program defects or faults)
Life-cycle causes		
Analysis	Incorrect customer understanding	Incorrect customer understanding
Feasibility	Incorrect user requirements	Incorrect user requirements
Design	Incorrect physical design	Incorrect program design
Development	Quality control problems	Incorrect program coding
Operation	Degradation and failure	Program errors (or remaining defects or faults)
Use effects	Hardware wears out and then fails	Software does not wear out, but fails from unknown defects (or faults)
Function of design	Physics of failure	Programmer skill
Domains	Time (t)	Time and data
Time relationships	Bathtub curve	Decreasing function
Math models	Theory well-established and accepted	Theory well-established, but not well-accepted
Time domain	$R = f(\lambda, t)$, λ = failure rate	$R = f$(failures [or defects or faults], t)
Functions	Exponential (constant λ) Weibull (increasing λ)	No agreement among the various time function models that have been proposed
Data domain	No meaning	Failures = f(data tests)
Growth models	Several models exist	Several models exist
Metric	λ, MTBF (mean time between failures) MTTF (mean time to failure)	Failure rate, number of defects (or faults) detected or remaining
Growth application	Design, prediction	Prediction
Prediction techniques	Block diagram, fault trees	Path analysis (actual analysis of all paths is an unsolvable problem, i.e. the number of possible dynamic paths for even simple programs can be shown to be infinite), complexity, simulation
Test and evaluation	Design and production acceptance	Design acceptance
Design	MIL-STD-781C (exponential) Other methods (nonexponential)	Path testing, simulation, error, seeding Bayesian
Operation	MIL-STD-781C	None
Use of redundancy		
Parallel	Can improve reliability	Need to consider common cause
Standby	Automatic error detection and correction, automatic fault detection and switching	Automatic error detection and correction, automatic audit software and software reinitialization
Majority logic	m-out-of-n	Impractical

It is desirable to treat the concept of human error with caution and to avoid an approach in which the operator appears to be held solely responsible; in reality, error arises out of a quite specific combination of conditions in the human–machine system, and it is on the total system that attention should be centered. Attacking the problem of human error in system performance involves training the workers and explaining the methodologies which must be mastered and actively applied in system development programs in order to accomplish the program objectives. Even where the reliabilities of the hardware components of military or space systems are extraordinarily high, poor system effectiveness is often observed in the field due to the human effect.

1.3.3   Software

A system consisting of software and hardware may fail due to the incapability of the software executed by external instructions. Software may fail when it is used in an undesirable environment. A software failure is defined as a departure from the expected external result or as output of the program operation which differs from the requirement. In other words, the program must be run in order for a failure to occur. A failure may be caused by a software fault or by some other cause.

Software has become an essential part of many industrial, military, and commercial systems. In today's large systems, software LCC typically exceeds hardware cost, with 80–90 percent of these costs going into software maintenance to fix, adapt, and expand the delivered program to meet the changing and growing needs of the users.

The current system cost trend is approaching software domination rather than hardware domination. Unfortunately, the relative frequency could be as high as 100:1 software to hardware failures [298]. For more complicated chips the ratio may be even higher than for normal chips. Software reliability and hardware reliability have distinct characteristics. Some major differences are outlined in Table 1.1.

1.3.4   Physical and economic constraints

All hardware and software systems operate under specific physical and economic constraints. Both types of constraints play a major role in the design and operation of a system. Physical constraints are normally related to the physical parameters of the system such as length, volume, weight, density, etc. In addition, constraints on component arrangement, temperature, radiation, humidity, etc. must be taken into account for successful operation of a system. The scope of a system design is generally limited by physical constraints. All systems require certain resources, and the consumption of resources varies with the system's design. Resource constraints include available budget, human power, material, time, etc. It is essential to satisfy the resource constraints in the design and operation of a reliability system.

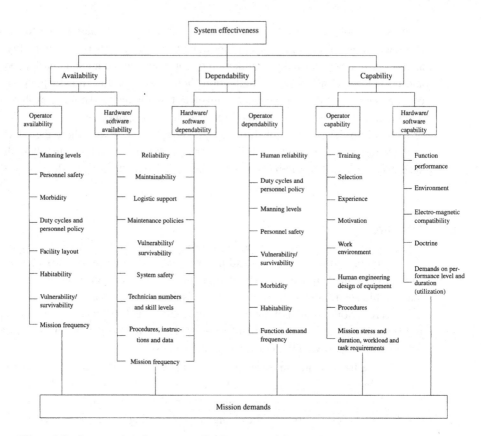

Figure 1.2. A comprehensive system effectiveness model

1.4   System effectiveness models

Due to various demands, a comprehensive system effectiveness model includes attributes as shown in Figure 1.2. System effectiveness was introduced in the late 1950s and early 1960s to describe the overall capability of a system for accomplishing its intended mission. The mission (to perform some intended function) was often referred to as the ultimate output of any system. Various system effectiveness definitions have been presented. The definition in ARINC [4] is: "the probability that the system can successfully meet an operational demand within a given time when operated under specified conditions." To define system effectiveness for a one-shot device such as a missile, the definition was modified to: "the probability that the system [missile] will operate successfully [kill the target] when called upon to do so under specified conditions." Effectiveness is obviously influenced by equipment design and construction. However, just as critical are the equipment's use and maintenance. Another famous and widely used definition of system effectiveness is from MIL-STD-721B [218]: "a measure of the degree to which an item can be expected to achieve a set of specific mission requirements and which may be expressed as a function of availability, dependability and capability." Several other definitions are used for other systems of interest. System effectiveness was originally confined to military and space systems, the ultimate customer being a branch of the US Government, such as the Department of Defense (DOD) or NASA. Notice that in Figure 1.2, performance of operator, hardware, and software all play important roles in the overall system effectiveness.

1.4.1   Attributes of system effectiveness

A number of different system effectiveness models are described by Tillman et al. [307] and each has somewhat different attributes and terminologies. The following are some of the terms used in these models. MIL-STD-721B definitions quoted in the following are used by the US Navy, Air Force, and Army.

1. *Reliability*: the probability that a system will perform satisfactorily for a specified mission under specified conditions for at least a given period of time. If the success of a system for a short period of time is of primary concern, the probability of success is termed the mission reliability.

2. *Availability*: a measure of the degree to which an item is in the operable and committable state at the start of the mission, when the mission is called for at an unknown (random) point in time [174]; operable is defined as the state of being able to perform the intended function.

3. *Repairability*: the probability that a failed system will be restored to operable condition in a given active repair time [4].

4. *Maintainability*: the probability that a failed system is restored to an operable condition in a specified down time. Maintainability is analogous to repairability. However, maintainability is based on total down time, while repairability is restricted solely to active repair time [4].

5. *Serviceability*: a measure of the degree to which servicing of an item will be accomplished within a given time under specified conditions. In this definition, servicing is referred to as the replenishment of consumables needed to keep an item in operating condition, not including any other preventive or corrective maintenance.

6. *Design adequacy*: the probability that a system will successfully accomplish its mission given that the system is operating within its design specifications throughout the mission [4].

7. *Capability*: a measure of the ability of an item to achieve mission objectives given the conditions during the mission. This definition is hard to implement in practice. An extended description of the MIL-STD-721B definition is: "the probability that the system's designed performance level will allow it to meet mission demands successfully provided that the system is available and dependable." This term

accounts for the adequacy of system elements to carry out the mission when operating in accordance with the system-design specifications as affected by the environment. Machine, software, and human modules of the operable system are included.

8. *Dependability*: a measure of the item operating condition at one or more points during the mission, including the effects of reliability, maintainability, and survivability, given the item condition(s) at the start of the mission.

9. *Human performance*: the severity of human-initiated malfunctions is a cause of failures in missile system operations. It is well-known that the high incident rate of human-initiated malfunctions is a symptom of inadequate system-engineering in system development programs.

10. *Environmental effect*: the environmental conditions encountered during a mission affect the performance of a system. Severe environmental conditions can make the actual mission duration longer than the ideal mission duration, thereby affecting mission reliability. System effectiveness can be influenced by environmental conditions in other ways: (1) mission availability can be severely decreased because of bad weather, e.g., shut-down of the transportation system; (2) operator performance will usually be affected by the working environment; (3) hardware reliability can be affected by humidity, atmospheric pressure, temperature, etc. Constraints are usually not included in the attributes of system effectiveness, although cost effectiveness as an important constraint is widely discussed.

1.4.2   Human factors in system effectiveness

As human beings become involved in systems, their abilities and limitations are manifested in their performance of mission tasks. Since humans are essential to the operation of such systems, it is important to measure the effect of human performance on the system reliability. There is evidence that the human component is responsible for 20–90 percent of the failures in many systems depending upon the degree of human involvement in the system [188], [191]. Human-factors specialists usually provide only qualitative analysis of human factors in human–machine systems. A better approach to study human factors in system effectiveness is to combine the human and hardware performance measures into a meaningful index taking into account the interaction of human and hardware components of the system. Comparison between human and hardware reliability is provided in Table 1.2. Many methods have been proposed in the literature to analyze the human function within the system. A survey of these methods is provided in Lee et al. [190]. The methods that are most widely used for studying human performance and human–machine system effectiveness are: (1) analytical methods, and (2) computer simulation methods. For several reasons, computer simulation modeling is considered to be more appropriate for modeling human–machine system effectiveness.

Table 1.2. A comparison between hardware and human reliability

Category	Hardware reliability	Human reliability	
		Discrete task	Continuous task
System definition	A set of components which perform their intended functions	A task which consists of several human behavioral units	Continuous control task such as vigilance, tracking, and stabilizing
System configuration	Functional relationships of components	Relationships of behavior units for a given task (task taxonomy)	Not necessary to define functional relationships between the task units
System failure analysis	Fault-tree analysis	Human error categorization: derivation of a mutually exclusive and exhaustive set of human errors for a given task	Binary error logic for continuous system response
Nature of failure	Mostly binary failure logic Multi-dimensionality of failure Common-cause failure	Sometimes hard to apply binary error logic to human action Multi-dimensionality of error Common-cause error Error correction	Same as discrete task
Cause of failure	Most hardware failures are explained by the laws of physics and chemistry	No well-codified laws which are generally accepted as explanations of human errors	Same as discrete task
System reliability evaluation	With probabilistic treatments of failure logic and statistical independence assumption between components, mathe- matical models are derived In cases of network reliability and phased mission reliability, which require statistical dependency between components, it is hard to evaluate exact system reliability	Very difficult because of problems in depicting the functional relationships between human behavioral units	With probabilistic treatments of binary error logic for system response, stochastic models are derived
Data	The data base for most types of machines is relatively large and robust compared to human reliability	No trustworthy and useful data base exists for human behavior units Largely depends on the judgment of experts	Same as discrete task

1.4.3   Mission effectiveness

Mission effectiveness, the probability of achieving a specific mission goal, was introduced to describe the capability of a system to accomplish its intended mission. The mission (to perform some intended function) is often referred to as the ultimate goal of any system. A state-of-the-art survey on mission effectiveness models is presented in [307]. Several mission effectiveness models, such as WSEIAC [325] are available in the literature. They express mission effectiveness as a combined probability measure of reliability and maintainability. Each model can be nominally expressed in a variety of ways – reflecting the disciplinary background of its modeler. Recognizing a lack of theoretical discussion of existing mission effectiveness models, Tillman et al. [308], [310] have established a stochastic model for mission effectiveness. Mission effectiveness is calculated by Tillman et al. [312] as the product of the availability function evaluated at task arrival time t and the reliability function at time t evaluated for a fixed time period x from this instant.

As an extension of Tillman [310], who considers only one task arrival during the mission, Lee et al. [189] and Lie et al. [196] have considered a more general mission effectiveness model with several randomly arriving tasks during the mission time. The mission effectiveness is defined as a combined measure of availability and reliability at each task arrival time. An analytical model for mission effectiveness is derived under a general set of assumptions and illustrated by numerical examples.

1.5   Fundamental system configurations and reliability functions

A system in many cases is not confined to a single component. What we really want to evaluate is the reliability of simple and complex systems. Consider a reliability system consisting of n components. These components can be hardware or human or even software. If some components are software products, then the modeling requires special attention. Let $\Pr(A_i), 1 \le i \le n$, denote the probability of event A_i that component i operates successfully during the intended period of time. Then the reliability of component i is $r_i = \Pr(A_i)$. Similarly, let $\Pr(\overline{A}_i)$ denote the probability of event $\overline{A}_i$ that component i fails during the intended period of time. Assume that the failure of any component is independent of that of other components unless stated otherwise. In the following subsections, we describe several important reliability configurations and present their reliability functions. We denote the system reliability for time t by $R_s(t)$. When t is considered as a fixed time, the system reliability is simply denoted by R_s.

1.5.1   Series configuration

The series configuration is the simplest and perhaps one of the most common structures. The block diagram in Figure 1.3 represents a series system consisting of n components.

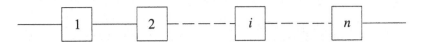

Figure 1.3. A series configuration

In this configuration, all n components must be operating to ensure system operation. In other words, the system fails when any one of the n components fails.

Thus, the reliability of a series system is

$$R_s = \text{Pr(all components operate successfully)}$$
$$= \text{Pr}(A_1 \cap A_2 \cap \cdots \cap A_n)$$
$$= \prod_{i=1}^{n} \text{Pr}(A_i).$$

The last equality holds since all components operate independently. Therefore, the reliability of a series system is

$$R_s = \prod_{i=1}^{n} r_i. \tag{1.1}$$

The lifetime of a series system is the minimum of component lifetimes. Suppose the failure time of component i, $1 \le i \le n$, follows a negative exponential distribution with failure rate λ_i and the system is required to operate for a specified time t. Then the reliability of component i is $r_i = e^{-\lambda_i t}$ and thus eq. (1.1) gives

$$R_s = \exp\left[-\left(\sum_{i=1}^{n} \lambda_i\right)t\right]. \tag{1.2}$$

It may be noted from eq. (1.2) that the failure time of the system follows a negative exponential distribution with failure rate $\sum_{i=1}^{n} \lambda_i$. Therefore, the mean time to failure of the system is $1/(\sum_{i=1}^{n} \lambda_i)$.

1.5.2 Parallel configuration

In many systems, several paths perform the same operation simultaneously forming a parallel configuration. A block diagram for this system is shown in Figure 1.4. There are m paths connecting input to output, and the system fails if all of the m components fail. This is sometimes called a redundant configuration. The word "redundant" is used only when the system configuration is deliberately changed to produce additional parallel paths in order to improve the system reliability. Thus, a parallel system may occur as a result of the basic system structure or may be produced by using redundancy in a reliability design or redesign of the system. In this book, we use synonymously "number of parallel redundant components" and "redundancy level."

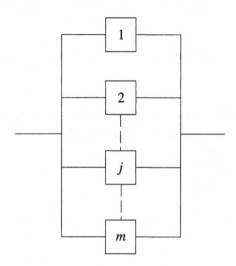

Figure 1.4. A parallel configuration

In a parallel configuration consisting of m components, the system is successful if any one of the m components is successful. Thus, the reliability of a parallel system is the probability of the union of the m events $A_1, A_2, \ldots, A_m$, which can be written as

$$R_s = \Pr(A_1 \cup A_2 \cup \cdots \cup A_m)$$
$$= 1 - \Pr(\overline{A}_1 \cap \overline{A}_2 \cap \cdots \cap \overline{A}_m)$$
$$= 1 - \prod_{j=1}^{m} \Pr(\overline{A}_j)$$
$$= 1 - \prod_{j=1}^{m} [1 - \Pr(A_j)].$$

The third equality holds since the component failures are independent. Therefore, the reliability of a parallel system is

$$R_s = 1 - \prod_{j=1}^{m} (1 - r_j). \tag{1.3}$$

If all components in the parallel configuration follow the negative exponential failure law, eq. (1.3) becomes

$$R_s = 1 - \prod_{j=1}^{m} \left(1 - e^{-\lambda_j t}\right), \tag{1.4}$$

where λ_j is the failure rate of component j. The lifetime of a parallel system is the maximum of component lifetimes.

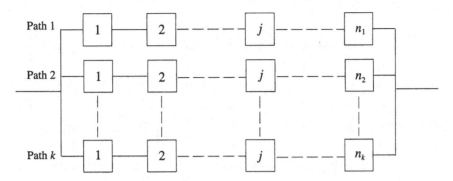

Figure 1.5. A series-parallel system

1.5.3 Series-parallel configuration

Consider a system which consists of k subsystems connected in parallel, with subsystem i consisting of n_i components in series for $i = 1, \ldots, k$. Such a system is called a *series-parallel* system. A series-parallel system is shown in Figure 1.5. Let R_i be the reliability of subsystem i and r_{ij} the reliability of component $j, 1 \leq j \leq n_i$, in subsystem i. Then

$$R_i = \prod_{j=1}^{n_i} r_{ij},$$

and the system reliability is

$$R_s = 1 - \prod_{i=1}^{k} (1 - R_i).$$

The above two equations together imply

$$R_s = 1 - \prod_{i=1}^{k} \left(1 - \prod_{j=1}^{n_i} r_{ij}\right). \tag{1.5}$$

When the components are identical in each subsystem the system reliability can be written as

$$R_s = 1 - \prod_{i=1}^{k} (1 - r_i^{n_i}),$$

where r_i is the reliability of each component in the subsystem, $i = 1, \ldots, k$.

1.5.4 Parallel-series configuration

Suppose a system consisting of k subsystems in series and subsystem i, $1 \leq i \leq k$, in turn consists of n_i components in parallel. Such a system is called a *parallel-series*

Stage   1   2 i k

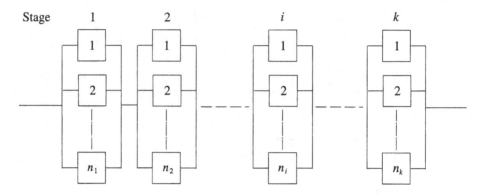

Figure 1.6. A parallel-series system

system. A parallel-series system is shown in Figure 1.6. Let R_i be the reliability of subsystem i and r_{ij} the reliability of component j, $1 \leq j \leq n_i$, in subsystem i. Then

$$R_i = 1 - \prod_{j=1}^{n_i}(1 - r_{ij}), \tag{1.6}$$

and the system reliability is

$$R_s = \prod_{i=1}^{k} R_i. \tag{1.7}$$

Equations (1.6) and (1.7) follow from eqs. (1.3) and (1.1), respectively. The two equations together imply

$$R_s = \prod_{i=1}^{k}\left[1 - \prod_{j=1}^{n_i}(1 - r_{ij})\right]. \tag{1.8}$$

If all components in subsystem i are identical, then r_{ij} is the same for $j = 1, \ldots, n_i$. Let $r_{ij} = r_i$ for $i = 1, \ldots, k$ and $j = 1, \ldots, n_i$. Then the system reliability is

$$R_s = \prod_{i=1}^{k}(1 - q_i^{n_i}),$$

where $q_i = 1 - r_i$ is the failure probability of a component in subsystem i. When r_i is large, the system reliability is approximated by

$$R_s \simeq 1 - \sum_{i=1}^{k} q_i^{n_i}.$$

The terminology used to name the structures in Figures 1.5 and 1.6 is not the same throughout the literature. Some authors have called the structures in Figures 1.5 and 1.6 parallel-series and series-parallel configurations, respectively. However, we follow our nomenclature for these systems throughout this book.

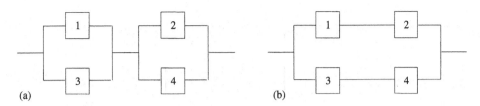

Figure 1.7. A comparison of parallel-series and series-parallel systems

In general, the reliability of a parallel-series configuration is greater than that of a series-parallel configuration, if the same set of components are used in both systems. Suppose components 1, 2, 3, and 4 are arranged in two kinds of configurations (a) and (b) as shown in Figure 1.7. The reliability of configuration (a) is

$$R_a = [1 - (1 - r_1)(1 - r_3)][1 - (1 - r_2)(1 - r_4)]$$

$$= r_1r_2 + r_1r_4 - r_1r_2r_4 + r_2r_3 + r_3r_4$$

$$- r_2r_3r_4 - r_1r_2r_3 - r_1r_3r_4 + r_1r_2r_3r_4,$$

where r_1, r_2, r_3, r_4 are the reliabilities of components 1, 2, 3, and 4, respectively.
The reliability of configuration (b) is

$$R_b = 1 - (1 - r_1r_2)(1 - r_3r_4)$$

$$= r_1r_2 + r_3r_4 - r_1r_2r_3r_4.$$

We then calculate

$$R_a - R_b = r_1r_4 - r_1r_2r_4 - r_1r_3r_4 + r_1r_2r_3r_4$$

$$+ r_2r_3 - r_2r_3r_4 - r_1r_2r_3 + r_1r_2r_3r_4$$

$$= r_1r_4(1 - r_3 - r_2 + r_2r_3) + r_2r_3(1 - r_1 - r_4 + r_1r_4)$$

$$= r_1r_4(1 - r_2)(1 - r_3) + r_2r_3(1 - r_1)(1 - r_4).$$

Since all r_i are smaller than 1, $R_a - R_b > 0$. Therefore, the reliability of configuration (a) is greater than that of configuration (b).

1.5.5   Hierarchical series-parallel systems

A reliability system is called a *hierarchical series-parallel system* (HSP) if the system can be viewed as a set of subsystems arranged in series or parallel; each subsystem has a similar configuration; subsystems of each subsystem have a similar configuration and so on. For example, consider the system in Figure 1.8. The system with components 1, 2, 3, 4, and 5 consists of two subsystems {1, 2, 3} and {4, 5} in parallel. The first subsystem has a series arrangement of component 3 and a subsystem which has components 1 and 2 in parallel, whereas the second one has components 4 and 5 in series. This is a HSP system.

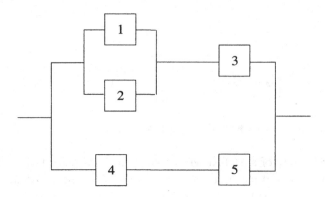

Figure 1.8. Five-component hierarchical series-parallel system

The reliability of each subsystem can be easily computed from those of its subsystems using eq. (1.1) or (1.3). Thus, the reliability of a HSP system can be evaluated in a recursive fashion starting with the reliability of the lowest level parallel or series subsystems. For example, let r_j denote the reliability of component j, $j = 1, \ldots, 5$, in the system shown in Figure 1.8. The reliabilities of subsystems $\{1, 2\}$ and $\{4, 5\}$ are

$$R(\{1, 2\}) = r_1 + r_2 - r_1 r_2$$

and

$$R(\{4, 5\}) = r_4 r_5,$$

respectively. Now, the reliability of subsystem $\{1, 2, 3\}$ is

$$R(\{1, 2, 3\}) = R(\{1, 2\}) R(\{3\})$$
$$= (r_1 + r_2 - r_1 r_2) r_3.$$

Similarly, the reliability of the entire system $\{1, 2, 3, 4, 5\}$ is

$$R(\{1, 2, 3, 4, 5\}) = 1 - [1 - R(\{1, 2, 3\})][1 - R(\{4, 5\})]$$
$$= 1 - [1 - (r_1 + r_2 - r_1 r_2) r_3](1 - r_4 r_5).$$

1.5.6   k-out-of-n systems

A k-out-of-n system is an n-component system which functions when at least k of its n components function. This redundant system is sometimes used in place of a pure parallel system. It is also referred to as a k-out-of-n: G system. An n-component series system is a n-out-of-n: G system whereas a parallel system with n components is a 1-out-of-n: G system. When all of the components are independent and identical, the reliability of a general k-out-of-n system can be written as

$$R_s = \sum_{j=k}^{n} \binom{n}{j} r^j (1 - r)^{n-j},$$

where r is the component reliability.

For example, the reliability of a 3-out-of-5: G system with component reliability r is

$$R_s = \binom{5}{3} r^3 (1-r)^{5-3} + \binom{5}{4} r^4 (1-r)^{5-4} + \binom{5}{5} r^5 (1-r)^{5-5}$$

$$= 10r^3 (1-r)^2 + 5r^4 (1-r) + r^5$$

$$= r^3 (6r^2 - 15r + 10).$$

A k-out-of-n: F system is an n-component system which fails when any k of its n components fail. A consecutive k-out-of-n: F system is an n-component system which fails when any consecutive k components in the system fail. See Appendix 4 for more details of consecutive k-out-of-n systems. The consecutive k-out-of-n: G and F systems can be defined similarly.

1.5.7 Complex configuration

In addition to reliability systems which can be reduced to series and parallel configurations, there exist combinations of components which are neither. Such systems are called complex or non-parallel-series systems (Figures 1.9(a) and 1.10); both are introduced here.

Non-parallel-series system

Two equal paths (series subsystems), 1–2 and 3–4 operate in parallel so that if at least one of them is good, the output is assured. However, because units 1 and 3 are not reliable enough, a third equal unit, 5, is inserted into the circuit so that it can supply either 2 or 4 with the necessary signal. Therefore, each of paths 1–2, 5–2, 5–4, and 3–4 ensures system success. The schematic reliability block diagram shown in Figure 1.9(a) is also different from the five-component complex system shown in Figure 1.9(b).

The five-component system in Figure 1.9(b), appears to be similar to the configuration in Figure 1.9(a). However, the additional paths 1–4 and 3–2 also ensure system success. It can be easily seen that the system in Figure 1.9(b) is a parallel-series configuration with two parallel systems, $\{1, 3, 5\}$ and $\{2, 4\}$, connected in series.

The failure probability Q_s of the complex system under consideration can be expressed as

$$Q_s = \Pr(\text{system failure given that } i \text{ is good})r_i$$
$$+ \Pr(\text{system failure given that } i \text{ is bad})q_i, \tag{1.9}$$

where q_i is the probability that component i is bad and r_i is the probability that component i is good. This holds for any component of the system. The system reliability then is

$$R_s = 1 - Q_s. \tag{1.10}$$

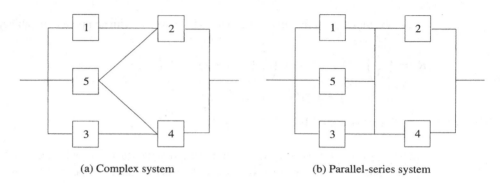

(a) Complex system   (b) Parallel-series system

Figure 1.9. A comparison of parallel-series and complex systems

Using eq. (1.9) with $i = 5$, the failure probability of the system in Figure 1.9(a) can be written as

$$Q_s = \text{Pr(system failure given that 5 is good)}r_5$$
$$+ \text{Pr(system failure given that 5 is bad)}q_5.$$

If component 5 is good, the system can fail only if both 2 and 4 fail. Since 2 and 4 are in parallel, the system failure probability, when 5 is good, amounts to

$$\text{Pr(system failure given that 5 is good)} = (1 - r_2)(1 - r_4).$$

Similarly, when 5 is bad, the system failure probability is

$$\text{Pr(system failure given that 5 is bad)} = (1 - r_1 r_2)(1 - r_3 r_4),$$

where $(1 - r_1 r_2)$ and $(1 - r_3 r_4)$ are the unreliabilities of paths 1–2 and 3–4, respectively. Therefore, the unreliability of the whole system can now be written as

$$Q_s = (1 - r_2)(1 - r_4)r_5 + (1 - r_1 r_2)(1 - r_3 r_4)q_5. \tag{1.11}$$

If the failure times of components 1, 2, 3, 4, and 5 are exponentially distributed with failure rates $\lambda_1, \lambda_2, \lambda_3, \lambda_4$, and λ_5, respectively, then

$$Q_s = (1 - e^{-\lambda_2 t})(1 - e^{-\lambda_4 t})e^{-\lambda_5 t}$$
$$+ (1 - e^{-(\lambda_1+\lambda_2)t})(1 - e^{-(\lambda_3+\lambda_4)t})(1 - e^{-\lambda_5 t}). \tag{1.12}$$

Thus, the reliability at time t of the complex system shown in Figure 1.9(a) is

$$R_s = 1 - Q_s$$
$$= 1 - (1 - e^{-\lambda_2 t})(1 - e^{-\lambda_4 t})(e^{-\lambda_5 t})$$
$$- (1 - e^{-(\lambda_1+\lambda_2)t})(1 - e^{-(\lambda_3+\lambda_4)t})(1 - e^{-\lambda_5 t}).$$

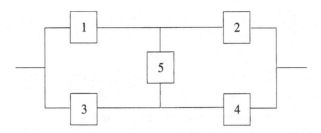

Figure 1.10. A bridge network

Complex system

This complex system is called a *bridge network*. It can operate successfully as long as one of the paths, 1–2 or 3–4, is good, whether unit 5, inserted in the middle, is good or not. However, if the pair $(1, 4)$ or the pair $(3, 2)$ fail, then unit 5 plays a vital role in system operation. There are two efficient methods to evaluate the reliability of the system shown in Figure 1.10.

Method 1

This method is based on a simple probability theorem

$$\Pr(X \cup Y) = \Pr(X) + \Pr(Y) - \Pr(X \cap Y), \tag{1.13}$$

where X and Y are two events.

To use eq. (1.13), it is required that all possible paths from the input node to the output node be found. Consider the sets of components $\{1, 2\}$, $\{3, 4\}$, $\{1, 5, 4\}$, and $\{3, 5, 2\}$. The system will operate if the components in any one of these sets operate. Moreover, when the system operates, the components in at least one of these sets operate. Thus, the system reliability is $\Pr(1\text{–}2 \cup 3\text{–}4 \cup 1\text{–}5\text{–}4 \cup 3\text{–}5\text{–}2)$, which can be evaluated by repeatedly using eq. (1.13). Since all the components operate independently, we have

$$\Pr(\{1, 2\}) = r_1 r_2,$$

$$\Pr(\{3, 4\}) = r_3 r_4,$$

$$\Pr(\{1, 5, 4\}) = r_1 r_5 r_4,$$

$$\Pr(\{3, 5, 2\}) = r_3 r_5 r_2.$$

Now, using eq. (1.13), we can write

$$\Pr(\{1, 2\} \cup \{3, 4\}) = \Pr(\{1, 2\}) + \Pr(\{3, 4\}) - \Pr(\{1, 2\} \cap \{3, 4\})$$

$$= r_1 r_2 + r_3 r_4 - r_1 r_2 r_3 r_4.$$

Similarly, we have

$$\Pr[(\{1, 2\} \cup \{3, 4\}) \cup \{1, 5, 4\}] = r_1 r_2 + r_3 r_4 - r_1 r_2 r_3 r_4 + r_1 r_5 r_4$$

$$- r_1 r_2 r_5 r_4 - r_1 r_3 r_4 r_5 + r_1 r_2 r_3 r_4 r_5,$$

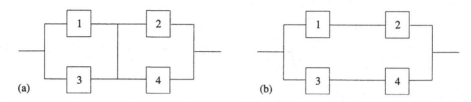

Figure 1.11. Configurations based on state of component 5 in Figure 1.10

and

$$\Pr[(\{1, 2\} \cup \{3, 4\} \cup \{1, 5, 4\}) \cup \{3, 5, 2\}] = r_1 r_2 + r_3 r_4 + r_1 r_5 r_4 + r_2 r_3 r_5$$

$$- r_1 r_2 r_3 r_4 - r_1 r_2 r_4 r_5 - r_1 r_3 r_4 r_5$$

$$- r_1 r_2 r_3 r_5 - r_2 r_3 r_4 r_5 + 2 r_1 r_2 r_3 r_4 r_5. \tag{1.14}$$

Therefore, the system reliability is given by the expression on the right-hand side of eq. (1.14). If $r_i = r$ for $i = 1, \ldots, 5$, then (1.14) reduces to

$$r_s = r^2 (2 + 2r - 5r^2 + 2r^3). \tag{1.15}$$

Method 2
This method is based on conditional probability. We can write the reliability R_s of the bridge network in Figure 1.10 as

$$R_s = \Pr(\text{system does not fail given that 5 is good}) r_5$$

$$+ \Pr(\text{system does not fail given that 5 is bad})(1 - r_5). \tag{1.16}$$

If 5 is good, then the original network is reduced to Figure 1.11(a), which is a parallel-series system with a reliability of

$$\Pr(\text{system does not fail given that 5 is good}) = (r_1 + r_3 - r_1 r_3)(r_2 + r_4 - r_2 r_4). \tag{1.17}$$

If 5 is bad, then the original network is reduced to Figure 1.11(b), which is a series-parallel system with a reliability of

$$\Pr(\text{system does not fail given that 5 is bad}) = (r_1 r_2 + r_3 r_4 - r_1 r_2 r_3 r_4). \tag{1.18}$$

Substitution of eqs. (1.17) and (1.18) into eq. (1.16) yields the same system reliability as given by eq. (1.14).

1.5.8   Coherent systems

In nonseries systems, it is not necessary that all components operate to ensure system operation. In such systems, we can also find subsets of components such that the failure of all components in the subset leads to system failure irrespective of the states of other components. The theory of coherent systems deals with the deterministic functional relationship between the system and its components. Such a relationship is useful for finding the reliability of large and complex systems.

Suppose a system consists of n components, $1, 2, \ldots, n$. Assume that the system and its components have two states each: "operating state" and "failure state." Define a binary variable a_j to represent the state of component j for $j = 1, 2, \ldots, n$. Let

$$a_j = \begin{cases} 1, & \text{if the component } j \text{ is in operating state,} \\ 0, & \text{if the component } j \text{ is in failure state,} \end{cases}$$

for $j = 1, 2, \ldots, n$. Let $\mathbf{a} = (a_1, \ldots, a_n)$. Similarly, define a binary variable ϕ to represent the state of the system: $\phi = 1$ if the system is in an operating state and 0, otherwise. The vector $\mathbf{a}$ can be any element in the set of 2^n binary vectors of dimension n. The value of ϕ is 1 for some possibilities of $\mathbf{a}$ and 0 for other possibilities. For a majority of reliability systems, we can explicitly write ϕ as a function of $\mathbf{a}$, say $\phi(\mathbf{a})$, so that the system state ϕ corresponding to vector $\mathbf{a}$ is given by

$$\phi = \phi(\mathbf{a}).$$

The function $\phi(\mathbf{a})$ is called the *structure function* of the system. The form of $\phi(\mathbf{a})$ depends on the system configuration. Consider a system with structure function $\phi(\mathbf{a})$. A component j is said to be *relevant* to function $\phi(\mathbf{a})$ if

$$\phi(a_1, \ldots, a_{j-1}, 0, a_{j+1}, \ldots, a_n) = 0,$$

and

$$\phi(a_1, \ldots, a_{j-1}, 1, a_{j+1}, \ldots, a_n) = 1$$

for some states $a_1, \ldots, a_{j-1}, a_{j+1}, \ldots, a_n$ of other components $1, 2, \ldots, j-1, j+1, \ldots, n$. The failure of a relevant component leads to system failure in at least one situation (a combination of states of other $n-1$ components). The structure function $\phi(\mathbf{a})$ is said to be *coherent* when:

1. $\phi(\mathbf{a})$ is nondecreasing in each a_j, and

2. each component is relevant to $\phi(\mathbf{a})$.

If a system has a coherent structure function, it is called a *coherent system*. Series and parallel systems are two simple examples of coherent systems.

For a series system with n components, $1, 2, \ldots, n$, the structure function is

$$\phi(\mathbf{a}) = \prod_{j=1}^{n} a_j.$$

This is true since $\phi = 1$ if and only if $a_j = 1$ for all j. Similarly, for a parallel system with n components $1, 2, \ldots, n$, the structure function is

$$\phi(\mathbf{a}) = 1 - \prod_{j=1}^{n} (1 - a_j).$$

It can be verified by observing that $\phi = 0$ if and only if $a_j = 0$ for all j.

Suppose $\phi(\mathbf{a}) = 1$ for a vector $\mathbf{a}$ of component states. Then the set $P = \{j: a_j = 1\}$ is called a *path set* and vector $\mathbf{a}$ is called a *path vector*. The system operates as long as all components in P operate. If $\phi(\mathbf{a}) = 0$ for vector $\mathbf{a}$, then the set $C = \{j: a_j = 0\}$ is called a *cut set* and vector $\mathbf{a}$ is called a *cut vector*.

Let P be a path set and $\mathbf{a}$ the corresponding path vector (that is, $a_j = 1$ for $j \in P$ and 0 for $j \in \overline{P}$). The set P is called a *minimal path set* if

$$\phi(\mathbf{b}) = 0 \text{ for any binary vector } \mathbf{b} \le \mathbf{a} \text{ and } \mathbf{b} \ne \mathbf{a}.$$

If P is a minimal path set, then no proper subset of P is a path set. Let C be a cut set and $\mathbf{a}$ the corresponding cut vector (that is, $a_j = 0$ for $j \in C$ and 1 for $j \in \overline{C}$). The set C is called a *minimal cut set* if

$$\phi(\mathbf{b}) = 1 \text{ for any } \mathbf{b} \ge \mathbf{a} \text{ and } \mathbf{b} \ne \mathbf{a}.$$

If P is a minimal cut set, then no proper subset of C is a cut set.

For the complex system shown in Figure 1.9, $P = \{1, 4, 5\}$ is a path set, whereas $P' = \{3, 4\}$ is a minimal path set. $C = \{1, 2, 3, 4\}$ is a cut set, while $C' = \{2, 4\}$ is a minimal cut set.

Let $\{P_1, P_2, \ldots, P_r\}$ be the collection of all minimal path sets of a coherent structure $\phi(\mathbf{a})$. Then

$$\phi(\mathbf{a}) = 1 - \prod_{i=1}^{r}\left(1 - \prod_{j \in P_i} a_j\right). \tag{1.19}$$

The structure function of a reliability system can be obtained by finding all minimal path sets and using eq. (1.19). For example, consider the series-parallel system shown in Figure 1.5. Let

$$a_{ij} = \begin{cases} 1, & \text{if } j\text{th component in } i\text{th series subsystem is in operating state,} \\ 0, & \text{otherwise,} \end{cases}$$

for $i = 1, \ldots, k$ and $j = 1, \ldots, n_i$. The path sets of this system are $P_1, \ldots, P_k$, where

$$P_i = \{\text{all components in } i\text{th series subsystem}\}$$

for $i = 1, \ldots, k$. Therefore, by eq. (1.19), the structure function of a series-parallel system is

$$\phi(\mathbf{a}) = 1 - \prod_{i=1}^{k}\left(1 - \prod_{j=1}^{n_i} a_{ij}\right),$$

where $\mathbf{a} = (a_{11}, \ldots, a_{1n_1}, \ldots, a_{k1}, \ldots, a_{kn_k})$.

Suppose $\{C_1, C_2, \ldots, C_s\}$ is the collection of all minimal cut sets of $\phi(\mathbf{a})$. Then

$$\phi(\mathbf{a}) = \prod_{i=1}^{s} \left[1 - \prod_{j \in C_i} (1 - a_j) \right] \qquad (1.20)$$

and $\phi(\mathbf{a})$ can be obtained by finding all minimal cut sets and using eq. (1.20). For example, consider the parallel-series system shown in Figure 1.6. Let

$$a_{ij} = \begin{cases} 1, & \text{if } j\text{th component in } i\text{th parallel subsystem is in operating state,} \\ 0, & \text{otherwise,} \end{cases}$$

for $i = 1, \ldots, k$ and $j = 1, \ldots, n_i$. The cut sets of this system are $C_1, \ldots, C_k$ where

$$C_i = \{\text{all components in } i\text{th parallel subsystem}\}$$

for $i = 1, \ldots, k$. By eq. (1.19), the structure function of parallel-series system can be written as

$$\phi(\mathbf{a}) = \prod_{i=1}^{k} \left[1 - \prod_{j=1}^{n_i} (1 - a_{ij}) \right].$$

The structure function $\phi(\mathbf{a})$ of a reliability system can also be obtained iteratively using the pivotal decomposition rule

$$\phi(\mathbf{a}) = a_j \phi(a_1, \ldots, a_{j-1}, 1, a_{j+1}, \ldots, a_n)$$
$$+ (1 - a_j) \phi(a_1, \ldots, a_{j-1}, 0, a_{j+1}, \ldots, a_n).$$

This method is quite useful for deriving the structure function of complex systems. For any zero–one variable y, we have $y = y^m$ for $m = 2, 3, \ldots$. Using this property, we can express $\phi(\mathbf{a})$ as a multi-linear function of $a_1, \ldots, a_n$ (that is, linear in each a_j). The system reliability can be obtained by substituting reliability p_j of component j for a_j in the multi-linear form of $\phi(\mathbf{a})$. Related theorems about the coherent structure can be seen in Barlow and Proschan [23]. The relationship between the reliability function of a coherent system and its minimum path (cut) sets can be found in El-Neweihi [85]. Structure analyses of multi-coherent systems are given in Andrzejczak [12] and Zettwitz [334].

1.5.9 Cold standby redundancy in a single-component system

Consider a system consisting of a single component. To increase the system reliability, the designer may provide several components of a similar type as redundancies. The redundant components perform the same operation as the single components of the system. There are three types of redundancy: (1) *cold standby*, (2) *warm standby*, and (3) *hot standby*. In cold standby redundancy, a component does not fail before it is put into operation. In warm standby redundancy, the component is more prone to failures before operation than the cold standby components. The failure pattern of a hot standby redundant component does not depend on whether the component is idle or in operation.

If all redundant components operate simultaneously from time zero, even though the system needs only one of them at any time, such an arrangement is called *parallel (or active)* redundancy. This arrangement is essential when switching or starting a good component following a component failure is ruled out. Parallel redundancy is considered in all of the following chapters. The mathematical models for hot standby and parallel redundancy arrangements are equivalent. In the cold standby redundancy arrangement, the redundant components are sequentially used in the system at component failure times. Each redundant component in the cold standby arrangement can operate only when it is switched on. When the component in operation fails, one of the redundant ones is switched on to continue the operation. The cold standby redundancies are termed spares. The system fails when the last spare fails. Cold standby redundancy provides longer system life than hot standby redundancy. Suppose m components are provided as spares to a single-component system. Let X_0 denote the lifetime of the system component and X_i denote that of the ith spare for $i = 1, \ldots, m$. Assume that all X_is are statistically independent. The system component fails at time X_0 and the ith spare fails at time $X_0 + X_1 + \cdots + X_i$ for $i = 1, 2, \ldots, m$. The system reliability at time t is

$$R_s(t) = \text{Pr (system operates successfully until time } t)$$
$$= \text{Pr}(X_0 + X_1 + \cdots + X_m > t).$$

Let $F_i(t)$ denote the probability distribution function of lifetime X_i and $f_i(t)$ the corresponding density function. Let $R_i(t) = 1 - F_i(t)$ for $i = 0, 1, \ldots, m$. For $m = 1$, the system reliability is

$$R_s(t) = \text{Pr}(X_0 + X_1 > t)$$
$$= \int_0^t \text{Pr}(X_1 > t - x_0) f_0(x_0)\, dx_0 + R_0(t).$$

Similarly, for $m = 2$, the system reliability is

$$R_s(t) = \text{Pr}(X_0 + X_1 + X_2 > t)$$
$$= \int_0^t \text{Pr}(X_1 + X_2 > t - x_0) f_0(x_0)\, dx_0 + R_0(t)$$
$$= \int_0^t \left[\int_0^{t-x_0} \text{Pr}(X_2 > t - x_0 - x_1) f_1(x_1)\, dx_1 + R_1(t - x_0) \right] f_0(x_0)\, dx_0$$
$$+ R_0(t).$$

The system reliability $R_s(t)$ can be expressed in this fashion for any m. However, the expression becomes quite cumbersome for general m. The value of $R_s(t)$ can also be found by evaluating the system failure probability $\text{Pr}(X_0 + X_1 + \cdots + X_m \le t)$.

This probability can be obtained when the distribution of the sum $X_0 + X_1 + \cdots + X_m$ can be derived. Laplace transformation techniques are sometimes useful to derive such distributions. For the intractable cases, one can adopt Monte Carlo simulation techniques to estimate the failure probability. Suppose the system component and the redundant ones are identical, each having a negative exponential life distribution with mean μ. Then,

$$f_i(t) = \frac{1}{\mu} e^{-t/\mu}, \quad t \geq 0, \qquad \text{for } i = 0, 1, \ldots, m,$$

and the system life $X_0 + X_1 + \cdots + X_m$ follows a gamma distribution with density function

$$f(t) = \frac{1/\mu}{m!} (t/\mu)^m e^{-t/\mu}, \quad t \geq 0.$$

Therefore, the probability of system failure by time t is

$$\Pr(X_0 + X_1 + \cdots + X_m \leq t) = 1 - \sum_{i=0}^{m} \frac{(t/\mu)^i}{i!} e^{-t/\mu},$$

and the system reliability is

$$R_s(t) = \sum_{i=0}^{m} \frac{(t/\mu)^i}{i!} e^{-t/\mu}. \tag{1.21}$$

The mean and variance of system life are $(m+1)\mu$ and $(m+1)\mu^2$, respectively.

The system reliability as given in eq. (1.21) can also be obtained using the theory of Poisson processes. If the number of redundant components is unlimited, the component failure times form a Poisson process with rate $1/\mu$. Then, the number of failures in time t follows a Poisson distribution with mean t/μ and therefore the right-hand side of eq. (1.21) is the probability that the number of failures in time t does not exceed m. This probability is the same as the required system reliability.

1.5.10 Redundancy with imperfect switching system

As mentioned earlier, the reliability of a system can be enhanced by providing component redundancy. When cold standby redundancy is feasible, it is preferable to parallel redundancy as it improves system life more effectively. The scope of cold standby redundancy increases when highly reliable switching devices (with high quality sensors) are available. The redundancy model considered in the previous section implicitly assumes that the switching devices and sensors are perfect (with a reliability of one). However, a switching system may fail or sensors may give wrong signals, thereby causing system failure. In fact, switching devices have multiple failure modes: false shorting of a circuit, opening of a circuit in the absence of failures, nonopening of a circuit when failures occur, etc. In a cold standby redundancy arrangement with imperfect switching, the system reliability depends not only on the failure patterns of the redundant components but also on those of switching devices and sensors.

It is very difficult to analyze cold standby redundancy models when the switching devices have multiple failure modes. Thus, let us assume that it is enough to know whether a switching device is in good or failed condition in order to determine the state of the system. To simplify the problem, let us assume further that the sensors are perfect.

Redundant system with m cold standby redundant components

Consider a system with one original component C_0 and m cold standby components $C_1, \ldots, C_m$. A switch S is used to switch on a good component whenever the component in operation fails. The switch is required to be in operating condition only for switching purposes (but not for the actual operation of any component). Let X_j represent the lifetime of C_j for $j = 0, 1, \ldots, m$ and Y represent the lifetime of the switch S. Assume that:

1. C_0 and S start operating at time zero.

2. S can fail at any time, even before the first redundant component C_1 is put into operation.

3. If S fails at time y, it is impossible to switch on any component after y.

4. C_j is required to operate when $C_0, C_1, C_2, \ldots, C_{j-1}$ fail and the switch S is in operating condition.

5. Lifetimes of components $X_0, X_1, \ldots, X_m$ follow an exponential distribution with failure rate λ.

6. Lifetime Y of the switch S follows an exponential distribution with failure rate μ.

7. Failures of the components and the switch are independent.

The system is in operating condition at time t when

- C_0 operates at t with or without failure of S, or

- a redundant component C_j operates at t following the failures of $C_0, C_1, \ldots, C_{j-1}$ (before t) and S does not fail until the failure of C_{j-1}.

The probability that C_0 operates at t is

$$P_0 = \Pr(X_0 > t) = e^{-\lambda t}. \tag{1.22}$$

Now we derive the probability P_j of the event that C_j operates at t. This event occurs when: (1) C_0 fails at time $x < t$; (2) C_{j-1} fails at time $(x + y) < t$, that is, $\sum_{k=1}^{j-1} X_k = y$; (3) S does not fail until $(x + y)$; and (4) C_j does not fail within the time $t - (x + y)$. Since $\sum_{k=1}^{j-1} X_k$ follows a gamma distribution and all lifetimes are mutually independent, we have

$\Pr(C_j$ operates at $t|$ C_0 has failed at $x < t)$

$$= \int_0^{t-x} [e^{-\lambda(t-x-y)}e^{-\mu(x+y)}]\frac{\lambda^j}{(j-1)!}y^{j-1}e^{-\lambda y}\,dy$$

$$= (\lambda/\mu)^j e^{-\lambda(t-x)-\mu x}\left[1 - \sum_{k=0}^{j-1}\frac{\mu^k(t-x)^k}{k!}e^{-\mu(t-x)}\right].$$

Therefore

$$P_j = \int_0^t \Pr(C_j \text{ operates at } t|\ C_0 \text{ has failed at } x)\lambda e^{-\lambda x}\,dx$$

$$= \int_0^t (\lambda/\mu)^j e^{-\lambda(t-x)-\mu x}\left[1 - \sum_{k=0}^{j-1}\frac{\mu^k(t-x)^k}{k!}e^{-\mu(t-x)}\right]\lambda e^{-\lambda x}\,dx.$$

By simplifying the last term, we get

$$P_j = (\lambda/\mu)^{j+1}e^{-\lambda t}\left[1 - \sum_{k=0}^{j}\frac{\mu^k t^k}{k!}e^{-\mu t}\right]. \tag{1.23}$$

Since the system reliability at time t is $R_s(t) = P_0 + P_1 + \cdots + P_m$, from eqs. (1.22) and (1.23), we now have

$$R_s(t) = e^{-\lambda t}\left\langle 1 + \sum_{j=1}^{m}(\lambda/\mu)^{j+1}\left[1 - \sum_{k=0}^{j}\frac{\mu^k t^k}{k!}e^{-\mu t}\right]\right\rangle.$$

Redundant system with m hot standby redundant components

Suppose all m redundant components are hot standby, that is, each one of them has a failure rate λ even before being put into operation. Assume that the system fails when a switch fails during the operation of a redundant component. The system operates when one of the following two events occur:

- Event E_1: C_0 does not fail until t, or

- Event E_2: C_0 fails before t, and the switch and at least one of the m redundant components survive until t.

We have

$$\Pr(E_1) = e^{-\lambda t},$$

$$\Pr(E_2) = \Pr(X_0 \le t, Y > t, \max\{X_1, \ldots, X_m\} > t)$$

$$= \Pr(X_0 \le t)\Pr(Y > t)\Pr(\max\{X_1, \ldots, X_m\} > t)$$

$$= (1 - e^{-\lambda t})e^{-\mu t}[1 - \Pr(X_1 \le t, X_2 \le t, \ldots, X_m \le t)]$$

$$= (1 - e^{-\lambda t})e^{-\mu t}[1 - (1 - e^{-\lambda t})^m].$$

Therefore, the system reliability at time t is

$$R_s(t) = e^{-\lambda t} + (1 - e^{-\lambda t})e^{-\mu t}[1 - (1 - e^{-\lambda t})^m]. \tag{1.24}$$

When $\mu = 0$, that is, the switch is perfect, (1.24) reduces to

$$R_s(t) = 1 - (1 - e^{-\lambda t})^{m+1}.$$

The derivation of $R_s(t)$ is difficult when the switch is required only for the purpose of switching on redundant components at component failure times.

1.5.11 Multi-cause failure model

A redundant system consists of m components which are subjected to some common causes of failure. Common-cause failure is a way to handle a dependent situation that occurs during critical system operation. Examples include the operation of nuclear reactors and software systems. Let $M = \{1, 2, \ldots, m\}$ denote the set of components. The system is subjected to n causes of failure $1, 2, \ldots, n$, which occur independently. The rate of occurrence of cause j is λ_j for $j = 1, \ldots, n$; that is, the time at which cause j occurs follows an exponential distribution with mean $1/\lambda_j$. Without loss of generality, we assume that $n \leq 2^m - 1$. Let C_j denote the set of components which fail due to cause j for $j = 1, \ldots, n$. A component may fail due to one or several causes. Let $a_{ij} = 1$ if component i fails due to cause j, and 0 otherwise. Let $A = [(a_{ij})]_{mn}$. Matrix A is called the *incidence matrix*.

A subset F of causes is a *cover* if occurrence of all causes in F leads to failure of all components of the system, that is, F is a cover if $\cup_{j \in F} C_j = M$. A subset F is *minimal cover* if no proper subset of F is a cover.

Let $\{F_1, F_2, \ldots, F_v\}$ be the collection of all minimal covers. Define event E_i for $i = 1, \ldots, v$ such that E_i occurs if, and only if, all causes in F_i occur. Let $V = \{1, 2, \ldots, v\}$ and for any $G = \{i_1, \ldots, i_r\} \subseteq V$ let $\Pr(\cup_{i \in G} E_i)$ denote the probability that the events $E_{i_1}, E_{i_2}, \ldots, E_{i_r}$ occur. Then the probability of system failure can be written as

$$F_s = [\Pr(E_1) + \cdots + \Pr(E_v)] - \sum_{i=1}^{v} \sum_{\substack{j=1 \\ j>i}}^{v} \Pr(E_i E_j) + \cdots + (-1)^{r+1} \sum_{\substack{G \subseteq V \\ |G|=r}} \Pr(\cap_{i \in G} E_i)$$

$$+ \cdots + (-1)^{v+1} \Pr(E_1 \cap E_2 \cap \cdots \cap E_v).$$

In the following examples, we use the notation: $a = 1 - e^{-\lambda t}$ and $b = 1 - e^{-\mu t}$.

Example 1-1

Let the number of components $m = 2$ and the number of causes $n = 3$ and let the incidence matrix be

		Causes		
		1	2	3
Components	1	1	0	1
	2	0	1	1

The set $F = \{1, 3\}$ is a cover since the occurrence of causes 1 and 3 leads to the failure of both the components. Similarly, set $\{3\}$ is also a cover. In fact, it is a minimal cover. Note that the set $F' = \{1\}$ is a not cover.

Here, the minimal covers are $F_1 = \{1, 2\}$ and $F_2 = \{3\}$. Let the rate of occurrence of causes 1 and 2 be λ and that of cause 3 be μ. In this case, occurrence of cause 3 leads to failure of both components 1 and 2. The probability of system failure by time t is

$$F(t) = \Pr(E_1 \cup E_2)$$
$$= \Pr(E_1) + \Pr(E_2) - \Pr(E_1 E_2)$$
$$= a^2 + b - a^2 b.$$

Thus, the system reliability by time t is $R_s(t) = 1 - a^2 - b + a^2 b$.

Example 1-2
Let $M = \{1, 2, 3\}$, $N = \{1, 2, 3, 4\}$ and the incidence matrix A be

		Causes			
		1	2	3	4
	1	1	0	0	1
Components	2	0	1	0	1
	3	0	0	1	0

Let the rate of occurrence of causes 1, 2, and 3 be λ, and that of cause 4 be μ.

The minimal covers are $F_1 = \{1, 2, 3\}$ and $F_2 = \{3, 4\}$. In this case, the probability of system failure by time t is given by $F(t) = a^3 + ab - a^3 b$ and the system reliability by time t is $R_s(t) = 1 - a^3 - ab + a^3 b$.

EXERCISES

1.1 What are the measures used to indicate the performance of equipment? Define and describe three performance measures.

1.2 Which process is highly related to the reliability of products? How can we reduce the human error?

1.3 Show that $R_s = \prod_{i=1}^{k}(1 - q_i^{n_i})$ is approximated by $R_s \simeq 1 - \sum_{i=1}^{k} q_i^{n_i}$ when q_i is small.

1.4   Consider the series-parallel and parallel-series systems shown in Figures 1.5 and 1.6, respectively. Let t_{ij} denote the failure time of component j in the ith subsystem. Show that the failure time of a series-parallel system is

$$T_{SP} = \max_{1 \le i \le k} \min_{1 \le j \le n_i} t_{ij}$$

and that of a parallel-series system is

$$T_{PS} = \min_{1 \le i \le k} \max_{1 \le j \le n_i} t_{ij}.$$

1.5   Let $\epsilon > 0$ be a small number. For the parallel-series system in Figure 1.6 with $1 - r_{ij} = v_i$ for all i and j, find a value $v_{\max}$ in terms of ϵ and k such that

$$R_s - \epsilon \le 1 - \sum_{i=1}^{k} v_i^{n_i} \le R_s + \epsilon, \qquad \text{for} \;\; v_i \in (0, v_{\max}), \;\; 1 \le i \le k.$$

1.6   Consider the hierarchical series-parallel system shown in Figure 1.8. Derive the system reliability by method 2 of Section 1.5.7 pivoting component 3.

1.7   Show that the reliability of the parallel-series system in Figure 1.9(b) is no less than that of the complex system in Figure 1.9(a).

1.8   Consider the pair of four-component systems shown in Figure 1.11. If all components have equal reliability, show that the parallel-series configuration gives higher reliability than the series-parallel configuration.

1.9   Suppose the reliability of all the components in the figure below are identical. Find the reliability of the system.

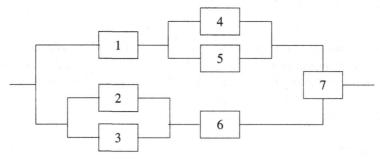

1.10 Calculate the reliability for a 3-out-of-5: F system. Is there any relationship between the 3-out-of-5: F system and the 3-out-of-5: G system? Assume that the reliability of each component is identical.

1.11 Let

$$a_i = \begin{cases} 1, & \text{if component } i \text{ is in operating state,} \\ 0, & \text{otherwise.} \end{cases}$$

Find the minimal path sets and minimal cut sets of a 3-out-of-4: G system.

1.12 Let $\mathbf{a} = (a_1, a_2, a_3, a_4, a_5)$.

(a) Find the structure function of a 3-out-of-5: G system using the minimal path sets.

(b) Find the structure function of a 3-out-of-5: G system using the minimal cut sets.

1.13 Using the network of Problem 1.9, find the structure function.

1.14 Derive the coherent structure function $\phi(\mathbf{a})$ for the complex system shown in Figure 1.10.

1.15 Consider an n-component system with a coherent structure function $\phi(\mathbf{a})$ and component reliabilities $r_1, \ldots, r_n$. Show that the system reliability is

$$R_s = \sum_{a_1, \ldots, a_n} \phi(a_1, \ldots, a_n) \prod_{j=1}^{n} r_j^{a_j} (1 - r_j)^{1-a_j}.$$

1.16 Consider a 2-out-of-5 system with component reliabilities $r_1, \ldots, r_5$. Derive the coherent structure function for this system and evaluate system reliability using the equation in Exercise 1.15.

1.17 A night guard carries three flashlights for 10 h everyday. He turns on a flashlight for 30 min a day. If a flashlight is out of order, he uses another one. Find the probability, R_s, that he will not use more than three flashlights in a given year. Assume that the redundant ones are identical, each having an exponential life distribution with mean $\mu = 50$ h.

1.18 A system is composed of one switch and one original and two standby components. If a component in operation fails, the switch makes a standby component work. However, the switch is not perfect. Assume that all components are identical, each having an exponential life distribution with mean $\mu = 5$ months, and the switch also has an exponential life distribution with mean 2 years. Find the probability, R_s, that this system will work for at least one year.

1.19 In a cold standby system with m components in standby where the failure rate of a component is λ, show that the mean time to failure is

$$\text{MTTF} = \frac{m + 1}{\lambda},$$

and find the variance of the MTTF.

1.20 A new component is to be designed. A stress analysis reveals that the component will be subjected to a tensile stress. Since there are variations in the load, the tensile stress is found to be normally distributed with a mean of 3500 kg/m² and a standard deviation of 400 kg/m². The manufacturing operations create a residual compressive stress that is normally distributed with a mean of 1000 kg/m² and a standard deviation of 150 kg/m². A strength analysis of the component shows that the mean value of the significant strength is 5000 kg/m². The variations introduced by various strength factors are not clear at the present time. The engineer wants to know the maximum value of the standard deviation for the strength that will insure that the component reliability will not drop below 0.999.

1.21 A two-state stochastic system follows an exponential distribution with failure rate λ and repair rate μ. Derive its instantaneous and steady-state availability.

1.22 Show that a system's failure rate can uniquely determine its reliability function.

1.23 List four hypotheses about program errors in developing software reliability modeling.

1.24 Let p_i be the probability of success of the ith component, $p_i \to 1.0$, q_i be $1 - p_i$, ϕ_i be q_i/p_i, and s denote "system".

(a) If there are n units in series, show that $\phi_s \approx \sum_{i=1}^{n} \phi_i$.

(b) If there are n units in parallel, show that $\phi_s \approx \prod_{i=1}^{n} \phi_i$.

(c) Given

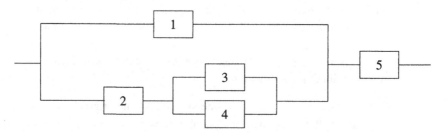

and $p_1 = 0.86$, $p_2 = 0.93$, $p_3 = 0.92$, $p_4 = 0.95$, and $p_5 = 0.98$. Find the system reliability using (a) and (b).

2 Analysis and classification of reliability optimization models

2.1 Introduction and notation

Notation

b_i	total amount of ith resource available
$f(\mathbf{x})$	system reliability for redundancy allocation $\mathbf{x}$
	(for fixed component reliabilities)
$h(R_1, \ldots, R_n)$	system reliability when stage reliabilities are $R_1, \ldots, R_n$
j	index for stage
ℓ_j	lower limit on x_j
m	number of resources
n	number of stages in the system
$R_j(x_j)$	$1 - (1 - r_j)^{x_j}$, the reliability of stage j
R_s	system reliability
r_j	reliability of a component at stage j with x_j redundant components
r_j^1	lower limit on r_j
r_j^u	upper limit on r_j
u_j	upper limit on x_j
$\mathbf{x}$	$(x_1, \ldots, x_n)$
x_j	number of redundant components at stage j

The reliability of a system can be improved in several ways, as explained in Chapter 1. However, improvement usually requires certain procedures/resources. To maximize system reliability, we consider the following options in Chapter 2:

1. enhancement of component reliability,

2. provision of redundant components in parallel,

3. a combination of options 1 and 2, and

4. reassignment of interchangeable components.

From Chapter 3 onwards, several system-reliability optimization models based on these options, as well as some applications of the models themselves are presented together with the corresponding solution methodologies. In Chapter 2 these options are assigned to optimization models, I–VIII. However, the reader is advised that: design for reliability allocation in a system, option 1, which was well-developed in the 1960s and 1970s, is best documented in Tillman et al. [306]; to understand enhancement of component reliability for modern semiconductor products see Kuo and Kim [176] and Kuo and Kuo [177]; many important approaches have been developed to solve problems in options 2–4 (option 4, in particular, requiring no resource can be very attractive).

Let us now begin by considering an n-stage reliability system. Each stage consists of exactly one component when there is no component redundancy at that stage. The total consumption of resource i is denoted by a function $g_i(\cdot)$. When the option is to select component reliabilities, it is denoted by $g_i(r_1, \ldots, r_n)$. Similarly, when the option is to select redundancy levels $x_1, \ldots, x_n$, the consumption of resource i is denoted by $g_i(x_1, \ldots, x_n)$. For the option of selecting both redundancy levels and reliabilities of components, it is represented by $g_i(x_1, \ldots, x_n; r_1, \ldots, r_n)$. If all of the constraints pertain to resource consumption, then all of them are separable in terms of consumption at individual stages. Let $g_{ij}(\cdot)$ denote the consumption of resource i at stage j. For the options 1–3 mentioned previously, we have

$$g_{ij}(\cdot) = g_{ij}(r_j),$$
$$g_{ij}(\cdot) = g_{ij}(x_j),$$
$$g_{ij}(\cdot) = g_{ij}(x_j, r_j),$$

respectively. The function $g_{ij}(\cdot)$ may represent cost, weight, volume, etc. Examples of these functions can be found in the following chapters. In all reliability optimization problems, we denote system reliability by $f(\cdot)$. We have

$$f(r_1, \ldots, r_n) = h(r_1, \ldots, r_n)$$

and

$$f(x_1, \ldots, x_n) = h[R_1(x_1), \ldots, R_n(x_n)]$$

for options 1 and 2, respectively. For the third option, we denote system reliability by $f(x_1, \ldots, x_n; r_1, \ldots, r_n)$. If lower limit ℓ_j on x_j is not specified, it can be taken as zero. Similarly, u_j can be taken as infinity when it is not specified.

2.2 Optimization models

The diversity of system structures, resource constraints, and options for reliability improvement lead to the construction and analysis of several optimization models. The vast majority of the reliability optimization models discussed in the literature can be presented in the following general framework.

Model I: Allocation of continuous component reliabilities

System reliability can be improved through the selection of component reliabilities at stages subject to resource constraints. The problem of maximizing system reliability through the selection of component reliabilities subject to resource constraints is of the form: maximize

$$R_s = f(r_1, \dots, r_n),$$

subject to

$$g_i(r_1, \dots, r_n) \le b_i, \qquad \text{for } i = 1, \dots, m,$$
$$r_j^\ell \le r_j \le r_j^u, \qquad \text{for } j = 1, \dots, n.$$

This problem is called the *reliability allocation problem*. Many researchers adopt the separability assumption that

$$g_i(r_1, \dots, r_n) = \sum_{j=1}^n g_{ij}(r_j).$$

This problem is a typical nonlinear programming problem.

Model II: Allocation of discrete and continuous component reliabilities

Suppose there are u_j discrete choices for component reliability at stage j for $j = 1, \dots, k$ $(\le n)$ and the choice for component reliability at stages $k+1, \dots, n$ is on a continuous scale. Let $r_j(1), r_j(2), \dots, r_j(u_j)$ denote the component reliability choices at stage j for $j = 1, \dots, k$. Then, the problem of selecting optimal component reliabilities that maximize system reliability can be written so as to maximize

$$R_s = h[r_1(x_1), \dots, r_k(x_k), r_{k+1}, \dots, r_n],$$

subject to

$$g_i[r_1(x_1), \dots, r_k(x_k), r_{k+1}, \dots, r_n] \le b_i, \qquad \text{for } i = 1, \dots, m,$$
$$x_j \in \{1, 2, \dots, u_j\}, \qquad \text{for } j = 1, \dots, k,$$
$$r_j^\ell \le r_j \le r_j^u, \qquad \text{for } j = k+1, \dots, n.$$

This problem can be viewed as a nonlinear mixed integer programming problem.

Model III: Redundancy allocation

System reliability can be improved through the selection of redundancy levels at stages subject to resource constraints. The problem of finding optimal redundancy levels $x_1, \dots, x_n$ for maximizing system reliability subject to resource constraints can be described so as to maximize

$$R_s = f(x_1, \dots, x_n),$$

subject to

$$g_i(x_1, \ldots, x_n) \le b_i, \qquad \text{for } i = 1, \ldots, m,$$

$$\ell_j \le x_j \le u_j, \qquad \text{for } j = 1, \ldots, n,$$

x_j being an integer.

The above problem, which is called the *redundancy allocation problem*, is a nonlinear integer programming problem. The vast majority of papers on redundancy allocation make a separability assumption for this case

$$g_i(x_1, \ldots, x_n) = \sum_{j=1}^{n} g_{ij}(x_j),$$

which holds most often in real life. This model is thoroughly discussed in the literature. A large number of solution methods including heuristics, exact algorithms, metaheuristic algorithms, approximation methods, etc., are described in Chapters 3–7.

Model IV: Reliability–redundancy allocation

System reliability can be improved through the selection of component reliabilities as well as redundancy levels at stages subject to resource constraints. The problem of finding simultaneously optimal redundancy levels $x_1, \ldots, x_n$ and optimal component reliabilities $r_1, \ldots, r_n$ that maximize system reliability subject to resource constraints can be expressed so as to maximize

$$R_s = f(x_1, \ldots, x_n; r_1, \ldots, r_n),$$

subject to

$$g_i(x_1, \ldots, x_n; r_1, \ldots, r_n) \le b_i, \qquad \text{for } i = 1, \ldots, m,$$

$$\ell_j \le x_j \le u_j, \qquad \text{for } j = 1, \ldots, n,$$

$$r_j^\ell \le r_j \le r_j^u, \qquad \text{for } j = 1, \ldots, n,$$

x_j being an integer.

This problem is called the *reliability–redundancy allocation problem*. Here also, it is common to assume that the constraints are separable, that is

$$g_i(x_1, \ldots, x_n; r_1, \ldots, r_n) = \sum_{j=1}^{n} g_{ij}(x_j, r_j), \qquad \text{for } i = 1, \ldots, m.$$

The above problem and its solution methodology are discussed in Chapter 8. As shown in Model II, it is possible to deal with discrete choices of component reliabilities in the above problem.

Model V: Allocation of discrete component reliabilities and redundancies

In Models III and IV, redundant components at stage j form a 1-out-of-x_j: G system, since each stage requires the operation of at least one component at any time. If a stage j requires the operation of at least ℓ_j components, then the parallel redundancy at stage j forms an ℓ_j-out-of-x_j: G system. Suppose component reliability is to be chosen from a set of discrete choices and no redundancy is allowed at stages $1, \dots, k$ ($\leq n$). Further, suppose for $j = k + 1, \dots, n$, redundancy (with fixed component reliability) is to be provided at stage j, which requires the simultaneous operation of ℓ_j (≥ 1) components at a time.

Let $\{r_j(1), r_j(2), \dots, r_j(u_j)\}$ be the set of choices for component reliability at stage j for $j = 1, \dots, k$. If stage j ($>k$) contains x_j components of reliability r_j in parallel, then the stage reliability is

$$R_j(x_j) = \sum_{w=\ell_j}^{x_j} \binom{x_j}{w} r_j^w (1 - r_j)^{x_j - w}.$$

The system-reliability optimization problem can be written so as to maximize

$$R_s = h[r_1(x_1), \dots, r_k(x_k), R_{k+1}(x_{k+1}), \dots, R_n(x_n)],$$

subject to

$$g_i[r_1(x_1), \dots, r_k(x_k), x_{k+1}, \dots, x_n] \leq b_i, \qquad \text{for } i = 1, \dots, m,$$

$$x_j \in \{1, 2, \dots, u_j\}, \qquad \text{for } j = 1, \dots, k,$$

$$x_j \in \{\ell_j, \ell_j + 1, \dots, u_j\}, \qquad \text{for } j = k + 1, \dots, n.$$

This problem is a generalization of Example 3-3. Note that all of the variables of this problem are integers. Thus, the problem can be viewed as a nonlinear integer programming problem. The solution methodologies described in Chapter 3 are useful for solving this class of problem.

Model VI: Redundancy allocation for cost minimization

Cost minimization is the objective of some redundancy allocation problems in reliability systems. Such problems can be stated so as to minimize

$$C_s = \sum_{j=1}^{n} c_j(x_j),$$

subject to

$$g_i(x_1, \dots, x_n) \geq b_i, \qquad \text{for } i = 1, 2, \dots, m,$$

$$\ell_j \leq x_j \leq u_j, \qquad \text{for } j = 1, 2, \dots, n,$$

where C_s is the total component cost of the system and $c_j(x_j)$ the cost of x_j components at stage j. A lower limit on system reliability usually appears as one of the constraints.

Cost can be interpreted as any parameter in either an analytical or an empirical function. For example, cost is most often treated as a function of the failure rate of a component or a system in the software industry (see Chapter 13 for illustrations). If our objective is to provide a design that will result in the lowest value of system cost, then we should appropriate our targeted system failure rate for each component such that C_s is minimized.

Model VII: Component assignment

Consider an n-stage reliability system in which the components at various stages are interchangeable. When the components have different reliabilities, the system reliability can be maximized by reassigning the components among the stages without consuming any resources. Suppose the components in stages $1, \ldots, n$ are interchangeable and numbered as $1, \ldots, n$, respectively. Let r_j denote the reliability of component j for $j = 1, \ldots, n$. Let $\pi = (\pi_1, \ldots, \pi_n)$ be a sequence of the components $1, \ldots, n$ and let $f(\pi_1, \ldots, \pi_n)$ represent the system reliability when component π_j is assigned to stage j for $j = 1, \ldots, n$. We have

$$f(\pi_1, \ldots, \pi_n) = h(r_{\pi_1}, \ldots, r_{\pi_n}).$$

Now, the problem is to maximize

$$R_s = f(\pi_1, \ldots, \pi_n),$$

subject to

$$(\pi_1, \ldots, \pi_n) \in P,$$

where P is the set of all $n!$ sequences of $1, \ldots, n$. If a sequence $(\pi_1, \ldots, \pi_n)$ is not feasible, that is a component π_j cannot be assigned to position j, then $f(\pi_1, \ldots, \pi_n) = 0$. The problem is also relevant when the system requires identical components at some stages, and the ages of existing components are not the same. It may be noted that, except for the exponential life distribution, the reliability of a component depends on the component's age. From a mathematical point of view, this problem is a nonlinear assignment problem.

Optimal component assignment is analytically derived for some special structures such as series-parallel and consecutive 2-out-of-n: F systems. Heuristics and exact methods that are based on the comparison of criticality of stages are available for general structures. The concept of majorization and the nature of Schur convex functions are used in the development of some solution procedures. The theory and solution methodologies for this problem class are provided in Chapter 9.

Model VIII: Multiple objective optimization

A design engineer is often required to consider, in addition to maximization of system reliability, other objectives such as minimization of cost, volume, weight, etc. It may not be easy to define limits on each objective in order to deal with them in the form of constraints. In such situations, the designer faces the problem of optimizing all objectives simultaneously. This is typically seen in aircraft design. Suppose the designer is considering only the option of providing redundancy for optimization purposes. Mathematically, this problem can be expressed so as to maximize vector

$$\mathbf{z} = [f_1(x_1, \ldots, x_n), f_2(x_1, \ldots, x_n), \ldots, f_s(x_1, \ldots, x_n)],$$

subject to

$$g_i(x_1, \ldots, x_n) \leq b_i, \qquad \text{for } i = 1, \ldots, m,$$
$$\ell_j \leq x_j \leq u_j, \qquad \text{for } j = 1, \ldots, n,$$

x_j being an integer and where $f_1(\mathbf{x}), f_2(\mathbf{x}), \ldots, f_s(\mathbf{x})$ are s objective functions to be optimized. A general approach for solving this multiple objective optimization problem is to find a set of nondominated feasible solutions and make interactive decisions based on the set. Multiple objective optimization problems in reliability systems are discussed in Chapter 10.

The objective in all of the above models, except Models VI and VIII, is to maximize the system reliability.

2.3   Problem reduction

Integer variables are involved in a vast majority of reliability optimization problems. The computational effort of the techniques developed for solving these problems usually depends on the number of variables, the number of constraints, and the type of the constraint functions. When the techniques involve partial or implicit enumeration, their computational effort depends also on the limits on the variables. It is possible to reduce the search process by tightening the limits on the integer variables when the system has coherent structure and constraint functions $g_i(\cdot)$s are nondecreasing. The upper limits can be reduced when $g_i(\cdot)$s are nondecreasing in each variable, whereas the lower limits can be increased for series systems using a feasible solution. We now describe a procedure for tightening the limits on the integer variables.

Consider Model III with a system-reliability function (for a series system)

$$f(x_1, \ldots, x_n) = \prod_{j=1}^{n} [1 - (1 - r_j)^{x_j}],$$

which is nondecreasing in each x_j. This problem involves only integer variables. Assume that $g_i(x_1, \ldots, x_n)$ is nondecreasing in each x_j. Let $(x_1^*, \ldots, x_n^*)$ denote an optimal solution and $R_s^* = f(x_1^*, \ldots, x_n^*)$.

New lower limits based on the objective function for a series system

Suppose a known feasible solution $\mathbf{x}^0$ gives system reliability R_s^0. Then R_s^0 ($\leq R_s^*$) is a lower bound on the optimal reliability R_s^*. If any feasible solution $\mathbf{x} = (x_1, \dots, x_n)$ is optimal, then the reliability R_j of each stage j must be at least R_s^*, that is,

$$R_j(x_j) = 1 - (1 - r_j)^{x_j} \geq R_s^* \geq R_s^0,$$

which implies

$$x_j \geq \left\lceil \frac{\ln(1 - R_s^0)}{\ln(1 - r_j)} \right\rceil,$$

where $\lceil a \rceil$ is the smallest integer greater than or equal to a. Let

$$\underline{x}_j = \max\left[\left\lceil \frac{\ln(1 - R_s^0)}{\ln(1 - r_j)} \right\rceil, \ell_j \right], \qquad \text{for } j = 1, \dots, n.$$

Now, a necessary condition for $\mathbf{x}$ to be optimal is

$$x_j \geq \underline{x}_j, \qquad \text{for } j = 1, \dots, n. \tag{2.1}$$

In eq. (2.1), $\underline{x}_j$ can be taken as a lower limit on x_j in order to reduce the set S of solutions to be explored. Note that the sharpness of a new lower limit depends on the magnitude of R_s^0. We can obtain a good feasible solution using any simple heuristic. To derive a good feasible solution, we can start with a trivial feasible solution, such as $(1, \dots, 1)$, and follow an iterative procedure in which each iteration causes an increment of one in some variable until a constraint is violated.

New upper limits based on constraints and lower limits

Suppose that $g_i(x_1, \dots, x_n)$ is nondecreasing in x_j for all i and j. In order to reduce the solution set S, the upper limits on the variables can be reduced to

$$\bar{x}_j = \max\{v : v \leq u_j, g_i(\ell_1, \dots, \ell_{j-1}, v, \ell_{j+1}, \dots, \ell_n) \leq b_i,$$
$$\text{for } i = 1, \dots, m, \text{ and } v \text{ is an integer}\}, \tag{2.2}$$

for $j = 1, \dots, n$. If

$$g_i(x_1, \dots, x_n) = \sum_{j=1}^{n} g_{ij}(x_j)$$

and $g_{ij}(x_j)$ is nondecreasing in x_j for all i and j, then

$$\bar{x}_j = \max\{v : v \leq u_j, g_{ij}(v) - g_{ij}(\ell_j) \leq s_i, \text{ for } i = 1, \dots, m, \text{ and } v \text{ is an integer}\},$$

where

$$s_i = b_i - \sum_{j=1}^{n} g_{ij}(\ell_j), \qquad \text{for } i = 1, \dots, m.$$

The upper limit u_j can be replaced by $\bar{x}_j$ in order to reduce the search over the set S. The effectiveness of this new limit depends on n, b_i, and the values of $g_{ij}(x_j)$. Reduction of the solution set S in this fashion may not be effective for large n. Hochbaum [127] presents lower and upper bounds for the allocation problem and other nonlinear optimization problems.

2.4   Classification of system-reliability optimization

Although the qualitative concepts of reliability are not new, its quantitative aspects began attracting attention in the 1960s. This interest resulted from the increasing need for highly reliable systems and components which were both safer and cheaper. As described in Chapter 1, there are several ways to improve system reliability. However, design-reliability experts have focused a great deal of their efforts on allocation of reliability and redundancy of components for maximizing system reliability. This approach is essential when there is no possibility for replacement or repair of failed components during system operation. Development of efficient solution methods for maximizing system reliability by allocation still remains a challenging task in design for reliability.

An overview of system-reliability optimization is presented in this chapter. The overview is based on Chapter 2 of Tillman et al. [306] and a recent review by Kuo and Prasad [180]. It includes genetic algorithms, the simulated annealing method, heuristics, and exact methods for optimal redundancy allocation; heuristics for optimal reliability–redundancy allocation; multi-criteria optimization in reliability systems; and optimal system assembly (the reassignment of interchangeable components or subsystems).

Note that, according to Chern [49], with multiple constraints it is quite often hard to find feasible solutions for redundancy allocation problems. For a good literature survey of the early work refer to Tillman et al. [305], [306], and Misra [227]. In their review, Tillman et al. [305] present a classification of papers on reliability optimization by system structure, problem type, and solution method. In Kuo and Prasad [180], the contributions that have been made to the literature since the publication of Tillman et al. [305] are discussed.

All of the articles on optimization methods for system reliability can be classified using three criteria:

1. system configuration,

2. problem type, and

3. optimization technique employed.

Classification by system configuration
System configuration quite often influences the approach used to solve a reliability optimization problem. For this reason, all of the articles on reliability optimization

Table 2.1. Reference classification by system configuration

System configuration	References
Series	[7], [20], [37], [38], [40], [41], [91], [107], [114], [123], [193], [240], [248], [249], [254], [285], [286], [301], [309], [320], [321]
Parallel	[7], [8], [114], [123], [144], [161], [201], [240], [248], [285], [301], [309], [320], [321]
Series-parallel	[123], [144], [148], [222], [240], [252], [285], [286], [320], [321]
Standby	[37], [123], [132], [146], [193], [234], [239], [240], [248], [274], [285], [320], [321]
Parallel-series	[5], [20], [24], [30], [37], [38], [43], [49]–[52], [60]–[62], [80], [86], [87], [89]–[91], [96], [101], [107], [112]–[115], [123], [136], [142]–[144], [146], [148], [150], [152], [164], [171], [173], [175], [178], [184], [201], [204], [213]–[215], [221], [223]–[225], [228]–[232], [234], [240], [245]–[248], [252], [258], [261], [262], [267], [271]–[274], [277], [282], [285], [286], [290], [295], [297], [301], [302], [304], [309], [311], [320], [321], [326], [328], [330]
General network including complex systems	[6]–[8], [21], [41], [43], [53], [54], [70], [73], [74], [126], [133], [153]–[155], [163], [175], [194], [197], [235], [248], [255], [275], [286], [292], [296], [303], [309], [316], [317], [333]
k-out-of-n: G(F) system	[19], [79], [82], [83], [87], [140], [141], [206], [322], [337]
Unspecified system	[28], [34], [55], [68], [263], [270]

are grouped by system configuration in Table 2.1. The reliability optimization methods that have emerged for several system structures include:

- series systems, parallel systems, series-parallel systems, and standby systems, which are defined in Chapter 1;

- parallel-series systems in which a number of stages exist in series and where redundant components can be added to enhance the reliability of the stages and hence the reliability of the systems (redundancy can be considered at both the stage level and the component level);

- general network systems, which include bridge networks, non-series-non-parallel structures, and other complex system configurations;

- k-out-of-n: G(F) systems, which include both consecutive and nonconsecutive systems; and

- unspecified systems where the structure of the systems are not explicit or the modules of the systems are not necessarily physically connected.

Table 2.2. Reference classification by problem type

Type of problem	References
Model I: separable constraints for series systems	[33], [168], [171], [202], [222], [254]
Model III: optimum redundancy allocation, maximization of systems reliability subject to cost constraints	[7], [20], [30], [37], [42], [43], [89], [91], [96], [114], [123], [132], [143], [146], [148], [152], [164], [171], [204], [213]–[215], [221]–[225], [229], [234], [239], [240], [242], [248], [249], [285], [286], [290], [295], [301], [302], [309], [311], [320], [321], [324], [332]
Models I and III: separable constraints and system-reliability maximization for a coherent system	
Reliability allocation	[43], [133], [171], [303]
Redundancy allocation	[5], [8], [248], [309]
Model VI: "Cost" minimization problems subject to the minimum requirement of system reliability	[20], [38], [132], [133], [146], [171], [221], [240], [248], [250], [277], [288], [301], [311]
Model VII	[24], [40], [79], [82], [83], [86], [87], [140], [141], [197], [198], [206], [266], [267], [272], [273], [322], [337]
Model VIII	[80], [101], [103], [126], [145], [194], [195], [231], [232], [246], [282], [283]
Others	
Maximization of the system profit	[90], [277]
Maximization of the ratio of system reliability to the power demand of the system	[201]

Each of the above five system structures can be further classified using the fundamental analysis of Chapter 1 as a general framework. Each system considered in Table 2.1 has one of seven configurations: series, parallel, series-parallel, standby, parallel-series, k-out-of-n: G(F), and unspecified system models.

Classification by problem type

Models I–VIII, described earlier, represent the vast majority of the mathematical formulations of reliability optimization problems. The methods developed for solving these problems generally exploit the mathematical structure of the problem. Classification of the work by problem type is shown in Table 2.2.

Classification by optimization techniques employed

The major focus of recent work is on the development of heuristic methods and metaheuristic algorithms for redundancy allocation problems. Little work has been directed toward exact solution methodologies for such problems. To the best of our knowledge, all of the reliability systems considered in this area belong to the class of coherent systems. In the 1960–1980s, many well-developed mathematical programming techniques were applied to solving some of the problems in Table 2.2. Recent developments are based on the following methods:

1. heuristics for redundancy allocation, special techniques developed for reliability problems;

2. metaheuristic algorithms for redundancy allocation, perhaps the most attractive development in the last ten years;

3. exact algorithms for redundancy allocation or reliability allocation (most are based on mathematical programming techniques, e.g. the reduced gradient methods presented in Hwang et al. [136]);

4. heuristics for reliability–redundancy allocation, a difficult but realistic situation in reliability optimization;

5. multiple objective optimization in system reliability, an important but not widely studied problem in reliability optimization;

6. optimal assignment of interchangeable components in reliability systems, a unique scheme that often takes no effort; and

7. others including decomposition, fuzzy apportionment, and effort function minimization.

Most of the system-reliability optimization problems listed are nonlinear integer programming problems. They are more difficult to solve than general nonlinear programming problems because their solutions must be integers. Several optimization methods are available in the literature, Table 2.3, for solving such problems: each of the techniques listed has had some success in solving particular reliability optimization problems. However, it is almost impossible to select one method to solve all reliability optimization problems.

Dynamic programming has dimensionality difficulties which increase with increasing number of state variables, and it is hard to solve problems with more than three constraints. Although integer programming methods yield integer solutions, transforming nonlinear objective functions and constraints into linear forms so that integer programming methods can be applied is a difficult task. In addition, the various integer programming techniques do not guarantee that optimal solutions can be obtained in a reasonable time. Branch-and-bound and other implicit enumeration methods require much computational effort to determine an exact optimal solution.

Table 2.3. Reference classification by optimization technique

Optimization technique	References
Dynamic programming	[30], [43], [96], [148], [152], [168], [184], [201], [215], [274]
Exact algorithms for redundancy allocation or reliability allocation	[19], [24], [50]–[52], [115], [136], [149], [220], [228], [230], [235], [246], [270], [292], [301]
Generalized reduced gradient method	[171]
Heuristics for redundancy allocation	[5], [6], [47], [71], [112], [113], [150], [153], [163], [171], [175], [178], [245], [247], [255], [261], [267], [290], [292], [296]
Heuristics for reliability–redundancy allocation	[43], [55], [80], [113], [126], [178], [304], [326]
Integer programming	[43], [104], [107], [132], [143], [164], [186], [204], [214], [215], [234], [291], [301], [311]
Linear programming	[164], [288]
The maximum principle	[90], [222], [277], [294], [302], [324]
Metaheuristic algorithms for redundancy allocation	[44], [58], [60]–[63], [71], [73], [98], [99], [102], [144], [205], [258], [275], [300], [318], [328]–[330]
The method of Lagrange multipliers and the Kuhn–Tucker conditions	[37], [38], [89], [171], [215], [222], [229], [250], [294]
Modified sequential simplex pattern search	[8], [225]
Multiple objective reliability optimization	[80], [101], [103], [126], [145], [194], [195], [231], [232], [246], [282], [283]
Optimal assignment of interchangeable components	[24], [40], [79], [82], [83], [86], [87], [140], [141], [197], [198], [206], [266], [267], [272], [273], [280], [281], [322], [337]
Parametric approach	[8], [20], [21], [107]
Pseudo-Boolean programming	[146]
Sequential unconstrained minimization technique	[53], [133], [295], [302], [309]
Other methods	[33], [37], [38], [68], [80], [91], [114], [123], [142], [161], [171], [195], [223], [224], [239], [240], [248], [254], [260], [271], [285], [286], [320], [321]

Discrete reliability optimization problems are sometimes solved by continuous versions and rounding off the optimal values. Although many algorithms have been proposed for nonlinear programming problems, only a few, such as the sequential unconstrained minimization technique (SUMT), the generalized reduced gradient method (GRG), the modified sequential simplex pattern search, and the generalized Lagrangian function method, have proved to be effective when applied to large-scale nonlinear programming problems. Of the algorithms available, the maximum principle has difficulty solving problems with more than three constraints. Likewise geometric programming is restricted to problems that can be formulated by polynomial functions.

Genetic algorithms and simulated annealing methods can be used to solve complex discrete optimization problems. These methods provide more flexibility and require fewer assumptions on the objective as well as the constraint functions. They can be effective particularly when the objective function is not available in a closed form and the underlying mathematical model is very complex. Moreover, they are relatively easier to implement on computers. However, they exhibit three drawbacks: (1) they involve a lot of computational effort; (2) they provide heuristic solutions; and (3) one needs more ingenuity to develop them. The tabu search method, Glover and Laguna [110], is demonstrated to be quite useful in solving discrete optimization problems. Nevertheless, development of an effective tabu search method requires great ingenuity and thorough understanding of the problem.

2.5   Recent developments in reliability optimization

Tillman et al. [306] reviewed the system-reliability optimization literature pre 1980, while Misra [227] presented a survey of the literature on system-reliability design pre 1986. Several interesting papers and, more recently, books on reliability optimization have been published thereafter.

The major focus of recent work based on Kuo and Prasad [180], is on the development of heuristic methods and genetic algorithms for redundancy allocation problems. Researchers in general have directed very little attention toward exact solution methodologies for such problems. To the best of our knowledge, all of the reliability systems considered previously belong to the class of coherent systems. Recent developments in system-reliability optimization can be classified into seven categories:

1. heuristics for redundancy allocation;

2. metaheuristic algorithms for redundancy allocation, including genetic algorithms, simulated annealing, and tabu search;

3. exact methods for redundancy allocation;

4. heuristics for reliability–redundancy allocation;

5. multiple objective optimization in reliability systems;

6. optimal assignment of interchangeable components in reliability systems; and

7. effort function optimization.

2.5.1   Heuristics for redundancy allocation

Almost all of the heuristics developed for redundancy allocation, Model III (Section 2.2), before 1980 have the common feature that, in any iteration, a solution is obtained from the solution of a previous iteration by increasing one of the variables by 1. Selection of the variable for the increment is based on a sensitivity factor. Nakagawa and Miyazaki [246] numerically compared the heuristic methods of Nakagawa and Nakashima [247], Kuo et al. [175], Gopal et al. [112], and Sharma and Venkateswaran [290] for a redundancy allocation problem with nonlinear constraints. They carried out extensive numerical investigation and reported the computational time and relative errors of solutions for these various methods.

Interestingly, the heuristics presented after 1980 are based on distinct approaches. Dinghua Shi [296] developed a heuristic method for Model III with separable, monotonic nondecreasing constraint functions following the approach of adjusting unit-increment with time. Dinghua's method, as it is known, requires determination of all minimal path sets of the reliability system. In every iteration of this method, a two-step stage is selected as a feasible redundancy increment. Based on different sensitivity factors, a minimal path set is selected in step one and a chosen path set in step two.

For Model III, Kohda and Inoue [163] developed a heuristic method in which the solutions of two successive iterations may differ on one or two variables. This method is applicable even when the constraints do not involve all the nondecreasing functions. If $\mathbf{x}$ is a solution obtained in some iteration and $\mathbf{x}'$ is the solution obtained from $\mathbf{x}$ in the following iteration, then one of the following cases will hold:

1. A variable is increased by 1.

2. Two variables are simultaneously increased by 1.

3. One variable is increased by 1 and some other variable is reduced by 1.

Cases 1 and 3 use a single-stage sensitivity factor $S(i)$, while case 2 is based on a two-stage sensitivity factor $S(i, j)$.

Kim and Yum [153] developed a heuristic algorithm for Model III with separable, monotonic nondecreasing constraint functions. The algorithm makes excursions to a bounded subset of infeasible solutions while improving a feasible solution. They assumed that the system is coherent and the constraint function $g_{ij}(x_j)$ increases with x_j for each j. The algorithm starts with a feasible solution and improves it as much as possible by giving increments to the variables. Later it goes to another feasible solution, passing through a sequence of solutions in a predetermined infeasible region B, where in each move a single variable increases or decreases by 1. The resulting feasible solution is improved as much as possible through increments. The cycle is repeated until it reaches an infeasible solution outside B.

Kuo et al. [178] presented a heuristic method for Model III based on a branch-and-bound strategy and Lagrange multiplier method, where:

- the initial node is associated with a relaxed version of Model III;
- each successive node is associated with a nonlinear programming problem, that is, a relaxed version of Model III with some variables fixed at integer values.

The bound associated with any node is the optimal value of the corresponding optimization problem, and the nonlinear programming problem associated with each node is solved by the Lagrangian multipliers method.

For Model III, Jianping [150] recently developed the *bounded heuristic method* for optimal redundancy allocation. It is assumed that the constraint functions are increasing in each variable. A feasible solution $(x_1, \ldots, x_n)$ is called a *bound point* if no feasible increment can be given to any x_j. In each iteration, the method moves from one bound point to another through an increase of 1 in a selected variable and changes in some other variables. It has some similarity with the method of Kohda and Inoue [163], in the sense that an addition and a subtraction are simultaneously done at two stages in some iterations.

2.5.2   Metaheuristic algorithms for redundancy allocation

An abundance of optimization methods have been used to solve various reliability optimization problems as described in Tillman et al. [306]. Methods including the maximum principle, once applied to reliability optimization problems, are not recommended because they are not only difficult to understand but often produce bad solutions. Most of these methods were generated from diverse research ideas, and reflect the state-of-the-art in the 1990s.

In recent years, metaheuristics have been selected and successfully applied to handle a number of reliability optimization problems. In this section, however, emphasis is placed on solving the redundancy allocation problem. These metaheuristics, based more on artificial reasoning than classical mathematics-based optimization, include genetic algorithms, simulated annealing, and tabu search. Genetic algorithms seek to imitate the biological phenomenon of evolutionary production through the parent–children relationship. Simulated annealing is based on a physical process in metallurgy. Tabu search derives from and exploits a collection of principles involved in intelligent problem solving.

Genetic algorithms

A genetic algorithm is a probabilistic search method for solving optimization problems. Holland [129] made pioneering contributions to the development of genetic algorithms, and there has been significant progress in the application of these methods during the 1980s and 1990s. The development of a genetic algorithm can be viewed as an adaptation of a probabilistic approach based on principles of natural evolution.

The genetic algorithm approach can be effectively adopted for complex combinatorial problems. However, it only provides a heuristic solution. This approach was adopted in the 1990s by several researchers, see Table 2.3, to solve reliability optimization problems (see below). For a detailed description of applications of genetic algorithms to combinatorial problems, including reliability optimization problems, one may refer to Gen and Cheng [99].

Painton and Campbell [258] adopted a genetic algorithm approach to solve a reliability optimization problem concerning the design of a personal computer (PC). The functional block diagram of the PC has a series-parallel configuration. There are three choices for each component: the first choice is the existing option, and the other two are reliability increments with additional costs. The component failure rate for each choice is random, following a known triangular distribution. Due to the randomness in the input, the mean time between failures (MTBF) of the system is also random. The problem considered by Painton and Campbell [258] is to maximize the fifth percentile of the statistical distribution of MTBF over the choices of components subject to a budget constraint. The problem has both combinatorial and stochastic elements: the combinatorial element is the choice of components, whereas the stochastic element is the randomness of input (component failure rates) and output (MTBF). In the genetic algorithm design, Painton and Campbell have taken the indices representing the component choices as the genes in the chromosome representation of the solution. The fitness of an infeasible solution is zero, and the fitness of feasible solutions linearly increases as the difference between the fifth percentile of MTBF and its predetermined goal decreases. Since the percentile is difficult to determine analytically, it is estimated through Monte Carlo simulation using 200 samples. Through several genetic algorithm runs, Painton and Campbell identified three best solutions, which were very similar.

Tillman [301] considered the problem of finding optimal redundancy allocation in a series system in which the components of each subsystem were subject to two classes of failure modes. He adopted an implicit enumeration method for solving the problem. Ida et al. [144] and Yokota et al. [328] designed a genetic algorithm, Problem 2.1, for optimal redundancy allocation in a series system in which the components of each subsystem were also subject to two classes of failure modes.

Problem 2.1

A series system consists of n subsystems. Subsystem j has $x_j + 1$ components in parallel, which are subject to two classes of failure modes: O and A. The subsystem j fails if a class-O failure mode occurs in at least one of the $x_j + 1$ components. In contrast, subsystem j fails when a class-A failure mode occurs in all $x_j + 1$ components. In general, subsystem j is subject to s_j failure modes $1, 2, \ldots, s_j$, among which the first h_j modes $1, 2, \ldots, h_j$ belong to class O and the others belong to class A. Let q_{ju} be the probability that a component in subsystem j fails resulting in failure mode u for $1 \le u \le s_j$ and $1 \le j \le n$.

Let $Q_j^O(x_j)$ be the probability that subsystem j, consisting of $x_j + 1$ redundant components, fails due to the occurrence of one of the h_j class-O failure modes.

Similarly, let $Q_j^A(x_j)$ be the probability that the subsystem j fails due to occurrence of one of the $s_j - h_j$ class-A failure modes. Then $Q_j(x_j) = Q_j^O(x_j) + Q_j^A(x_j)$ is the failure probability of subsystem j, and the system reliability to be maximized is

$$R_s = \prod_{j=1}^{n} [1 - Q_j(x_j)],$$

subject to

$$\sum_{j=1}^{n} g_{ij}(x_j) \le b_i, \qquad \text{for } i = 1, \dots, m,$$

$$\ell_j \le x_j \le u_j, \qquad \text{for } j = 1, \dots, n.$$

We have

$$Q_j^O(x_j) \approx \sum_{u=1}^{h_j} [1 - (1 - q_{ju})^{x_j+1}]$$

and

$$Q_j^A(x_j) \approx \sum_{u=h_j+1}^{s_j} (q_{ju})^{x_j+1},$$

x_j being a nonnegative integer.

Gen and Cheng [99], Gen et al. [101], [102], Ida et al. [144], and Yokota et al. [328] designed a genetic algorithm for Problem 2.1, namely, maximize

$$f(x_1, \dots, x_n) = \prod_{j=1}^{n} \left\langle 1 - \sum_{u=1}^{h_j} [1 - (1 - q_{ju})^{x_j+1}] - \sum_{u=h_j+1}^{s_j} (q_{ju})^{x_j+1} \right\rangle,$$

subject to

$$\sum_{j=1}^{n} g_{ij}(x_j) \le b_i, \qquad \text{for } i = 1, 2, \dots, m,$$

$$\ell_j \le x_j \le u_j, \qquad \text{for } j = 1, 2, \dots, n,$$

x_j being a nonnegative integer, where q_{ju} is the known failure probability.

In the genetic algorithm design, using the lower and upper bounds, the decision variables are converted into binary strings which are used in the chromosome representation of the solutions. A large penalty is included in the fitness of infeasible solutions.

Problem 2.2

Coit and Smith [58], [60], and [61] developed genetic algorithms for a series-parallel system in which each subsystem is a k-out-of-n: G system and the components to be used in the subsystem are dissimilar. Their models are described as follows.

A reliability system consists of n subsystems $P_1, P_2, \ldots, P_n$ in a series, where subsystem P_i requires operation of at least ℓ_i components in order to function. There are m_i types of components available to be used in P_i and the reliability of jth type components in subsystem P_i is r_{ij}. Let x_{ij} denote the number of jth type components to be used in subsystem P_i. Then vector $\mathbf{x}_i = (x_{i1}, x_{i2}, \ldots, x_{im_i})$ represents an allocation of components of m_i types to P_i. Let $c_i(\mathbf{x}_i)$ and $w_i(\mathbf{x}_i)$ denote, respectively, the total cost and total weight of P_i for the allocation $\mathbf{x}_i$. The problem considered by Coit and Smith [58], [60] is to maximize

$$R_s = \prod_{i=1}^{n} R_i(\mathbf{x}_i | \ell_i), \tag{2.3}$$

subject to

$$\sum_{i=1}^{n} c_i(\mathbf{x}_i) \le C,$$

$$\sum_{i=1}^{n} w_i(\mathbf{x}_i) \le W,$$

$$\ell_i \le \sum_{j=1}^{m_i} x_{ij} \le u_i, \qquad \text{for } i = 1, 2, \ldots, n,$$

x_{ij} being a nonnegative integer; where $R_i(\mathbf{x}_i | \ell_i)$ is the reliability of subsystem P_i; ℓ_i and u_i are the lower and upper limits on the total number of components in P_i; and C and W are the limits on the total cost and total weight of the system, respectively.

Coit and Smith [60] also considered the problem of minimizing the total cost, subject to a minimum requirement of system reliability and other constraints such as weight. Their objective function involves a quadratic penalty function for solution infeasibility, with the penalty depending on the extent of infeasibility. Later, Coit and Smith [58] introduced a robust adoptive penalty function to penalize the infeasible solutions. This function is based on a *near-feasibility threshold* (NFT) for all constraints. The NFT-based penalty encourages the genetic algorithm to explore the feasible region and the infeasible region closest to the boundary of the feasible region. In the penalty function, they also used a dynamic NFT which depends on the generation number. On the basis of extensive numerical investigations, they reported that genetic algorithms with dynamic NFT in the penalty function are superior to genetic algorithms with several penalty strategies, including genetic algorithms that consider only feasible solutions. Based on numerical experimentation, they have also reported that the genetic algorithm method can give better results than the surrogate constraint method of Nakagawa and Miyazaki [245].

Coit and Smith [61] have also designed a genetic algorithm for an optimal redundancy allocation problem, which is the same as Problem 2.2 except that the cost is minimized subject to a minimum requirement of system reliability. The total cost involves a quadratic penalty for solution infeasibility. The system reliability is evaluated using a neural network approach. Assuming that component reliabilities are random following known probability distributions, Coit and Smith [62] developed a genetic algorithm for Problem 2.2, where the objective is replaced by maximization of a percentile of the statistical distribution of system reliability. The genetic algorithm can also be used for maximization of a percentile of the distribution of system failure time. The objective function is evaluated using a Newton–Raphson search for both system reliability and system failure time.

Majety and Rajagopal [205] developed an evolution strategy based on an adoptive penalty function to solve some reliability optimization problems. They demonstrated their method for parallel-series and series-parallel systems. A striking feature of the method is that a penalty is imposed not only on infeasible solutions, but also on feasible solutions which are far from the optimal solution. As the method continues, solutions closer and closer to the optimum are penalized.

Genetic algorithms have also been developed for cost-optimal network designs. Suppose a communication network has nodes $1, 2, \dots, n$ and a set of h_{ij} links are available to directly connect a pair of nodes i and j for $i = 1, \dots, n$, and $j = 1, \dots, n$ and $i \neq j$. The links have different reliabilities and costs associated with them, and only one of the h_{ij} links is used when nodes i and j are directly connected. The network is in good condition as long as all nodes remain connected, that is, the operating links form a graph that contains a spanning tree. Let x_{ij} denote the index of the link used to connect the pair (i, j). If the pair (i, j) is not directly connected, then $x_{ij} = 0$. The problem under consideration is to minimize $Z = C(\mathbf{x})$, subject to $R(\mathbf{x}) \geq R_0$, where $C(\mathbf{x})$ and $R(\mathbf{x})$ are total cost and network reliability, respectively, for network configuration $\mathbf{x}$, and R_0 is the minimum required network reliability.

Dengiz et al. [74] have designed a genetic algorithm for cost-optimal network design when $h_{ij} = 1$ and $C(\mathbf{x}) = \sum \sum c_{ij} x_{ij}$. In this case, only one link is available to connect any particular pair of nodes. The evaluation of exact network reliability requires a large amount of computational effort. To avoid extensive computation, each network generated by the algorithm is first screened using a connectivity check for a spanning tree and a 2-connectivity measure. If the network passes the screening, then an upper bound on network reliability is computed and used in the calculation of the objective function (fitness of solution). For network designs for which the upper bound is at least R_0 and total cost is the lowest, Monte Carlo simulations are used to estimate the reliability. The penalty for not meeting the minimum reliability requirement is proportional to $[R(\mathbf{x}) - R_0]^2$. Deeter and Smith [71] developed a genetic algorithm for cost-optimal network design without any assumption on h_{ij}. The penalty considered by them involves the difference $R_0 - R(\mathbf{x})$, the population size, and the generation number.

Simulated annealing method

The simulated annealing algorithm is a general method used to solve combinatorial optimization problems. It involves probabilistic transitions among the solutions to the problem. Unlike iterative improvement algorithms, which improve the objective value continuously, a simulated annealing may encounter some adverse changes in objective value in the course of its progress. Such changes are intended to lead to a global optimal solution instead of a local one.

Annealing is a physical process in which a solid is heated up to a high temperature and then allowed to cool slowly and gradually. In this process, all the particles arrange themselves gradually in a low energy ground state level. The ultimate energy level depends on the temperature and the rate of cooling. The annealing process can be described by the following stochastic model. At each temperature T, the solid undergoes a large number of random transitions among different energy level states until it attains a thermal equilibrium in which the probability of the solid appearing in a state with energy level E is given by the *Boltzmann distribution*. As the temperature T decreases, equilibrium probabilities associated with higher energy level states decrease. When the temperature approaches zero, only the states with the lowest energy levels will have nonzero probability. If cooling is not sufficiently slow, thermal equilibrium will not be attained at any temperature and consequently the solid will finally have a metastable condition.

To simulate the random transitions among the states and the attainment of thermal equilibrium at a fixed temperature T, Metropolis et al. [216] developed a method in which a transition from one state to another occurs due to a random perturbation in the state. If the perturbation results in a reduction of energy level, transition to the new state is accepted. If, instead, the perturbation increases the energy level by ΔE (>0), then transition to the new state is accepted with a given probability governed by the Boltzmann distribution. This method is called the *Metropolis algorithm*. The criterion for acceptance of the transition is called the *Metropolis criterion*.

Based on the simulation of the annealing process, Kirkpatrick et al. [158] developed a simulated annealing algorithm for solving combinatorial optimization problems. Although simulated annealing gives satisfactory solutions for combinatorial optimization problems, its major disadvantage is the amount of computational effort involved. In order to improve the rate of convergence and reduce the computational time, Cardoso et al. [44] introduced the nonequilibrium simulated annealing algorithm (NESA) by modifying the algorithms of Metropolis et al. [216]. In NESA, there is no need to reach an equilibrium condition through a large number of transitions at any fixed temperature. The temperature is reduced as soon as an improved solution is obtained. Ravi et al. [275] have recently improved NESA by incorporating a simplex-like heuristic in the method. They applied this variant of NESA, denoted I-NESA, to reliability optimization problems. It consists of two phases: phase I uses a NESA and collects solutions obtained at regular NESA progress intervals; phase II starts with the set of solutions obtained in phase I and uses a heuristic procedure to improve the best solution further.

Tabu search method

Tabu search is another metaheuristic that guides a heuristic method to expand its search beyond local optimality. It is an artificial intelligence technique which utilizes memory (information about the solutions visited) at every stage to provide an efficient search for optimality. It is based on ideas proposed by Fred Glover. An excellent description of tabu search methodology can be found in Glover and Laguna [110].

Tabu search for any complex optimization problem combines the merits of artificial intelligence with those of optimization procedures. Tabu search allows the heuristic to cross boundaries of feasibility or local optimality which are major impediments in any local search procedure. The most prominent feature of tabu search is the design and use of memory-based strategies for exploration in the neighborhood of a solution at every stage. Tabu search ensures responsive exploration by imposing restrictions on the search at every stage based on memory structures. It is useful for solving large complex optimization problems that are very difficult to solve by exact methods. To solve redundancy allocation problems, we recommend that tabu search be used alone, or in conjunction with the heuristics presented in Section 2.1, to improve the quality of the heuristics.

2.5.3   Exact methods for redundancy allocation

The purpose of exact methods is to obtain an exact optimal solution to a problem. It is generally difficult to develop exact methods for reliability optimization problems, which are equivalent to methods for nonlinear integer programming problems. Such methods involve more computational effort and usually require larger computer memory. For these reasons, researchers in reliability optimization have placed more emphasis on heuristic approaches. However, development of good exact methods always poses a challenge. Such methods are particularly advantageous when the problem is not large. Moreover, the exact solutions provided by such methods can be used to measure performance of heuristic methods. Many exact methods were developed before 1980 and documented in Tillman et al. [305].

Nakagawa and Miyazaki [245] adopted the surrogate constraints method to solve Model III when there are exactly two constraints (pertaining to resources), and the objective as well as the constraint functions are separable.

Problem 2.3

Here one can apply dynamic programming (DP) by either using the Lagrange multipliers or defining the state space with respect to both of the constraints. Of course, there is no guarantee that DP with Lagrangian multipliers will yield an exact optimal solution. With the surrogate constraints method, Nakagawa and Miyazaki [246] solve the surrogate problem: maximize

$$z(w) = \sum_{j=1}^{n} f_j(x_j),$$

subject to

$$\sum_{j=1}^{n}[(1 - w)g_{1j}(x_j) + wg_{2j}(x_j)] \leq (1 - w)b_1 + wb_2,$$

$$\ell_j \leq x_j \leq u_j, \qquad \text{for } j = 1, \ldots, n,$$

x_i being a nonnegative integer, for several values of w in the interval $[0, 1]$. If the optimal solution of Problem 2.3 satisfies the separable constraints of Model III with $m = 2$ for some $w, 0 \leq w \leq 1$, then it is optimal for the original problem. If such w does not exist, a stopping rule [245] enables us to terminate the method after solving a few surrogate problems. Nakagawa and Miyazaki [245] have reported that the performance of their method is superior to DP with Lagrangian multipliers for the problem under consideration. They have also indicated that it is possible, although remotely, for this method to fail to yield an exact optimal solution.

Misra [223] has proposed an exact algorithm for optimal redundancy allocation, Model III, based on a search near the boundary of the feasible region. This method was later implemented by Misra and Sharma [230], Sharma et al. [292], and Misra and Misra [220] for solving various redundancy allocation problems. This algorithm does not always give an exact optimal solution. Prasad and Kuo [270] recently developed a partial enumeration method based on a lexicographic search with an upper bound on system reliability. The method is demonstrated for both small and large problems. For large systems with a good modular structure, Li and Haimes [195] proposed a three-level decomposition method for reliability optimization subject to resource constraints. At level 1, a nonlinear programming problem is solved for each module. At level 2, the problem is transformed into a multiple objective optimization problem, which is solved by the ϵ-constraint method of Chankong and Haimes [46]. This approach involves optimization at three levels. At level 3 (the highest level), the lower limits, ϵ_i on multiple objective functions, are chosen, while Kuhn–Tucker multipliers are chosen at level 2 for fixed ϵ. For fixed Kuhn–Tucker multipliers and fixed ϵ, a nonlinear programming problem is solved for each module of the system at level 1. Later Li and Haimes [195] also proposed a parametric DP method for optimizing a nonseparable objective function subject to resource constraints. This can be adopted for large-scale reliability optimizations with good modular structure.

Mohan and Shanker [235] adopted a random search technique for finding a global optimal solution to the problem of maximizing system reliability through the selection of only component reliabilities subject to cost constraints. Bai et al. [19] considered a k-out-of-n: G system with common-cause failures. The components are subjected not only to intrinsic failures but also to a common failure cause following independent exponential distributions. If there is no inspection, the system is restored upon failure to its initial condition through necessary component replacements. If there is inspection, failed components are replaced during the inspection. For both of the cases, that is, with and without inspection, Bai et al. [19] derived, using renewal theory, an optimal n that minimizes the mean cost-rate. They also demonstrated their procedure with numerical examples.

2.5.4   Heuristics for reliability–redundancy allocation

The reliability of a system can be enhanced by either providing redundancy at the component level or increasing component reliabilities or both. Redundancy and component reliability enhancement, however, lead to increase in system cost. Thus, a tradeoff between these two options is necessary for budget-constrained reliability optimization. The problem of maximizing system reliability through redundancy and component reliability choices is referred to as a reliability–redundancy allocation problem. Mathematically, it can be described by Model IV. Model IV is a nonlinear mixed integer programming problem, which is more difficult than a pure redundancy allocation problem. Some interesting heuristic methods were developed for this model during the 1980s and 1990s.

Tillman et al. [304] were among the first to solve the problem using a heuristic and search technique. Gopal et al. [113] developed a heuristic method that starts with 0.5 as the component reliability at each stage of the system, and increases component reliability at one of the stages by a specified value h in every iteration. The selection of a stage for improving a component's reliability is based on a *stage sensitivity factor*. For any particular choice of component reliabilities $r_1, \ldots, r_n$, an optimal redundancy allocation $x_1, \ldots, x_n$ is derived by a heuristic method. Any heuristic redundancy allocation method can be used for this purpose. When such increments in component reliabilities do not give any higher system reliability, the increment h is reduced and the procedure is repeated with the new increment h. This process is discontinued when h falls below a specified limit h_0.

The branch-and-bound method of Kuo et al. [178] described in Section 8.5 is useful for solving Model IV. The bound associated with a node is the optimal value of the relaxed version of Model IV with some integer variables fixed at integral values. The method requires the assumption that all functions of the problem are differentiable. The relaxed version, which is a nonlinear programming problem, is solved by the Lagrangian multipliers method. Kuo et al. [178] demonstrated the method for a series system with five subsystems.

Later, assuming that the functions f and g_{ij} are differentiable and monotonic nondecreasing functions, Xu et al. [326] offered an iterative heuristic method for Model IV with separable constraints. In each iteration, a solution is derived from the previous solution in one of the following two ways:

1. One x_j is increased by 1 and an optimal $(r_1, \ldots, r_n)$ vector is obtained corresponding to the new $(x_1, \ldots, x_n)$ by solving a nonlinear programming problem.

2. One x_j is increased by 1 and another reduced by 1, and an optimal $(r_1, \ldots, r_n)$ vector is obtained for the new $(x_1, \ldots, x_n)$ by solving a nonlinear programming problem.

Although this method can be applied to Model IV even when the constraints are not separable, it may not be computationally viable in such a case.

Hikita et al. [126] developed a surrogate constraints method to solve Model IV with separable constraints. The method is based on the theory developed by Luenberger [203] for minimizing a quasi-convex function subject to convex constraints. In this method, a series of surrogate optimization problems, each consisting of a single constraint, is solved. In each surrogate problem, the objective is the same as that in Model IV, but the single constraint is obtained by taking a convex linear combination of the m constraints. The surrogate constraint approach to Model IV (with separable constraints) is to find a convex linear combination that gives the least optimal objective value of the surrogate problem and takes the corresponding surrogate optimal solution as the required solution.

Hikita et al. [126] use the dynamic programming approach to solve the single-constraint surrogate problem. For this method, it is required that either the objective function f is separable or that the surrogate problem can be formulated as a multi-stage decision-making problem. The surrogate constraint method is useful for special structures, including parallel-series and series-parallel designs.

Reliability–redundancy allocation problems arise in software reliability optimization also. The redundant components of software may be programs developed by different groups of people for given specifications. The reliability of any software component can be enhanced by additional testing, which requires various resources. Another feature of software systems is that the components are not necessarily completely independent. Chi and Kuo [55] formulated mixed integer nonlinear programming problems for reliability–redundancy allocation in software systems and systems involving both software and hardware.

2.5.5   Multiple objective optimization in reliability systems

In single-objective optimization problems relating to system design, either the system reliability is maximized subject to limits on resource consumption, or the consumption of one resource is minimized subject to a lower limit on system reliability and other resource constraints. While designing a reliability system, it is always desirable to simultaneously maximize system reliability and minimize resource consumption. When the limits on resource consumption are flexible or they cannot be determined properly and precisely, it is better to adopt a multiple objective approach to system design. In this approach, there may not exist a single solution that is optimal with respect to each objective. The approach usually involves determination of all Pareto optimal (nondominated) solutions.

A design engineer is often required to consider, in addition to the maximization of system reliability, other objectives such as minimization of cost, volume, weight, etc. It may not be easy to define limits on each objective in order to deal with them in the form of constraints. In such situations, the designer faces the problem of optimizing all objectives simultaneously. This is typically seen in aircraft design.

Problem 2.4

Suppose a designer is considering the option of providing redundancy for optimization purposes. Mathematically, this can be expressed so as to maximize vector

$$\mathbf{z} = [f_1(x_1, \dots, x_n), f_2(x_1, \dots, x_n), \dots, f_s(x_1, \dots, x_n)],$$

subject to

$$g_i(x_1, \dots, x_n) \le b_i, \qquad \text{for } i = 1, \dots, m,$$

$$\ell_j \le x_j \le u_j, \qquad \text{for } j = 1, \dots, n,$$

x_j being an integer, where $f_1(\mathbf{x})$, $f_2(\mathbf{x})$, $\dots$, $f_s(\mathbf{x})$, $\mathbf{x} = (x_1, x_2, \dots, x_n)$ are s objective functions to be optimized. A general approach for solving this multiple objective optimization problem is to find a set of nondominated feasible solutions and make interactive decisions based on this set.

Sakawa [282] adopts a large-scale multiple objective optimization method to deal with the problem of determining optimal levels of component reliabilities and redundancies. He considers a large-scale series system with four objectives: maximization of system reliability, and minimization of cost, weight, and volume. In this approach, he derives Pareto optimal solutions by optimizing composite objective functions, which are obtained as linear combinations of the four objective functions. The Lagrangian function for each composite problem is decomposed into parts and optimized by applying both the dual decomposition method and the surrogate worth tradeoff method, treating redundancy levels as continuous variables. Later, the resulting redundancy levels are rounded off and the Lagrangian function is optimized with respect to component reliabilities by the dual decomposition method in order to obtain an approximate Pareto solution. Sakawa [283] provides a theoretical framework for the sequential proxy optimization technique (SPOT); which is an interactive, multiple objective decision-making technique for selection among a set of Pareto optimal solutions. He applies SPOT to optimize system reliability, cost, weight, volume, and product of weight and volume for series-parallel systems subject to some constraints.

To solve multiple objective redundancy allocation problems in reliability systems, Misra and Sharma [231] adopt an approach which involves the Misra integer programming algorithm (discussed in Section 3.3.1) and a multi-criteria optimization method based on the min–max concept for obtaining Pareto optimal solutions. Misra and Sharma [232] also presented a similar approach to solve multiple objective reliability–redundancy allocation problems in reliability systems. Their methods take into account two objectives: maximization of system reliability and minimization of total cost subject to resource constraints.

Dhingra [80] adopts another multiple objective approach to maximize system reliability and minimize consumption of resources: cost, weight, and volume. He uses the goal programming formulation and the goal attainment method to generate Pareto optimal solutions. For system designs in which the problem parameters and goals are not formulated precisely, he suggests the multiple objective fuzzy optimization approach. He demonstrates the multiple objective approach for a four-stage series system with constraints on cost, weight, and volume. Similarly, Gen et al. [100] also solve reliability optimization problems using goal programming.

2.5.6   Optimal assignment of interchangeable components in coherent systems

If a system has interchangeable components with different reliabilities, then the system reliability depends on the assignment of such components to required positions. Considerable work has been done on optimal assignment of interchangeable components in coherent systems, particularly consecutive k-out-of-n systems. The component assignment problem is described by some researchers as the problem of optimal assembly of systems. Before summarizing this work, we first give a brief description of the problem.

Consider a coherent system with n interchangeable components $1, 2, \ldots, n$. Let $f(a_1, \ldots, a_n)$ denote the system reliability when the component in position i has reliability a_i, $1 \leq i \leq n$. The reliability of a component may also depend on the position at which it is fixed. Suppose the reliability of component j is p_{ij} when it is assigned to position i. If component v_i is assigned to position i for $i = 1, \ldots, n$, then the system reliability is $f(p_{1v_1}, \ldots, p_{nv_n})$. Now, the problem is to find a component assignment $(v_1, \ldots, v_n)$ that maximizes $f(p_{1v_1}, \ldots, p_{nv_n})$. When component reliability p_{ij} is independent of position i, some systems admit optimal assignments which depend only on the increasing (or decreasing) order of component reliabilities but not on their actual magnitudes. Such assignments are called *invariant optimal assignments* in the literature.

El-Neweihi et al. [86] solved the problem analytically for series-parallel structures assuming that component reliabilities were invariant of position, that is, $p_{ij} = r_j$ for all (i, j). They suggested a 0–1 linear programming approach for parallel-series structures. They elegantly used the theory of majorization in their approach to solve the problem for both structures. Later, Prasad et al. [272], assuming that $p_{ij} = g_i r_j$ (a separability condition), developed an algorithm to solve the problem for series-parallel structures. They also provided two greedy algorithms for this problem. If both algorithms yield the same solution, then that solution is optimal. They presented a simple optimal solution for parallel-series structures when $p_{ij} = r_j$ for all (i, j) and each parallel subsystem consists of two components.

Prasad et al. [267] developed a heuristic method to solve the problem for series-parallel structures without any assumption on p_{ij}. This method involves some of the classical assignment problems. Baxter and Harche [24] have presented a heuristic called a *top–down heuristic* for optimal component assignment in parallel-series systems under the assumption that $p_{ij} = r_j$ for all (i, j). They have derived upper bounds on the absolute and relative errors of heuristic optimal system reliability. Assuming that r_j is an independent and identical random variable following uniform distribution over the interval $[0, 1]$, they have shown that the system reliability obtained by their heuristic converges to the optimal value as the number of components n and subsystem sizes tend to infinity. Prasad and Raghavachari [273] developed a heuristic method for parallel-series structures under the assumption $p_{ij} = r_j$. In this paper, they use some important results of El-Neweihi et al. [86], and approximate the problem as a mixed integer linear programming problem. They derive an optimal solution of the relaxed problem; round it off to obtain a feasible solution; and improve the latter iteratively by a switching rule until there is no further improvement. Prasad et al. [266] have used their bicriteria optimization methods for solving component assignment problems in parallel-series and series-parallel systems consisting of two subsystems. These methods give exact optimal assignments without any assumption on p_{ij}.

El-Neweihi et al. [87] considered the problem of allocating m types of components to a general assembly of n series systems. Under certain conditions, they derived an allocation that stochastically maximizes the number of functioning systems. As a consequence, this allocation also maximizes the probability that at least k-out-of-n systems function. They also analyzed a situation where each system (need not be a series) requires one component of each type. Malon [208] presented a greedy rule to assemble modules of a coherent system out of a collection of available components. The greedy rule assembles modules one by one using best available components. If the modules have series structure, the rule gives an optimal assignment of components to the entire system provided that the modules are arranged in the system in an optimal manner. Boland et al. [40] have suggested a procedure using a pairwise interchange of components for obtaining optimal component assignment in coherent systems. They have introduced the notion of comparison of criticality of two positions in the system and used it to improve system reliability through pairwise interchange of components.

Lin and Kuo [197] present a greedy method for optimal component assignment in a general coherent structure when the component reliabilities are invariant of positions. Some interesting work has been done on the optimal allocation of a single component in parallel/standby redundancy to a coherent system in a general framework. Through the relationship to the Fibonacci sequence, with order k, a closed-form solution of structure importance for each component of the consecutive k-out-of-n system is also obtained by Lin et al. [198].

Some work has been done particularly on optimal component assignment in consecutive k-out-of-n: F systems in which the components are arranged in linear or cyclical fashion. For consecutive 2-out-of-n: F linear systems, Derman et al. [79] made a conjecture that when $r_1 \leq r_2 \leq \cdots \leq r_n$, the assignment

$$u^* = (1, n, 3, n - 2, \ldots, n - 3, 4, n - 1, 2)$$

maximizes system reliability. Wei et al. [322] have proved the conjecture for two special cases. Extending this conjecture to a cyclical system, Hwang [139] conjectured that

$$v^* = (n, 1, n - 1, 3, n - 3, \ldots, n - 4, 4, n - 2, 2, n)$$

maximizes the reliability of a consecutive 2-out-of-n: F cyclical system. Du and Hwang [82] proved Hwang's conjecture and argued that the assignment problem in a linear system with n components is equivalent to that of a cyclical system which has $(n + 1)$ components with $r_{n+1} = 1$. Thus, the conjecture of Derman et al. [79] relating to a linear consecutive 2-out-of-n: F system holds true as a special case. Note that the assignments u^* and v^* are invariant optimal for linear and cyclical consecutive 2-out-of-n: F systems, respectively. Malon [206] directly and independently proved the conjecture for consecutive 2-out-of-n: F linear systems. Later, Malon [207] showed that consecutive k-out-of-n: F linear systems admit invariant optimal component assignment if, and only if, $k \in \{1, 2, n - 2, n - 1, n\}$. Through relationship to the Fibonacci sequence, with order k, a closed-form solution of structure importance for each component of the consecutive k-out-of-n systems is also obtained by Lin et al. [198].

Hwang [140] has considered the component assignment problem in two-stage consecutive k-out-of-n: F systems in which each of n subsystems is a consecutive h_i-out-of-m_i: F system. He has derived optimal assignment when: (1) $h_i = h$ for all i, and (2) all n subsystems are arranged in parallel, or each one of them has a series structure. He has also proposed a heuristic method for the general case. Hwang and Dinghua Shi [141] have considered redundant consecutive k-out-of-n: F systems in which each of n stages consists of h components in parallel. They have also identified some systems which do not admit the invariant optimal assignment. Zuo and Kuo [337] have summarized the results available for the invariant optimal design of consecutive k-out-of-n systems. They have also identified invariant optimal designs for some consecutive k-out-of-n systems and proved that invariant optimal designs for other consecutive k-out-of-n systems do not exist. Zhang et al. [335] have applied the invariant optimal design concept to a railway management system. Shen and Zuo [293] have studied the optimal design of a series system that consists of linear consecutive k-out-of-n: G subsystems. Invariant optimal designs are identified when $k < n \leq 2k$ for each subsystem.

2.5.7   Effort function optimization

One of the standard approaches for enhancing system reliability is to increase the reliability of the components. However, an increase in component reliability requires some effort, which may be cost, volume, weight, power consumption, etc., and thus system-reliability enhancement also requires such effort. Assume that the effort to increase the reliability of any component from one level to another is measurable by a mathematical function. Such functions, called *effort functions*, are not necessarily explicit. Reliability engineers usually formulate the effort functions based on knowledge of the development process. The problem under consideration is to minimize the total effort required to increase the reliability of a general coherent system from an existing level to a desired level through incremental increases in component reliabilities. Albert [9] solves this problem for series systems when the effort functions are the same for all components. For a good description of this method, refer to Lloyd and Lipow [202]. Dale and Winterbottom [68] provide a solution approach for a general coherent structure.

Table 2.4. Applications of system-reliability optimization methods

Applications	References
Software development	[15], [28], [34], [55], [263], [284], [327]
Systems with common-cause failures	[19], [55]
Assembly	[24], [87], [142], [208]
Maintenance policy	[63]
Burn-in	[53], [173], [174]
Network	[17], [147], [160], [165], [169], [170], [318]

2.6   Applications

Reliability optimization techniques have been developed as design tools for various system applications. The most commonly documented applications of these tools are summarized in Table 2.4. Because of the increasing complexity and criticality of software development, it is expected that more applications will be seen in software development in the near future. As documented in Kuo et al. [174], both the philosophy and reliability optimization techniques are extensively implemented in semiconductor burn-in. In addition, more applications to network optimization are expected to fulfill the increasing needs for network communications.

2.7   Discussion

We have reviewed the work on reliability optimization published since 1977. A major part of this work is devoted to the development and application of heuristic methods and metaheuristic algorithms to redundancy allocation problems, which can be extended to optimal reliability–redundancy allocation problems. It is interesting to note that these heuristics were developed on the basis of a very distinct perspective. However, the extent to which they are superior to previous methods is still not clear. While developing heuristics, it is relevant to seek answers to two important questions: (1) under what conditions does a heuristic give an optimal solution, and (2) what are the favorable conditions for a heuristic to give a satisfactory solution? The answers to these questions enhance the importance and applicability of a heuristic. We can understand the merit of a newly developed heuristic only when it is compared with existing ones for a large number of numerical problems. For the redundancy allocation problem, Nakagawa and Miyazaki [245] carried out numerical experimentation to compare several heuristics. For the reliability–redundancy allocation problem, Xu et al. [326] made a thorough comparison of a number of algorithms.

Genetic algorithms, treated as probabilistic heuristic methods, are metaheuristic methods which imitate the natural evolutionary process. They are very useful for solving complex discrete optimization problems and do not require sophisticated mathematical treatment. They can be easily designed and implemented on a computer for a wide spectrum of discrete problems. Genetic algorithms have been designed for solving redundancy allocation problems in reliability systems. The chromosome definition and selection of the genetic algorithm parameters provide a lot of flexibility in adopting genetic algorithms for a particular type of problem. However, there is some difficulty in determining appropriate values for the parameters and a penalty for infeasibility. If these values are not selected properly, a genetic algorithm may rapidly converge to a local optimum or slowly converge to the global optimum. A large population size and the number of generations enhance the solution quality while increasing the computational effort. Experiments are usually recommended to obtain appropriate genetic algorithm parameters for solving a specific type of problem. An important advantage of the genetic algorithm method is its presentation of several good solutions (mostly optimal or near optimal). The multiple solutions yielded by the genetic algorithm method provide a great deal of flexibility in decision making for reliability design.

Simulated annealing is a global optimization technique that can be used for solving large size combinatorial optimization problems. It may be noted that unlike many discrete optimization methods, simulated annealing does not exploit any special structure that exists in the objective function or in the constraints. However, simulated annealing is relatively more effective when a problem is highly complex without having any special structure. Redundancy allocation problems in reliability systems are nonlinear integer programming problems of this type. Thus, simulated annealing can be quite useful in solving complex reliability optimization problems. Although several approaches are available in the literature for designing a simulated annealing method, the design still requires ingenuity and sometimes considerable experimentation. A major disadvantage of simulated annealing is that it requires a large amount of computational effort (with a large number of function evaluations and tests for solution feasibility). However, it has great potential for yielding an optimal or near-optimal solution.

Tabu search is very useful for solving large-scale complex optimization problems. The salient feature of this method is the utilization of memory (information about previous solutions) to guide the search beyond local optimality. There is no fixed sequence of operations in tabu search and its implementation is problem-specific. Thus, tabu search can be described as a metaheuristic rather than a method. A simple tabu search method which uses only short-term memory is quite easy to implement. Usually such methods yield good solutions when attributes, tabu tenure, and aspiration criteria are appropriately defined. A simple tabu search can be implemented for solving redundancy allocation and reliability–redundancy allocation problems. One major disadvantage of tabu search is the difficulty involved in defining effective memory structures and memory-based strategies which are problem-dependent. This task really requires good knowledge of the nature of the problem, ingenuity, and some numerical experimentation. A well-designed tabu search can offer excellent solutions in large-scale system-reliability optimization.

To derive an exact optimal redundancy allocation in reliability systems, Misra [228] has presented a search method which has been used in several papers to solve a variety of reliability optimization problems including some multiple objective optimization problems. Unfortunately, this method is invalid, as shown in Section 2.3. Very little progress has been made on multiple objective optimization in reliability systems although such work could provide the system designer with an interactive environment. These problems belong to the class of nonlinear integer multiple objective optimization problems. A fuzzy optimization approach has also been adopted by Park [260] and Dhingra [80] to solve reliability optimization problems in a fuzzy environment.

Optimal assignment of interchangeable components in a reliability system is a nonlinear assignment problem. From a mathematical point of view, the nature of this problem is much different from that of redundancy allocation problems because no resource is needed for an optimal assignment. Heuristic methods are developed for general structures, while exact methods are available for special cases. Some special structures are shown to admit invariant optimal assignments. Also, some structures are shown to have no such assignments. Occasionally, an optimal arrangement can be obtained at the minimum resource (effort) when the resource function does not have to be explicit. Dale and Winterbottom presented such an application in [68].

Notice that many of these optimization techniques can be very powerful if they are utilized together. For example, Lagrange multipliers, branch-and-bound, lexicographic search, and other search techniques should be considered along with other applicable optimization methods. In addition, the problem reduction techniques outlined in Section 2.3 should be applied frequently in order to reduce the search space. To enhance the computational efficiency, heuristics based on the physical meaning of the reliability systems can also be extremely useful information.

EXERCISES

2.1   Describe a few reliability optimization problems which can be formulated as Models III, IV, and VI.

2.2   Discuss the similarities and differences between Models III and VI.

2.3   Show that problems formulated as Models IV and V can also be modeled in more general terms as Model II.

2.4   Formulate Model VII as Model III, and explain the advantages and disadvantages of such formulation.

2.5   Find invariant optimal component assignments for the linear consecutive 2-out-of-n: F system and the circular consecutive 2-out-of-n: F system. Discuss which sequence is optimal, and why?

2.6   What are some exact optimization algorithms? Describe each exact algorithm. What are the drawbacks of the branch-and-bound method and dynamic programming?

2.7 Genetic algorithms, simulated annealing, and tabu search are used to solve complex discrete optimization problems. Compare each method. What are the merits and drawbacks of each?

2.8 When do we use multiple objective optimization methods? Show one real example and describe SPOT and goal programming.

2.9 What are component reliability allocation problems? Show the general formulation of these. Are these nonlinear programming problems? What kinds of approaches are used to solve them?

2.10 Describe the merits of problem reduction. What kinds of approaches are used?

2.11 If nonlinear programming functions are not differentiable, what kinds of approaches will you use to seek the optimal solution?

2.12 What is separable programming? Discuss its merits and show the function types.

2.13 Discuss methods that can be used to solve mixed integer problems.

2.14 Find a coherent system for which the component assignment problem can be solved by a standard assignment technique.

3 Redundancy allocation by heuristic methods

3.1 Introduction

Maximization of the system reliability through redundancy at each stage of operation is known to be a nonlinear integer programming problem involving nonlinear constraints. Such system-reliability optimization through component redundancy was initially considered by Moskowitz and McLean [240] and Mine [219]. Although several procedures are available in the literature to obtain exact optimal solutions to the problem, they require a lot of computational effort and time and are not easy to solve. Chern [49] has shown that even a simple redundancy allocation problem in series systems with linear constraints is NP-hard. For NP-hard problems, which from a computational point of view are classified as highly complex, it is very unlikely that simple and efficient algorithms can be developed to obtain exact optimal solutions: see Garey and Johnson [97] for a good description of this class of problems. Therefore, it is rational to depend on a heuristic method which gives a reasonably good solution, if not the exact one, with less computational effort. In many real-life problems, it is not essential to obtain an exact optimal solution. Any solution that gives a system reliability very close to the optimal value is satisfactory. In fact, an exact optimal solution has less significance when the estimation of component reliability and resource consumption includes errors and approximations. Therefore, in practice, any simple and computationally efficient heuristic method may be useful for solving large-scale reliability optimization problems.

Several heuristic methods have been developed in the last four decades for solving reliability optimization problems. Almost all are methods that improve the solution iteratively, starting with a feasible answer. We describe some of the important methods in this chapter. Section 3.2 contains some notation, definitions, and examples. Section 3.3 gives heuristic methods in which only one of the variables is increased by 1 in each iteration; and the selection of the variable to be increased is based on a sensitivity factor. Section 3.4 presents heuristics in which changes may take place in more than one variable in every iteration.

3.2   Definitions and examples

The general redundancy optimization problem is to maximize

$$R_s = f(x_1, \ldots, x_n), \tag{3.1}$$

subject to

$$\sum_{j=1}^{n} g_{ij}(x_j) \le b_i, \qquad \text{for } i = 1, 2, \ldots, m, \tag{3.2}$$

$$\ell_j \le x_j \le u_j, \qquad \text{for } j = 1, 2, \ldots, n, \tag{3.3}$$

x_j being an integer, where $f(x_1, \ldots, x_n)$ is the system reliability and $g_{ij}(x_j)$ is the consumption of resource i by x_j components at the jth stage. This problem is a special case of Model III described in Chapter 2. The function $f(x_1, \ldots, x_n)$ is nondecreasing in each variable x_j. Since eq. (3.2) pertains to resource consumptions, it is assumed that $g_{ij}(x_j)$ is increasing in x_j. Let

$$\Delta g_{ij}(x_j) = g_{ij}(x_j + 1) - g_{ij}(x_j), \qquad \text{for } i = 1, 2, \ldots, m \text{ and } j = 1, 2, \ldots, n. \tag{3.4}$$

The quantity $\Delta g_{ij}(x_j)$ is the increase in consumption of the ith resource due to an increment of 1 in x_j. Let $\Delta R_s(\mathbf{x}, j)$ be the increase in system reliability due to the increment of 1 in x_j, that is,

$$\Delta R_s(\mathbf{x}, j) = f(x_1, x_2, \ldots, x_j + 1, \ldots, x_n) - f(x_1, x_2, \ldots, x_j, \ldots, x_n).$$

Let $s_i = b_i - \sum_{j=1}^{n} g_{ij}(x_j)$ for $i = 1, 2, \ldots, m$. For a feasible solution $\mathbf{x} = (x_1, \ldots, x_n)$, s_i gives the residual amount of resource i and is nonnegative.

k-neighborhood
Let $\mathbf{x} = (x_1, \ldots, x_n)$ be an arbitrary solution. Then a solution $\mathbf{y} = (y_1, \ldots, y_n)$ is said to be in k-neighborhood of $\mathbf{x}$ if the sum of absolute differences between x_j and y_j does not exceed k, that is,

$$\sum_{j=1}^{n} |y_j - x_j| \le k.$$

The k-neighborhood of $\mathbf{x}$ is

$$N_k(\mathbf{x}) = \left\{ (y_1, \ldots, y_n) : \sum_{j=1}^{n} |y_j - x_j| \le k \right\}.$$

If $y_j = x_j + 1$ and $y_h = x_h$ for all $h \ne j$, then $(y_1, \ldots, y_n)$ is in 1-neighborhood $N_1(\mathbf{x})$ of $\mathbf{x}$.

Let

$$S_k(\mathbf{x}) = \left\{ \mathbf{y} : \sum_{j=1}^{n} |y_j - x_j| \le k, \text{ and } \mathbf{y} \text{ satisfies (3.2) and (3.3)} \right\}.$$

The set $S_k(\mathbf{x})$ is called feasible k-neighborhood of $\mathbf{x}$. Let

$$S_1^+(\mathbf{x}) = \{\mathbf{y}: y_r = x_r + 1, \ y_j = x_j \text{ for all } j \neq r \text{ for some } r,$$

$$\text{and } \mathbf{y} \text{ satisfies (3.2) and (3.3)}\}.$$

Note that $S_1^+(\mathbf{x})$ is a subset of feasible 1-neighborhood $S_1(\mathbf{x})$ of $\mathbf{x}$ and the number of solutions in $S_1^+(\mathbf{x})$ is no more than n.

Saturated stage

A stage j is said to be saturated with respect to a solution $\mathbf{x}$ if

$$x_j = u_j \text{ or } \Delta g_{ij}(x_j) > s_i, \text{ for some } i. \tag{3.5}$$

It means that if the additional requirement $\Delta g_{ij}(x_j)$ of the ith resource, reaches its upper limit u_j (due to the increment of one unit in x_j exceeding the residual amount of the ith resource for some i or x_j) then stage j becomes saturated.

Let $\mathbf{x} = (x_1, x_2, \ldots, x_n)$ and $\mathbf{y} = (y_1, y_2, \ldots, y_n)$ be two feasible solutions and $y_j \geq x_j$ for all j. If a stage j is saturated with respect to $\mathbf{x}$, then it is saturated with respect to $\mathbf{y}$ also.

Example 3-1

Consider the following redundancy allocation problem in a five-stage series system. Maximize

$$R_s = \prod_{j=1}^{5} R_j(x_j),$$

subject to

$$g_1 = \sum_{j=1}^{5} p_j x_j^2 \leq P,$$

$$g_2 = \sum_{j=1}^{5} c_j [x_j + \exp(x_j/4)] \leq C,$$

$$g_3 = \sum_{j=1}^{5} w_j x_j \exp\left(\frac{x_j}{4}\right) \leq W,$$

x_j being a positive integer, where $P, C,$ and W are given resources; and $R_j(x_j) = 1 - (1 - r_j)^{x_j}$ is the reliability of stage j when x_j redundant components, each with reliability r_j, are arranged at stage j. The constants in the above constraints are given in Table 3.1.

This problem was presented originally by Tillman and Littschwager [311], and used by Tillman et al. [302], Sharma and Venkateswaran [290], and others to demonstrate a number of optimization techniques.

Table 3.1. Coefficients used in Example 3-1

j	r_j	p_j	P	c_j	C	w_j	W
1	0.80	1		7		7	
2	0.85	2		7		8	
3	0.90	3	110	5	175	8	200
4	0.65	4		9		6	
5	0.75	2		4		9	

Table 3.2. Coefficients used in Example 3-2

Stage j	1	2	3	4	5
r_j	0.70	0.85	0.75	0.80	0.90
c_j	2	3	2	3	1

Example 3-2

Consider a five-stage complex system as shown in Figure 1.10. The reliabilities and costs of components at the five stages are given in Table 3.2.

The total cost of all components must not exceed 20 units. The problem is to find an optimal redundant allocation that maximizes system reliability subject to the cost constraint. Mathematically, the objective is to minimize the system unreliability. Thus,

$$Q_s = Q_1 Q_3 + Q_2 Q_4 + Q_1 Q_4 Q_5 + Q_2 Q_3 Q_5 - Q_1 Q_2 Q_3 Q_4$$
$$- Q_1 Q_3 Q_4 Q_5 - Q_1 Q_2 Q_3 Q_5 - Q_1 Q_2 Q_4 Q_5 - Q_2 Q_3 Q_4 Q_5$$
$$+ 2 Q_1 Q_2 Q_3 Q_4 Q_5,$$

subject to

$$g = \sum_{j=1}^{5} c_j x_j \leq 20,$$

x_j being a positive integer, where $Q_j = (1 - r_j)^{x_j}$ is the failure probability of stage j with x_j redundant components.

Example 3-3

Consider a four-stage series system. Stage 1 does not allow component redundancy, but its reliability can be enhanced by choosing a more reliable component at that stage. There are six component choices with reliabilities $r_1(1), r_1(2), \ldots, r_1(6)$ at stage 1.

The reliabilities of stages 2 and 4 can only be increased by providing redundancy. However, stage 3 has a 2-out-of n: G configuration. The problem is to maximize

$$R_s = \prod_{j=1}^{4} R_j(x_j),$$

subject to

$$10 \exp\left[\frac{0.02}{1 - R_1(x_1)}\right] + 10x_2 + 6x_3 + 15x_4 \leq 150, \tag{3.6}$$

$$10 \exp\left(\frac{x_1}{2}\right) + 4 \exp(x_2) + 2\left[x_3 + \exp\left(\frac{x_3}{4}\right)\right] + 6x_4^2 \leq 200, \tag{3.7}$$

$$40x_1^2 + 6 \exp(x_2) + 3x_3 \exp\left(\frac{x_3}{4}\right) + 8x_4^3 \leq 750, \tag{3.8}$$

x_j being a positive integer, where stage reliabilities are

$R_1(x_1) = 0.94, 0.95, 0.96, 0.965, 0.97, 0.975,$ for $x_1 = 1, 2, \ldots, 6,$ respectively,

$R_2(x_2) = 1 - (1 - 0.75)^{x_2},$

$$R_3(x_3) = \sum_{k=2}^{x_3} \binom{x_3}{k} (0.90)^k (1 - 0.90)^{x_3-k},$$

$R_4(x_4) = 1 - (1 - 0.95)^{x_4}.$

This example is very similar to the one presented by Nakagawa and Nakashima [247], and is a special case of Model V.

Example 3-4

Consider a multi-function system consisting of n stages. We can use an arbitrary type of redundancy (cold/warm/hot) to increase the reliability of each stage. Let $R_j(x_j)$ denote the probability of failure-free operation of stage j when $(x_j - 1)$ redundant components are used to ensure its operability. Let c_j denote the cost of each component at stage j. The system is required to execute k functions. The execution of function $i, i = 1, 2, \ldots, k$, requires a set J_i of stages, and the probability that function i is executed must be at least $R_{\min}(i)$. The problem considered by Ushakov [314] is to minimize

$$C_s = \sum_{j=1}^{n} c_j x_j,$$

subject to

$$R(i) = \prod_{j \in J_i} R_j(x_j) \geq R_{\min}(i), \qquad i = 1, 2, \ldots, k,$$

x_j being a nonnegative integer, where J_i denotes the minimal set of stages required for the execution of function i for $i = 1, 2, \ldots, k$. Let $n = 3$, and let the reliabilities and costs of the components at the three stages be those shown in Table 3.3.

Let $J_1 = \{1, 2\}$, $J_2 = \{1, 3\}$, $R_{\min}(1) = 0.94$, and $R_{\min}(2) = 0.96$.

Table 3.3. Coefficients used in Example 3-4

j	1	2	3
r_j	0.8	0.75	0.85
c_j	3	2	4

3.3   Heuristic methods with 1-neighborhood solutions

The heuristic methods developed by Sharma and Venkateswaran [290], Aggarwal [5], Aggarwal et al. [7], Gopal et al. [112], Nakagawa and Nakashima [247], Kuo et al. [175], and Dinghua Shi [296] have the following similarities:

1. Starting with a feasible solution, system reliability is improved iteratively.

2. In every iteration, a solution is obtained from the preceding feasible 1-neighborhood solution.

3. A sensitivity factor $F_j(\mathbf{x})$ is calculated for all unsaturated stages in every iteration, where $\mathbf{x} = (x_1, \dots, x_n)$ is the solution obtained in the previous iteration.

4. If a stage v gives a maximum sensitivity factor in an iteration, then the corresponding variable x_v is increased by a value of 1 provided that such an increase does not violate the constraints.

5. The method stops when all the stages become saturated.

The following can serve as a generic algorithm for the above-mentioned class of heuristics. In every iteration of the algorithm, if the subset $S_1^+(\mathbf{x})$ of 1-neighborhood current feasible solutions $\mathbf{x}$ is nonempty, a solution in $S_1^+(\mathbf{x})$ is selected on the basis of sensitivity factor values. The algorithm stops when $S_1^+(\mathbf{x})$ becomes empty.

Generic algorithm
- Step 0: Take a feasible solution $\mathbf{x} = (x_1, \dots, x_n)$.
- Step 1: Determine the residual resource $s_i = b_i - \sum_{j=1}^{n} g_{ij}(x_j)$ for $i = 1, \dots, m$. If $s_i = 0$ for some i, go to step 3. Otherwise, find the subset of unsaturated stages

$$U(\mathbf{x}) = \{j: x_j < u_j \text{ and } \Delta g_{ij}(x_j) \le s_i \text{ for all } i\}.$$

If $U(\mathbf{x})$ is empty, go to step 3. Otherwise, compute the sensitivity factor $F_j(\mathbf{x})$ for each stage in the set $U(\mathbf{x})$ and find the stage v which gives the maximum sensitivity factor over the set $U(\mathbf{x})$.

- Step 2: Let $x_v = x_v + 1$, and go to step 1.
- Step 3: Take $\mathbf{x}$ as the heuristic solution and stop.

For the initial feasible solution, we can either take $(\ell_1, \dots, \ell_n)$ or obtain a better one by iteratively allocating a component to an arbitrarily selected unsaturated stage until all stages become saturated. All the heuristics described in this section start with the solution $(1, \dots, 1)$. The above-mentioned heuristics vary mainly in using different sensitivity factors. Each of the methods is implemented below in a numerical example giving a description of the corresponding sensitivity factor.

3.3.1   The Misra, Sharma, and Venkateswaran method

Sharma and Venkateswaran [290] developed a heuristic method for reliability optimization of series systems: this is the simplest of all heuristics for optimal redundancy allocation. Misra [223] also proposed the same method independently and simultaneously. Therefore, we refer to it as the MSV method. It is applicable only for redundancy optimization in a series system. The reliability of a series system with n stages is

$$f(x_1, \dots, x_n) = \prod_{j=1}^{n} (1 - q_j^{x_j}),$$

for an allocation $\mathbf{x} = (x_1, \dots, x_n)$, where q_j is the unreliability of any component at stage j. The value of $q_j^{x_j}$ gives the unreliability of stage j, and $q_j^{x_j}$ is assumed to be small enough so that the objective function can be approximated by

$$f(x_1, \dots, x_n) \approx 1 - \sum_{j=1}^{n} q_j^{x_j}.$$

In the MSV method, the sensitivity factor for stage j is the unreliability of stage j, that is,

$$F_j(\mathbf{x}) = q_j^{x_j}.$$

Note that in every iteration of the MSV method, a component is added to the stage with maximum stage unreliability. The initial feasible solution suggested by Misra [223] and Sharma and Venkateswaran [290] is $(1, \dots, 1)$.

Solution of Example 3-1 by the MSV method

The approximate objective function in this example is

$$f(x_1, x_2, \dots, x_n) = 1 - (q_1^{x_1} + q_2^{x_2} + q_3^{x_3} + q_4^{x_4} + q_5^{x_5}),$$

where $(q_1, q_2, q_3, q_4, q_5) = (0.20, 0.15, 0.10, 0.35, 0.25)$. The following steps describe the first iteration.

- Step 0: Take $\mathbf{x} = (1, 1, 1, 1, 1)$ as the initial feasible solution.

- Step 1: For the solution $\mathbf{x} = (1, 1, 1, 1, 1)$, the surplus amounts of resources are

$$s_1 = P - \sum_{j=1}^{5} p_j x_j^2$$

$$= 110 - \left[(1.0)(1)^2 + (2.0)(1)^2 + (3.0)(1)^2 + (4.0)(1)^2 + (2.0)(1)^2 \right]$$

$$= 98,$$

$$s_2 = C - \sum_{j=1}^{5} c_j \left[x_j + \exp\left(\frac{x_j}{4}\right) \right]$$

$$= 175 - \left[(7.0)(1 + e^{0.25}) + (7.0)(1 + e^{0.25}) + (5.0)(1 + e^{0.25}) \right.$$
$$\left. + (9.0)(1 + e^{0.25}) + (4.0)(1 + e^{0.25}) \right]$$

$$= 101.91,$$

$$s_3 = W - \sum_{j=1}^{5} w_j x_j \exp\left(\frac{x_j}{4}\right)$$

$$= 200 - \left[(7.0)(1)e^{0.25} + (8.0)(1)e^{0.25} + (8.0)(1)e^{0.25} \right.$$
$$\left. + (6.0)(1)e^{0.25} + (9.0)(1)e^{0.25} \right]$$

$$= 151.21,$$

and the set of unsaturated stages is $U(\mathbf{x}) = \{1, 2, 3, 4, 5\}$. The sensitivity factors for the unsaturated stages for the allocation $\mathbf{x} = (1, 1, 1, 1, 1)$ are

$$F_1(\mathbf{x}) = q_1^{x_1} = (0.20)^1 = 0.20,$$
$$F_2(\mathbf{x}) = q_2^{x_2} = (0.15)^1 = 0.15,$$
$$F_3(\mathbf{x}) = q_3^{x_3} = (0.10)^1 = 0.10,$$
$$F_4(\mathbf{x}) = q_4^{x_4} = (0.35)^1 = 0.35,$$
$$F_5(\mathbf{x}) = q_5^{x_5} = (0.25)^1 = 0.25.$$

- Step 2: Since stage 4 gives a maximum sensitivity factor, x_4 is increased from 1 to 2. The resulting feasible solution is $\mathbf{x} = (1, 1, 1, 2, 1)$.

The procedure is continued until either a constraint is satisfied as an equality or a constraint is violated. The algorithm finally gives the allocation $(3, 2, 2, 3, 3)$, with system reliability 0.9045. For every allocation $\mathbf{x}$ derived in the process, the corresponding resource consumption and sensitivity factors (only for stages where an additional redundancy can be provided without violating any constraint) are given in Table 3.4.

Table 3.4. Results of Example 3-1 by the MSV method

System reliability	Allocation $(x_1, \ldots, x_5)$	Residual resource			Sensitivity factors				
		s_1	s_2	s_3	1	2	3	4	5
0.298 35	$(1, 1, 1, 1, 1)$	98.00	101.91	151.21	0.200	0.150	0.100	0.350[a]	0.250
0.402 77	$(1, 1, 1, 2, 1)$	86.00	89.63	139.13	0.200	0.150	0.100	0.122	0.250[a]
0.503 47	$(1, 1, 1, 2, 2)$	80.00	84.17	121.01	0.200[a]	0.150	0.100	0.122	0.063
0.604 16	$(2, 1, 1, 2, 2)$	77.00	74.62	106.91	0.040	0.150[a]	0.100	0.122	0.063
0.694 78	$(2, 2, 1, 2, 2)$	71.00	65.06	90.80	0.040	0.023	0.100	0.122[a]	0.063
0.757 83	$(2, 2, 1, 3, 2)$	51.00	51.85	72.48	0.040	0.023	0.100[a]	0.043	0.063
0.833 61	$(2, 2, 2, 3, 2)$	42.00	45.03	56.38	0.040	0.023	0.010	0.043	0.063[a]
0.875 29	$(2, 2, 2, 3, 3)$	32.00	39.15	28.89	0.040[a]	0.023	0.010	[b]	[b]
0.904 47	$(3, 2, 2, 3, 3)$	27.00	28.88	7.52	[b]	[b]	[b]	[b]	[b]

[a] Maximum sensitivity factor.
[b] Saturated stage.

3.3.2 The Gopal, Aggarwal, and Gupta method

Aggarwal [5] proposed a heuristic method for optimal redundancy allocation in a general system using a sensitivity factor which can be expressed in the form

$$F_j(x_1, \ldots, x_n) = \frac{r_j(1 - r_j)^{x_j} \partial R_s / \partial R_j}{\displaystyle\prod_{i=1}^{m} \Delta g_{ij}(x_j)}, \tag{3.9}$$

where r_j is the component reliability at stage j, and R_j and R_s are the reliabilities of stage j and the system, respectively. The numerator in (3.9) is the increase in system reliability due to the unit-increment in x_j. It follows that the increment in reliability function $h(p_1, \ldots, p_n)$ of a coherent system, due to an increment Δ in a component reliability p_j, is

$$h(p_1, \ldots, p_j + \Delta, \ldots, p_n) - h(p_1, \ldots, p_n) = \Delta \frac{\partial h}{\partial p_j}.$$

The sensitivity factor takes into account not only the increase in system reliability, but also the entire increase in resource consumption. Noticing that the product term in eq. (3.9) reduces the effectiveness of the heuristic as the number of constraints increases, Gopal et al. [112] modified Aggarwal's sensitivity factor. The modified term can be written in the form

$$F_j(x_1, \ldots, x_n) = \frac{r_j(1 - r_j)^{x_j} \partial R_s / \partial R_j}{\max_i \Delta \bar{g}_{ij}(x_j)}, \tag{3.10}$$

where

$$\Delta \bar{g}_{ij}(x_j) = \frac{\Delta g_{ij}(x_j)}{\max_j \Delta g_{ij}(x_j)}.$$

The algorithm with the modified sensitivity factor is referred to as the GAG1 method.

Solution of Example 3-1 by the GAG1 method

The following are the computational details of the first iteration in the implementation of the GAG1 method for Example 3-1.

- Step 0: Take $\mathbf{x} = (1, 1, 1, 1, 1)$ as the initial feasible solution.

- Step 1: As shown in the implementation of the MSV method for the same example, the residual amounts of resources for the solution $\mathbf{x} = (1, 1, 1, 1, 1)$ are

$$s_1 = P - \sum_{j=1}^{5} p_j x_j = 98.00,$$

$$s_2 = C - \sum_{j=1}^{5} c_j \left[x_j + \exp\left(\frac{x_j}{4}\right) \right] = 101.91,$$

$$s_3 = W - \sum_{j=1}^{5} w_j x_j \exp\left(\frac{x_j}{4}\right) = 151.21,$$

and the set of unsaturated stages is $U(\mathbf{x}) = \{1, 2, 3, 4, 5\}$.

When $x_j = 1$ for all j, the values of $\Delta g_{ij}(x_j)$ for $i = 1, 2, 3$ and $j = 1, 2, 3, 4, 5$ are

i	j					$\max_j \Delta g_{ij}(x_j)$
	1	2	3	4	5	
1	3.00	6.00	9.00	12.00	6.00	12.00
2	9.55	9.55	6.82	12.28	5.46	12.28
3	14.09	16.11	16.11	12.08	18.12	18.12

and the values of $\Delta \bar{g}_{ij}(x_j)$ are

i	j				
	1	2	3	4	5
1	0.25	0.50	0.75	1.00	0.50
2	0.78	0.78	0.56	1.00	0.44
3	0.78	0.89	0.89	0.67	1.00

Since the system has series configuration,

$$\frac{\partial R_s}{\partial R_j} = \prod_{i \neq j} R_i(x_i) = \frac{R_s}{1 - (1 - r_j)^{x_j}}.$$

For $\mathbf{x} = (1, 1, 1, 1, 1)$, we have

$$[R_1(x_1), \ldots, R_5(x_5)] = (0.80, 0.85, 0.90, 0.65, 0.75),$$

$$R_s = 0.298\,35.$$

Table 3.5. Results of Example 3-1 by the GAG1 method

System reliability	Allocation $(x_1, \ldots, x_5)$	Residual resources			Sensitivity factors				
		s_1	s_2	s_3	F_1	F_2	F_3	F_4	F_5
0.298 35	(1, 1, 1, 1, 1)	98.00	101.91	151.21	0.077	0.050	0.034	0.104[a]	0.075
0.402 77	(1, 1, 1, 2, 1)	86.00	89.63	139.13	0.105[a]	0.069	0.046	0.037	0.102
0.483 33	(2, 1, 1, 2, 1)	83.00	80.08	125.03	0.016	0.096	0.064	0.044	0.143[a]
0.604 16	(2, 1, 1, 2, 2)	77.00	74.62	106.91	0.026	0.125[a]	0.103	0.055	0.030
0.694 78	(2, 2, 1, 2, 2)	71.00	65.06	90.80	0.030	0.015	0.119[a]	0.063	0.035
0.764 26	(2, 2, 2, 2, 2)	62.00	58.24	74.70	0.033	0.017	0.008	0.069[a]	0.038
0.833 61	(2, 2, 2, 3, 2)	42.00	45.03	56.38	0.036	0.018	0.009	0.024	0.042[a]
0.875 29	(2, 2, 2, 3, 3)	32.00	39.15	28.89	0.041[a]	0.024	0.013	[b]	[b]
0.904 47	(3, 2, 2, 3, 3)	27.00	28.88	7.52	[b]	[b]	[b]	[b]	[b]

[a] Maximum sensitivity factor.
[b] Saturated stage.

Now, the sensitivity factor of stage 1 is

$$F_1(\mathbf{x}) = \frac{r_1(1 - r_1)^{x_1} \partial R_s / \partial R_1}{\max_{1 \le i \le 3} \Delta \bar{g}_{i1}(x_1)} = \frac{(0.80)(1 - 0.80)(0.372\,94)}{\max(0.25, 0.78, 0.78)} = 0.077.$$

Similarly, the sensitivity factors for the other unsaturated stages are

$$F_2(\mathbf{x}) = 0.050,$$

$$F_3(\mathbf{x}) = 0.034,$$

$$F_4(\mathbf{x}) = 0.104,$$

$$F_5(\mathbf{x}) = 0.075.$$

- Step 2: Since stage 4 gives a maximum sensitivity factor, x_4 is increased from 1 to 2. The resulting feasible solution is $\mathbf{x} = (1, 1, 1, 2, 1)$.

The procedure is continued until the increment in any variable violates at least one constraint. The final allocation given by the algorithm is $(3, 2, 2, 3, 3)$ and the corresponding system reliability is 0.9045. For every allocation $\mathbf{x}$ obtained in the process, the corresponding resource consumption and sensitivity factor are also given in Table 3.5.

3.3.3   The Nakagawa–Nakashima method

Nakagawa and Nakashima (NN) [247] proposed a heuristic method for optimal redundancy allocation in series systems. Let $\mathbf{x} = (x_1, \ldots, x_n)$ be a feasible allocation and let

$$s_i = b_i - \sum_{j=1}^{n} g_{ij}(x_j)$$

and

$$\Delta x_j = \min_i \frac{s_i}{\Delta g_{ij}(x_j)}.$$

The sensitivity factor used in the NN method is

$$F_j(x_1, \ldots, x_n) = [\ln R_j(x_j + 1) - \ln R_j(x_j)]\left[\alpha \Delta x_j + (1 - \alpha) \min_{k \in L(\mathbf{x})} \Delta x_k\right],$$

where

$$L(\mathbf{x}) = \{j: \Delta x_j \geq 1\},$$

and α is a balancing coefficient.

Note that $L(\mathbf{x})$ is the set of unsaturated stages for allocation $\mathbf{x}$. In this case, it is implicit that $u_j = \infty$ for all j, and therefore, $L(\mathbf{x}) = U(\mathbf{x})$. In the NN method, the generic algorithm is applied separately for 14 values of α: $0, 0.1, 0.2,$ $\ldots, 0.9, 1.0, 1/0.9, 1/0.6, 1/0.3$. The best among the 14 solutions yielded by the algorithm is taken as the final solution.

Solution of Example 3-3 by the NN method

The results of Example 3-3 obtained by the NN method for $\alpha = 0.5$ are presented below along with details of the first iteration.

The initial allocation is $\mathbf{x} = (1, 1, 2, 1)$. The residual amounts of resources are

$$s_1 = 99.04, \ s_2 = 159.34, \ \text{and} \ s_3 = 675.80.$$

The values of $\Delta g_{ij}(\ell_j) = g_{ij}(\ell_j + 1) - g_{ij}(\ell_j)$ for $i = 1, 2, 3$ and $j = 1, 2, 3, 4$ are given below.

i	j			
	1	2	3	4
1	0.9621	10.0000	6.0000	15.0000
2	10.6956	18.6831	2.9366	18.0000
3	120.0000	28.0246	9.1607	56.0000

The set of unsaturated stages corresponding to the allocation $\mathbf{x} = (1, 1, 2, 1)$ is $L(\mathbf{x}) = \{1, 2, 3, 4\}$. To calculate the sensitivity factor $F_j(x_j)$ for $j \in L(\mathbf{x})$, we first require the values of Δx_j and $\ln R_j(x_j + 1) - \ln R_j(x_j)$ for all $j \in L(\mathbf{x})$. We have

$$\Delta x_1 = \min_{1 \leq i \leq 3} \frac{s_i}{\Delta g_{i1}(1)}$$

$$= \min\left\{\frac{99.04}{0.9621}, \frac{159.34}{10.6956}, \frac{675.80}{120.0000}\right\}$$

$$= 5.63.$$

Similarly,

$$\Delta x_2 = \min \left\{ \frac{99.04}{10.0000}, \frac{159.34}{18.6831}, \frac{675.80}{28.0246} \right\} = 8.53,$$

$$\Delta x_3 = \min \left\{ \frac{99.04}{6.0000}, \frac{159.34}{2.9366}, \frac{675.80}{9.1607} \right\} = 16.51,$$

$$\Delta x_4 = \min \left\{ \frac{99.04}{15.0000}, \frac{159.34}{18.0000}, \frac{675.80}{56.0000} \right\} = 6.60.$$

Now,

$$\min_{j \in L(\mathbf{x})} \Delta x_j = \min\{5.63, 8.53, 16.51, 6.60\} = 5.63.$$

The values of $[\ln R_j(x_j + 1) - \ln R_j(x_j)]$ for $j \in L(\mathbf{x})$ are

$$\ln R_1(2) - \ln R_1(1) = \ln 0.95 - \ln 0.94 = 0.0106,$$

$$\ln R_2(2) - \ln R_2(1) = \ln \left[1 - (1 - 0.75)^2 \right] - \ln \left[1 - (1 - 0.75)^1 \right] = 0.2232,$$

$$\ln R_3(3) - \ln R_3(2) = \ln \left[\binom{3}{2}(0.9)^2(0.1)^1 + \binom{3}{3}(0.9)^3(0.1)^0 \right]$$

$$- \ln \binom{2}{2}(0.9)^2(0.1)^0 = 0.1823,$$

$$\ln R_4(2) - \ln R_4(1) = \ln \left[1 - (1 - 0.95)^2 \right] - \ln \left[1 - (1 - 0.95)^1 \right] = 0.0488.$$

Therefore, the sensitivity factors for all unsaturated stages are

$$F_1(\mathbf{x}) = [\ln R_1(2) - \ln R_1(1)] \left[\alpha \Delta x_1 + (1 - \alpha) \min_{k \in L(\mathbf{x})} \Delta x_k \right]$$

$$= (0.0106)[(0.5)(5.63) + (0.5)(5.63)]$$

$$= 0.0597,$$

$$F_2(\mathbf{x}) = [\ln R_2(2) - \ln R_2(1)] \left[\alpha \Delta x_2 + (1 - \alpha) \min_{k \in L(\mathbf{x})} \Delta x_k \right]$$

$$= (0.2232)[(0.5)(8.53) + (0.5)(5.63)]$$

$$= 1.5803,$$

$$F_3(\mathbf{x}) = [\ln R_3(3) - \ln R_3(2)] \left[\alpha \Delta x_3 + (1 - \alpha) \min_{k \in L(\mathbf{x})} \Delta x_k \right]$$

$$= (0.1823)[(0.5)(16.51) + (0.5)(5.63)]$$

$$= 2.0181,$$

$$F_4(\mathbf{x}) = [\ln R_4(2) - \ln R_4(1)] \left[\alpha \Delta x_4 + (1 - \alpha) \min_{k \in L(\mathbf{x})} \Delta x_k \right]$$

$$= (0.0488)[(0.5)(6.60) + (0.5)(5.63)]$$

$$= 0.2984.$$

Table 3.6. Results of Example 3-3 by the NN method

Allocation	Residual resources			Sensitivity factors			
$(x_1, \ldots, x_4)$	s_1	s_2	s_3	F_1	F_2	F_3	F_4
$(1, 1, 2, 1)$	99.0439	159.3422	675.7980	0.0597	1.5803	2.0181[a]	0.2984
$(1, 1, 3, 1)$	98.0439	156.4057	666.6373	0.0588	1.5538[a]	0.2600	0.2868
$(1, 2, 3, 1)$	78.0439	137.7226	638.6127	0.0425	0.1323	0.2044[a]	0.2012
$(1, 2, 4, 1)$	77.0439	134.5200	625.0463	0.0416	0.1292	0.0251	0.1899[a]
$(1, 2, 4, 2)$	62.0439	116.5200	569.0463	0.0372	0.1119[a]	0.0205	0.0072
$(1, 3, 4, 2)$	52.0439	65.7341	492.8674	0.0333[a]	[b]	0.0176	0.0052
$(2, 3, 4, 2)$	51.0818	55.0385	372.8674	0.0194[a]	[b]	0.0168	0.0044
$(3, 3, 4, 2)$	49.5128	37.4044	172.8674	[b]	[b]	0.0152[a]	0.0027
$(3, 3, 5, 2)$	43.5128	33.8603	153.1316	[b]	[b]	0.0013	0.0024[a]
$(3, 3, 5, 3)$	28.5128	3.8603	1.1316	[b]	[b]	[b]	[b]

[a] Maximum sensitivity factor.
[b] Saturated stage.

Since stage 3 gives the highest sensitivity factor, x_3 is increased from 2 to 3. The results of other iterations for $\alpha = 0.5$ are given in Table 3.6.

The optimal allocation given by the NN method for $\alpha = 0.5$ is $(3, 3, 5, 3)$ and the corresponding system reliability is 0.9444. Incidentally, this allocation is obtained for all other values of α. However, the sequence of allocations is not the same for all α.

3.3.4   An extension of the NN method for complex systems

Kuo et al. [175] extended the NN method for general reliability systems by modifying the sensitivity factor of Nakagawa and Nakashima. Note that when the number of components at stage j is increased from x_j to $x_j + 1$, the increase in jth-stage reliability is $r_j(1 - r_j)^{x_j}$ and the corresponding increase in the system reliability is given by

$$\Delta R_s(\mathbf{x}, j) = r_j(1 - r_j)^{x_j} \frac{\partial R_s}{\partial R_j}.$$

Therefore, the modified sensitivity factor of Nakagawa and Nakashima becomes

$$F_j(x_1, \ldots, x_n) = \left[r_j(1 - r_j)^{x_j} \frac{\partial R_s}{\partial R_j} \right] \left[\alpha \Delta x_j + (1 - \alpha) \min_{k \in L(\mathbf{x})} \Delta x_k \right]. \qquad (3.11)$$

Also the term $\partial R_s / \partial R_j$ is replaced by $\Delta R_s / \Delta R_j$ if R_s is not differentiable with respect to R_j.

When the extended NN method is applied to Example 3-2, it indeed gives the exact optimal allocation $(3, 2, 2, 1, 1)$ with system reliability 0.9932 for $\alpha = 1/0.6$. The computational details for this α-value are given in Table 3.7.

Table 3.7. Results of Example 3-2 by the extended NN method

Allocation	Residual resource	Sensitivity factor				
$(x_1, \ldots, x_5)$	s_1	F_1	F_2	F_3	F_4	F_5
(1, 1, 1, 1, 1)	9.00	0.2914	0.0744	0.3035[a]	0.0745	0.0693
(1, 1, 2, 1, 1)	7.00	0.0681[a]	0.0578	0.0590	0.0636	0.0483
(2, 1, 2, 1, 1)	5.00	0.0146	0.0429[a]	0.0138	0.0415	0.0138
(2, 2, 2, 1, 1)	2.00	0.0051[a]	b		0.0042	b	0.0042
(3, 2, 2, 1, 1)	0	b	b	b	b	b

[a] Maximum sensitivity factor.
[b] Saturated stage.

3.3.5 The Dinghua method

Dinghua Shi [296] presented a heuristic method for redundancy optimization in complex systems. In every iteration of this method, an index is computed for each unsaturated minimal path set of the system, and the minimal path set with maximum index is selected. A minimal path set is unsaturated if at least one of its stages is unsaturated. Later, a sensitivity factor is computed for all unsaturated stages in the selected path set and a component is added to the stage with the maximum sensitivity factor.

Let $P_1, \ldots, P_k$ be the minimal path sets of the system and $\mathbf{x} = (x_1, \ldots, x_n)$ a feasible solution. The index of path set P_h for solution $\mathbf{x}$ is defined as

$$a_h(\mathbf{x}) = \frac{\prod_{j \in P_h} R_j(x_j)}{\sum_{j \in P_h} \sum_{i=1}^{m} g_{ij}(x_j)/(mb_i)},$$

for $h = 1, 2, \ldots, k$. The sensitivity factor of stage j is defined as

$$F_j(\mathbf{x}) = \frac{R_j(x_j)}{\sum_{i=1}^{m} g_{ij}(x_j)/(mb_i)}.$$

It is difficult to describe the computational procedure of Dinghua Shi [296] by the generic algorithm without sacrificing some clarity. For this reason, the procedure given by Dinghua is below.

- Step 0: Find all minimal path sets of the system and use them to calculate the index $a_h(\mathbf{x})$. Take $\mathbf{x} = (1, \ldots, 1)$ as the initial solution.

- Step 1: Determine the residual resource $s_i = b_i - \sum_{j=1}^{n} g_{ij}(x_j)$ for $i = 1, 2, \ldots, m$. If $s_i = 0$ for some i, go to step 4.

- Step 2: If there is no unsaturated minimal path set under consideration, go to step 4. Otherwise, compute the index $a_h(\mathbf{x})$ for each unsaturated minimal path set under consideration and find the minimal path set P_{h*} that gives the maximum index. Also, compute the sensitivity factor $F_j(\mathbf{x})$ for each unsaturated stage j in P_{h*} and find the stage v that gives the maximum sensitivity factor.

- Step 3: Let $x_v = x_v + 1$. If the new solution $\mathbf{x}$ violates any constraint, remove the minimal path set P_{h*} from further consideration, let $x_v = x_v - 1$, and go to step 2. If $\mathbf{x}$ satisfies all the constraints, go to step 1.

- Step 4: Take $\mathbf{x}$ as the optimal solution and stop.

Solution of Example 3-2 by the Dinghua method

The minimal path sets in this example are

$$P_1 = \{1, 2\},\ P_2 = \{3, 4\},\ P_3 = \{1, 4, 5\},\ P_4 = \{2, 3, 5\},$$

and the number of constraints is $m = 1$. The following are computational details of the first iteration.

- Step 0: Let $\mathbf{x} = (1, 1, 1, 1, 1)$.

- Step 1: The residual resource is $b_1 - \sum_{j=1}^{5} c_j x_j = 20 - (2 + 3 + 2 + 3 + 1) = 9$. Go to step 2 since the residual resource is positive.

- Step 2: The stage reliabilities are

$$R_1(x_1) = R_1(1) = 1 - (1 - 0.70)^1 = 0.70,$$
$$R_2(x_2) = R_2(1) = 1 - (1 - 0.85)^1 = 0.85,$$
$$R_3(x_3) = R_3(1) = 1 - (1 - 0.75)^1 = 0.75,$$
$$R_4(x_4) = R_4(1) = 1 - (1 - 0.80)^1 = 0.80,$$
$$R_5(x_5) = R_5(1) = 1 - (1 - 0.90)^1 = 0.90,$$

and the values of $\sum_{i=1}^{m} g_{ij}(x_j)/mb_i$ are

Stage j	$\sum_{i=1}^{m} g_{ij}(x_j)/mb_i = c_j x_j/20$
1	0.10
2	0.15
3	0.10
4	0.15
5	0.05

Table 3.8. Results of Example 3-2 by the Dinghua method

Allocation	Residual resource $20 - \sum_{j=1}^{5} c_j x_j$	Index of minimal path set				Sensitivity factor				
$(x_1, x_2, \ldots, x_5)$		P_1	P_2	P_3	P_4	F_1	F_2	F_3	F_4	F_5
$(1, 1, 1, 1, 1)$	9	2.38	2.40[a]	1.68	1.91				7.5[b]	5.3
$(1, 1, 2, 1, 1)$	7	2.38[a]	2.14	1.68	1.79	7.0[b]	5.7			
$(2, 1, 2, 1, 1)$	5	2.21[a]	2.14	1.64	1.79	4.6	5.7[b]			
$(2, 2, 2, 1, 1)$	2	1.78	2.14[a]	1.64	1.50				4.7	5.3[b]
$(2, 2, 2, 2, 1)$	-1[c]	1.78[a]	[d]	1.64	1.50	4.6[b]	3.3			
$(3, 2, 2, 1, 1)$	0									

[a] Maximum index of minimal path sets.
[b] Highest sensitivity factor.
[c] Constraint violation.
[d] Minimal path set removed from consideration.

Now, the index of the minimal path set $P_1 = \{1, 2\}$ is

$$a_1(\mathbf{x}) = \frac{\prod_{j \in P_1} R_j(x_j)}{\sum_{j \in P_1} \sum_{i=1}^{m} g_{ij}(x_j)/mb_i}$$

$$= \frac{R_1(x_1) R_2(x_2)}{\sum_{i=1}^{m} g_{i1}(x_1)/mb_i + \sum_{i=1}^{m} g_{i2}(x_2)/mb_i} = \frac{(0.70)(0.85)}{0.10 + 0.15} = 2.38.$$

Similarly, $a_2(\mathbf{x}) = 2.40$, $a_3(\mathbf{x}) = 1.68$, and $a_4(\mathbf{x}) = 1.91$. Since the minimal path set $P_2 = \{3, 4\}$ gives the maximum index value of 2.40, the sensitivity factor is computed for stages 3 and 4 as

$$F_3(\mathbf{x}) = \frac{R_3(x_3)}{\sum_{i=1}^{m} g_{i3}(x_3)/mb_i} = 7.5,$$

and

$$F_4(\mathbf{x}) = \frac{R_4(x_4)}{\sum_{i=1}^{m} g_{i4}(x_4)/mb_i} = 5.3.$$

- Step 3: Since stage 3 gives a higher sensitivity factor than stage 4, increase x_3 from 1 to 2. Go to step 1, as the new solution satisfies the constraint.

The results of the remaining iterations are provided in Table 3.8.

The final allocation derived by Dinghua's method is $(3, 2, 2, 1, 1)$, which gives a system reliability of 0.9932. This allocation is indeed the exact optimal solution.

3.4   Other heuristic methods

3.4.1   The Kohda–Inoue method

Kohda and Inoue (KI) [163] presented a heuristic method for redundancy optimization in general coherent systems. It is also an iterative algorithm starting with a feasible solution. It does not require the increasing nature of the constraint functions $g_{ij}(x_j)$.

In every iteration, if the subset $S_1^+(\mathbf{x})$ of the current feasible solution $\mathbf{x}$ is nonempty, the best solution in $S_1^+(\mathbf{x})$ is selected. Otherwise, the search for a better feasible solution is carried out in the feasible 2-neighborhood $S_2(\mathbf{x})$ of $\mathbf{x}$. Since no solution $\mathbf{y} \leq \mathbf{x}$ is better than $\mathbf{x}$, only feasible solutions of the type $(x_1, \ldots, x_i + 1, \ldots, x_j + 1, \ldots, x_n)$, $(x_1, \ldots, x_i - 1, \ldots, x_j + 1, \ldots, x_n)$, and $(x_1, \ldots, x_i + 1, \ldots, x_j - 1, \ldots, x_n)$ in $S_2(\mathbf{x})$ are considered in the search. It can easily be seen that in the KI method, two feasible solutions obtained in successive iterations are in 2-neighborhoods of each other, but not necessarily 1-neighborhood.

3.4.2   The Kim–Yum method

Kim and Yum (KY) [153] proposed a heuristic method for solving the redundancy allocation problem. Starting with a reasonably good feasible solution, their algorithm improves the solution iteratively. Between successive improvements, it collects some solutions in a predetermined bounded set of infeasible solutions. The method stops when it yields an infeasible solution outside the bounded set. In this method, two feasible solutions obtained in successive iterations need not even be in 2-neighborhoods of each other.

Let $\delta_1, \delta_2, \ldots, \delta_m$ be m nonnegative values and let B be the set of infeasible solutions satisfying

$$\left| b_i - \sum_{j=1}^{n} g_{ij}(x_j) \right| \leq \delta_i, \qquad \text{for } i = 1, \ldots, m.$$

The computational effort involved in the KY method increases with the size of set B, which in turn increases with the values of $\delta_1, \delta_2, \ldots, \delta_m$. These values can be chosen in several ways. For example, δ_i can be fixed as $\delta_i = \alpha_i b_i$ for some constants α_is. If the constraints are linear, say $g_{ij}(x_j) = a_{ij}x_j$, Kim and Yum [153] suggest $\delta_i = 2 \max_j a_{ij}$. To enhance the performance of the KY method, we modify set B as the set of solutions satisfying

$$\sum_{j=1}^{n} g_{ij}(x_j) \le b_i + \delta_i, \qquad \text{for } i = 1, \ldots, m$$

and

$$\sum_{j=1}^{n} g_{ij}(x_j) > b_i, \qquad \text{for at least one } i.$$

Consider a solution $\mathbf{x} = (x_1, \ldots, x_n)$. Select u, $1 \le u \le n$, such that

$$\frac{\triangle R_s(\mathbf{x}, u)}{\sum_{i=1}^{m} \triangle g_{iu}(x_u)/b_i} = \max_{1 \le j \le n} \frac{\triangle R_s(\mathbf{x}, j)}{\sum_{i=1}^{m} \triangle g_{ij}(x_j)/b_i}, \tag{3.12}$$

where $\triangle R_s(\mathbf{x}, j)$ is the increase in system reliability due to unit increase in x_j. Similarly, select v such that

$$\frac{\triangle R_s(\mathbf{x}, -v)}{\sum_{i=1}^{m} \triangle g_{iv}(x_v - 1)/b_i} = \max_{1 \le j \le n} \frac{\triangle R_s(\mathbf{x}, -j)}{\sum_{i=1}^{m} \triangle g_{ij}(x_j - 1)/b_i}, \tag{3.13}$$

where $\triangle R_s(\mathbf{x}, -j)$ is the decrease in system reliability due to unit reduction in x_j. Increasing x_u by 1 is called a *step-up operation*, whereas decreasing x_v by 1 is called a *step-down operation*. The right-hand side of (3.12) for any j can be used as a sensitivity factor of the jth stage. Selection of u and v can also be based on any one of the sensitivity factors given by Aggarwal [5], Nakagawa and Nakashima [247], and Gopal et al. [112].

Let S denote the set of all feasible solutions, and D the set of all infeasible solutions which are not in B. At any stage of the algorithm, there are two sets (E and F) of bounded infeasible solutions which are collected after derivation of the current best feasible solution. The nature of the algorithm can be described by a dynamic system using Figure 3.1.

If the current solution is in set S, the state of the system is said to be S. The same explanation also holds for E, F, and D. The system starts in state S, moves from one state to another due to the step-up or step-down operation, and finally stops after reaching state D. Transitions among the four states are as shown in Figure 3.1. The step-up operation is done in states S and F, while the step-down operation is done in state E. If the step-down operation in E does not cause a transition to S or F, the system is said to remain in E in that step. The sets E and F are initially made empty and are updated whenever a better feasible solution S is obtained. When a transition takes place from E to F, or when an inferior solution is obtained in a transition to S, set F is updated as $F = F \cup E$ and E is made empty. The KY method is presented below in stepwise fashion.

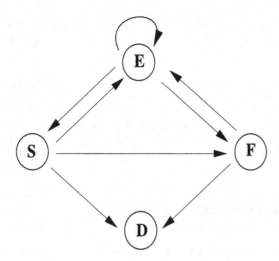

Figure 3.1. Transitions among four states

- Step 0: Determine δ_i for $i = 1, 2, \ldots, m$. Find an initial feasible solution $\mathbf{x} = (x_1, x_2, \ldots, x_n)$ and let $\mathbf{x}^* = \mathbf{x}$ and $R_s^* = f(\mathbf{x}^*)$.

- Step 1: Let $F = \phi$ and $E = \phi$.

- Step 2: Perform a step-up operation on $\mathbf{x}$ to derive a new solution $\mathbf{x}^+$. Let $\mathbf{x} = \mathbf{x}^+$. If $\mathbf{x} \in D$, take $\mathbf{x}^*$ as the heuristic solution and stop. If $\mathbf{x} \in F$, repeat step 2. If $\mathbf{x}$ is neither in D nor in F, let $E = \{\mathbf{x}\}$ and go to step 3.

- Step 3: Perform a step-down operation on $\mathbf{x}$ to obtain a new solution $\mathbf{x}^-$. If $\mathbf{x}^-$ is in S, let $\mathbf{x} = \mathbf{x}^-$ and go to step 4. If $\mathbf{x}^-$ is in F, let $F = F \cup E$, $E = \phi$, and $\mathbf{x} = \mathbf{x}^-$, and go to step 2. If $\mathbf{x}^-$ is neither feasible nor in F, let $E = E \cup \{\mathbf{x}^-\}$ and $\mathbf{x} = \mathbf{x}^-$; repeat step 3.

- Step 4: Improve the feasible solution $\mathbf{x}$ as much as possible (by adding increments to $x_1, x_2, \ldots, x_n$) without violating the constraints. Take the improved solution as $\mathbf{x}$ when there is any improvement. If $R_s^* < f(\mathbf{x})$, let $\mathbf{x}^* = \mathbf{x}$ and $R_s^* = f(\mathbf{x}^*)$, and go to step 1. Otherwise let $F = F \cup E$, $E = \phi$ and, $\mathbf{x} = \mathbf{x}^*$, and go to step 2.

Solution of Example 3-2 by the KY method

The following is a stepwise description of how the KY method is implemented to solve Example 3-2.

- Step 0: Take $\delta_1 = 6$ and initial solution $\mathbf{x} = (2, 2, 1, 2, 2)$. Let $\mathbf{x}^* = (2, 2, 1, 2, 2)$ and $R_s^* = f(\mathbf{x}^*) = 0.9765$.

- Step 1: Let $F = \phi$ and $E = \phi$.

- Step 2: Step-up operation on $\mathbf{x}$ yields $\mathbf{x}^+ = (2, 2, 2, 2, 2)$. Let $\mathbf{x} = \mathbf{x}^+$. Since $s_1 \geq -6$ and $\mathbf{x}$ is neither in D nor in F, let $E = \{(2, 2, 2, 2, 2)\}$ and go to step 3.

- Step 3: Step-down operation on $\mathbf{x}$ yields $\mathbf{x}^- = (2, 2, 2, 2, 1)$. Since $\mathbf{x}^-$ is neither feasible nor in F, update E as $E = \{(2, 2, 2, 2, 2), (2, 2, 2, 2, 1)\}$. Let $\mathbf{x} = \mathbf{x}^-$, and repeat step 3.

- Step 3: Step-down operation on $\mathbf{x}$ yields $\mathbf{x}^- = (2, 2, 2, 1, 1)$, which is feasible. Let $\mathbf{x} = \mathbf{x}^-$, and go to step 4.

- Step 4: The solution $\mathbf{x}$ is improved to $(3, 2, 2, 1, 1)$. Since $f(3, 2, 2, 1, 1) = 0.9932 > R_s^*$, let $\mathbf{x} = (3, 2, 2, 1, 1)$, $\mathbf{x}^* = (3, 2, 2, 1, 1)$, and $R_s^* = 0.9932$; go to step 1.

- Step 1: Let $F = \phi$ and $E = \phi$.

- Step 2: Step-up operation on $\mathbf{x}$ yields $\mathbf{x}^+ = (3, 2, 2, 2, 1)$. Let $\mathbf{x} = \mathbf{x}^+$. Since $s_1 \geq -6$ and $\mathbf{x}$ is neither in D nor in F, update E as $E = \{(3, 2, 2, 2, 1)\}$, and go to step 3.

- Step 3: Step-down operation on $\mathbf{x}$ yields $\mathbf{x}^- = (3, 2, 2, 1, 1)$, which is feasible. Let $\mathbf{x} = \mathbf{x}^-$, and go to step 4.

- Step 4: The solution $\mathbf{x}$ cannot be improved further and $f(\mathbf{x}) \not> R_s^*$. So, update F as $F = F \cup E$, that is, $F = \{(3, 2, 2, 2, 1)\}$, and let $E = \phi$, and $\mathbf{x} = \mathbf{x}^*$. Go to step 2.

- Step 2: Step-up operation on $\mathbf{x}$ yields $\mathbf{x}^+ = (3, 2, 2, 2, 1)$, which belongs to F. So, let $\mathbf{x} = \mathbf{x}^+$, and repeat step 2.

- Step 2: Step-up operation on $\mathbf{x}$ yields $\mathbf{x}^+ = (3, 2, 3, 2, 1)$. Since $s_1 \geq -6$, let $\mathbf{x} = \mathbf{x}^+$, and $E = \{(3, 2, 3, 2, 1)\}$; go to step 3.

- Step 3: Step-down operation on $\mathbf{x}$ yields $\mathbf{x}^- = (2, 2, 3, 2, 1)$ and $E = \{(3, 2, 3, 2, 1), (2, 2, 3, 2, 1)\}$; repeat step 3.

- Step 3: Step-down operation on $\mathbf{x}$ yields $\mathbf{x}^+ = (2, 2, 3, 1, 1)$, which is feasible. So, let $\mathbf{x} = \mathbf{x}^+$, go to step 4.

- Step 4: The solution $\mathbf{x}$ cannot be improved further. Therefore let $F = F \cup E$, that is, $F = \{(3, 2, 2, 2, 1), (3, 2, 3, 2, 1), (2, 2, 3, 2, 1)\}$, and let $E = \phi$ and $\mathbf{x} = \mathbf{x}^*$. Go to step 2.

- Step 2: Step-up operation on $\mathbf{x}$ yields $\mathbf{x}^+ = (3, 2, 2, 2, 1)$, which belongs to F. Let $\mathbf{x} = \mathbf{x}^+$, and repeat step 2.

- Step 2: Step-up operation on $\mathbf{x}$ yields $\mathbf{x}^+ = (3, 2, 3, 2, 1)$, which belongs to F. Let $\mathbf{x} = \mathbf{x}^+$, and repeat step 2.

- Step 2: Step-up operation on $\mathbf{x}$ yields $\mathbf{x}^+ = (3, 2, 3, 3, 1)$, which belongs to D. Therefore, take $\mathbf{x}^* = (3, 2, 2, 1, 1)$ as the heuristic solution.

The solution $(3, 2, 2, 1, 1)$ given by the KY method is indeed an exact optimal solution. However, we have noticed that performance of the algorithm depends on the initial feasible solution. The algorithm does not yield an exact optimal solution for some selections of an initial feasible solution.

3.4.3 Ushakov's heuristic method

Ushakov [314] developed an interesting heuristic method for solving Example 3-4:

- Step 1: Solve the problem of minimizing $\sum_{j \in J_i} c_j x_j$, subject to

$$\prod_{j \in J_i} R_j(x_j) \geq R_{\min}(i),$$

for each i. Let x_j^i, $j \in J_i$ denote the optimal solution for the ith problem.

- Step 2: For each stage j, find $x_j^* = \max_i \{x_j^i : J_i \text{ contains } j\}$.
- Step 3: For every function i, find the set J_i^{**} of stages which is necessary only for executing function i. Let $J_i^* = J_i - J_i^{**}$.
- Step 4: For each set J_i^*, find the value of $R_i^* = \prod_{j \in J_i^*} R_j(x_j^*)$. If J_i^* is empty, that is, if function i is executed with an independent group of elements, take $R_i^* = 1$.
- Step 5: For each function i, calculate $R_i^{**} = R_{\min}(i)/R_i^*$, which is the minimum joint probability requirement of failure-free operation of stages in J_i^{**}.
- Step 6: For each nonempty J_i^{**}, solve the problem of minimizing $\sum_{j \in J_i^{**}} c_j x_j$ subject to

$$\prod_{j \in J_i^{**}} R_j(x_j) \geq R_i^{**}.$$

(Note that J_i^{**} is a pairwise disjoint set.) Let x_j^{**} denote the optimal value for each stage j in $J_1^{**} \cup J_2^{**} \cup \cdots \cup J_k^{**}$.

- Step 7: Let $x_j^* = x_j^{**}$ for each stage j in $J_1^{**} \cup J_2^{**} \cup \cdots \cup J_k^{**}$, and take $(x_1^*, \ldots, x_n^*)$ as the heuristic solution.

Solution of Example 3-4 by the Ushakov method

Following the above described computational procedure, we obtain

- Step 1: $(x_1^1, x_2^1) = (2, 3)$ and $(x_1^2, x_3^2) = (3, 2)$.
- Step 2: $(x_1^*, x_2^*, x_3^*) = (3, 3, 2)$.
- Step 3: $J_1^{**} = \{2\}$, $J_2^{**} = \{3\}$, $J_1^* = \{1\}$, $J_2^* = \{1\}$.
- Step 4: $R_1^* = 1 - (1 - 0.80)^3 = 0.992$, and $R_2^* = 1 - (1 - 0.80)^3 = 0.992$.
- Step 5: $R_1^{**} = 0.94/0.992 = 0.9475$, and $R_2^{**} = 0.96/0.992 = 0.9677$.
- Step 6: $x_2^{**} = 3$ minimizes $2x_2$ subject to $1 - (1 - 0.75)^{x_2} \geq 0.9475$. Similarly, $x_3^{**} = 2$ minimizes $4x_3$ subject to $1 - (1 - 0.85)^{x_3} \geq 0.9677$.
- Step 7: Let $x_2^* = x_2^{**}(= 3)$ and $x_3^* = x_3^{**}(= 2)$. Now, the final solution is $(x_1^*, x_2^*, x_3^*) = (3, 3, 2)$.

For other multi-purpose systems, Genis and Ushakov [105] also present an optimization procedure.

3.4.4   The Misra method

Misra [223] presented a method, steps 1–5 below, for deriving optimal redundancy in a coherent system subject to linear constraints. In this method, the solution of redundancy optimization problems involving m linear constraints is obtained using the solution of m single-constraint optimization problems.

- Step 1: Obtain a feasible allocation following the MSV method and evaluate the corresponding system reliability R_s.

- Step 2: With respect to each of m constraints, obtain the allocation

$$x_j = \left\lfloor \frac{\ln(1 - R_s^{a_j})}{\ln q_j} \right\rfloor, \qquad \text{for } j = 1, 2, \ldots, n,$$

where

$$a_j = \frac{c_j / \ln q_j}{\sum\limits_{h=1}^{n} c_h / \ln q_h},$$

and $q_j = 1 - r_j$, and $\lfloor z \rfloor$ represents the largest integer less than or equal to z.

- Step 3: From the m allocations obtained in step 2, choose the one that gives maximum system reliability. Take the corresponding system reliability as a reference reliability index for comparison.

- Step 4: For each allocation with system reliability less than the reference reliability index, compute the desirability factor

$$F_j = \frac{\Delta R_s / R_s}{c_j / b_i}$$

for every stage and add a component to the stage that gives the highest desirability factor: $c_1, \ldots, c_n$ are the cost coefficients in the corresponding constraint and b_i the value on the right-hand side of the constraint. Now some allocation other than the previous one may provide the highest system reliability. This value becomes the reference reliability index and the allocation is not changed further. Go to step 3 if the allocation chosen in step 2 still has highest reliability. Otherwise repeat this step. This process continues until all m allocations give the same system reliability or a constraint is violated.

- Step 5: If all m allocations give the same system reliability, then the common allocation is optimal. Otherwise, the allocation with highest system reliability is taken as the near-optimal allocation.

3.5   Discussion

Several heuristic methods are available in the literature for solving redundancy allocation problems in reliability systems. Almost all of them are iterative algorithms yielding an allocation in each iteration. In the class of methods discussed in Section 3.3, allocations obtained in two successive iterations are in 1-neighborhood, that is, either of them can be obtained from the other by increasing or decreasing the allocation at a stage by one unit. Other methods which fall in this class are not discussed here due to space considerations. The salient feature of this class of methods is that in each iteration, a stage is selected on the basis of a stage sensitivity factor and the allocation at that stage is increased by one unit. The definition of this factor varies with the method.

The NN method developed by Nakagawa and Nakashima [247] for series systems involves application of such iterative algorithms for 14 values of parameter α, and it selects the best among 14 resulting allocations. Although the NN method takes more time than other heuristic methods, its overall performance appears to be much better. This method is extended by Kuo et al. [175] for general reliability systems by modifying the sensitivity factor. Nakagawa and Miyazaki [245] compared the performance of the heuristic methods MSV, GAG, and NN. In view of computational time, MSV is better than GAG, which is better than NN. However, in terms of optimality rate (that is, the proportion of times the exact optimal solution is obtained) and relative error, these methods rank in the reverse order. In fact, the NN method is superior to other heuristic methods in terms of optimality rate and relative error. The numerical experimentation of Nakagawa and Miyazaki shows that the NN method loses its advantage if it is applied to only a single α parameter value among the 14 given, that is, $0, 0.1, \ldots, 0.9$, $1.0, 1/0.9, 1/0.6$, and $1/0.3$.

The heuristic method presented by Dinghua Shi [296] is very similar to the above-mentioned class. However, in each iteration of the method, one of the minimal path sets is selected by any heuristics, and one of the stages in the selected path set is chosen on the basis of a sensitivity factor for a one-unit-increment in the allocation. In the Dinghua method [296], the allocation obtained in any iteration is in 1-neighborhood or 2-neighborhoods of the one obtained in the previous iteration.

Kim and Yum [153] have presented a heuristic method in which the allocation obtained in any iteration is either a feasible solution or an infeasible solution close to the boundary of the feasible region. A striking feature of the method is that a one-unit reduction in allocation may take place at some stage during transition from one allocation to another.

For a system with k different functions, each of which requires a subset of system components, Ushakov [314] developed a heuristic method for optimal allocation of redundancies. This method aims to give a cost-optimal redundancy allocation subject to a specified minimum k function system reliability requirement. To derive optimal redundancy in a coherent system subject to m linear constraints, the method of Misra [223] uses solutions of m single-constrained redundancy optimization problems.

Recently, Jianping [150] has proposed an iterative heuristic method in which each iteration involves application of an existing heuristic method for optimal redundancy.

Selection of a heuristic method always depends on the tradeoff between the quality of the solution and the computational effort. Since powerful PCs, such as the Pentium, are currently available at most work places, it may be advantageous to use the NN method for solving reliability optimization problems of moderate size.

EXERCISES

3.1 Discuss the advantages and disadvantages of heuristics for redundancy allocation problems.

3.2 Determine some classes of problems for which 1-neighborhood heuristics are more likely to yield optimal solutions.

3.3 Perform the second iteration of Example 3-1 by the MSV method.

3.4 Perform the second iteration of Example 3-1 by the GAG1 method.

3.5 Apply the GAG1 method to Example 3-2 for two iterations.   Start with $(1, 1, 1, 1, 1)$ as the initial solution.

3.6 Apply the extended NN method to Example 3-2 to find the optimal solution($\alpha = 0.8$). Start with $(1, 1, 1, 1, 1)$ as the initial solution. Compare this result with the result of $\alpha = 1/0.6$.

3.7 Perform the second iteration of Example 3-3 by the NN method ($\alpha = 0.5$).

3.8 Perform the second iteration of Example 3-2 by the Dinghua method.

3.9 Generate 50 random instances of Example 3-1 taking r_j as $(0.7, 1.0)$, P_j as $\{1, 2, 3, 4, 5\}$, c_j as $\{6, 7, 8, 9, 10\}$, and w_j as $\{6, 7, 8, 9, 10\}$, for $n = 20$, $P = 225$, $c = 350$, and $w = 400$. Using these instances, compare the performance of the methods of MSV, GAG1, and NN with respect to solution quality and execution time.

3.10 Compare the performance, with respect to solution quality and execution time, of the methods of Kuo et al. [175] and Dinghua using several randomly generated instances of Example 3-2.

3.11 Show that sensitivity factors are the same (but for a multiplier) in the GAG1 method and the method of Kuo et al. [175] when s_i is proportional to $\max_j \Delta g_{ij}(x_i)$ and $\alpha = 1$.

3.12 In the Dinghua method, an index is computed for each unsaturated minimal path set of the system and the minimal path set with maximum index is selected. Discuss whether it is possible to find the solution by using the minimal cut set. If possible, develop the algorithm for the minimal cut set.

3.13 Develop a simple heuristic based on the linearized approximation by suitably transforming the nonlinear integer programming formulation of redundancy allocation problems in a series system.

3.14 Develop a heuristic method based on Lagrangian relaxation for redundancy allocation problems.

3.15 Show the formulation of the general allocation problem with linear objective functions and constraints. Is it an NP complete problem? What are the computational characteristics of it?

3.16 Compare the 1-neighborhood algorithm with the 2-neighborhood algorithm. Describe the merits and disadvantages of each.

3.17 Apply the KY method on Example 3-1 for two iterations. Start with $(3, 3, 3, 2, 2)$ as the initial solution. Let $\delta_1 = \delta_2 = 24$ and $\delta_3 = 36$.

3.18 Explain how Ushakov's method differs from all other heuristics discussed in this chapter.

3.19 Consider a problem to minimize

$$C_s = \sum_{j=1}^{n} c_j x_j,$$

subject to

$$R(i) = \prod_{j \in J_i} R_j(x_j) \geq R_{\min}(i), \qquad i = 1, 2,$$

x_j being a nonnegative integer. Let $J_1 = 1, 2, 4$, $J_2 = 1, 3, 4$, $R_{\min}(1) = 0.93$, and $R_{\min}(2) = 0.95$. The reliabilities and costs of the components at four stages are

j	1	2	3	4
r_j	0.75	0.8	0.85	0.8
c_j	3	2	2	4

Let the initial solutions be $(x_1^1, x_2^1, x_4^1) = (3, 4, 2)$ and $(x_1^2, x_3^2, x_4^2) = (3, 2, 3)$. Apply Ushakov's heuristic method for one iteration.

4 Redundancy allocation by dynamic programming

4.1 Introduction

Dynamic programming (DP) is a solution methodology based on Richard Bellman's "principle of optimality" [29]. DP is an approach for solving a wide spectrum of decision-making problems, and it is highly effective in solving certain types of deterministic and stochastic optimization problems. In this approach, a decision-making problem involving n variables is reduced to a set of n single-variable problems, which are derived sequentially in such a way that the solution of each problem is obtained using the preceding answer. The DP approach transforms an n-variable optimization problem into a multi-stage decision-making process. Such a transformation is not always straight-forward, and it quite often requires a lot of ingenuity. For this reason, it is not easy to describe the general DP approach as an algorithm. An outline of DP is given for various optimization problems in Appendix 1.

Redundancy optimization in series systems subject to separable constraints can be viewed as a set of nonlinear integer programming problems with separable objective functions and separable constraints. The DP approach provides an excellent procedure for deriving an optimal solution when a problem involves one constraint, at most. However, the computational complexity of the approach increases exponentially with the number of constraints. There are several variations of the DP approach, which are classified in Table 4.1, to solve these kinds of problems.

The second approach in Table 4.1, where Lagrange multipliers were used for problems involving two or more constraints, was originally adopted by Bellman and Dreyfus [30]. The use of Lagrange multipliers eliminates the dimensionality problem that results from a multitude of constraints. However, the computational complexity of this approach also increases significantly with the number of constraints. The third approach listed in Table 4.1 utilizes the concept of dominating sequence: Kettelle [152], apparently, was the first to introduce this concept to solve a problem with a single linear constraint. This approach is also applicable to problems with several separable constraints: the lower bounds on the variables can be used to reduce the computation involved. Other DP approaches are also available for solving multi-constraint problems: (1) Morin and Marsten [237] presented a method which uses an embedded state space approach; (2) others, e.g. Morin and Marsten [238], Denardo and Fox [72], and Aust [18], adopted branch-and-bound strategies; Cooper [66] developed a method whereby

Table 4.1. Classification of approaches

DP approach	Application to examples	References
Basic	Examples 4-1, 4-2	[29]–[31], [43], [96], [168], [184], [201], [215], [277]
Lagrange multipliers	Examples 4-3–4-5	[30], [96], [120], [215]
Concept of dominating sequence	Examples 4-2–4-5	[148], [152], [324]
Other		[18], [66], [72], [195], [210], [237], [238]

the constraints were used only to test partial solutions for infeasibility, thus avoiding the "dimensionality curse" of the DP approach.

The basic DP approach, use of Lagrange multipliers, and Kettelle's approach based on dominated solutions are described in the following sections. The examples presented below are used to illustrate these methods. Throughout this chapter, we deal only with integer decision variables. For general information about dynamic programming, refer to Dreyfus [81].

Example 4-1

Consider an n-stage series reliability system. Let r_j and c_j denote, respectively, reliability and component cost at the jth stage for $j = 1, \ldots, n$. If x_j components are arranged in parallel at the jth stage for $j = 1, \ldots, n$, then the total cost of the components is $\sum_{j=1}^{n} c_j x_j$ and the system reliability is

$$R_s = f(x_1, \ldots, x_n) = \prod_{j=1}^{n} \left[1 - (1 - r_j)^{x_j} \right].$$ (4.1)

Let P' be the profit obtained when the system operates successfully. The system reliability R_s is the fraction of the trials that are successful, and hence the expected profit for the system is $P' f(x_1, \ldots, x_n)$. The net profit G for the entire system is the profit less the total cost, that is,

$$G(x_1, \ldots, x_n) = P' f(x_1, \ldots, x_n) - \sum_{j=1}^{n} c_j x_j.$$ (4.2)

Now the problem is to determine optimal redundancy levels that maximize the expected net profit at all stages. Mathematically, $G(x_1, \ldots, x_n)$ has to be maximized over x_j positive integer values. This problem is a nonlinear integer programming problem without any constraint. For illustrative purposes, let $n = 3$, $(r_1, r_2, r_3) = (0.75, 0.5, 0.333)$, $(c_1, c_2, c_3) = (1.0, 1.0, 0.2)$, and $P' = 10$.

Example 4-2

Consider a four-stage series system with reliabilities and component costs of

$$(r_1, r_2, r_3, r_4) = (0.60, 0.75, 0.50, 0.70),$$

$$(c_1, c_2, c_3, c_4) = (2, 3, 1, 2),$$

respectively. Suppose the system reliability can be improved by providing redundant components at each stage, but the total number of components at stage j must lie between $\ell_j = 1$ and $u_j = 6$, for $j = 1, 2, 3, 4$. Find the optimal number of parallel components at each stage such that the system reliability is maximized, subject to an upper limit $b = 30$ on the total cost. Mathematically, the problem is to maximize

$$f(x_1, \ldots, x_4) = \prod_{j=1}^{4} [1 - (1 - r_j)^{x_j}], \tag{4.3}$$

subject to

$$\sum_{j=1}^{4} c_j x_j \le b,$$

$$\ell_j \le x_j \le u_j, \qquad \text{for } j = 1, \ldots, 4,$$

x_j being a nonnegative integer.

This is a nonlinear integer programming problem with a linear constraint and limits on the variables.

Example 4-3

As for Example 3-1.

Example 4-4

Consider an n-stage series system in which there are d_j design alternatives at stage j, for $j = 1, \ldots, n$. Let $R'_j(v, d)$ denote the reliability of stage j when a design alternative indexed as d is chosen and v components of this design are used at stage j. Assume that $R'_j(v, d)$ is known for all possible values of v and d at the jth stage, for $j = 1, 2, \ldots, n$. If x_j components of design alternative a_j are used at stage j, for $j = 1, 2, \ldots, n$, then the system reliability R_s is $f(x_1, \ldots, x_n; a_1, \ldots, a_n) = \prod_{j=1}^{n} R'_j(x_j, a_j)$. Now the problem is to maximize

$$f(x_1, \ldots, x_n; a_1, \ldots, a_n) = \prod_{j=1}^{n} R'_j(x_j, a_j), \tag{4.4}$$

subject to

$$\sum_{j=1}^{n} g_j(x_j, a_j) \le C,$$

$$a_j = 1, 2, \ldots,$$

x_j being a nonnegative integer, where $g_j(x_j, a_j)$ is nonnegative and nondecreasing in x_j for each value of a_j.

This problem is a nonlinear integer programming problem with a single nonlinear constraint involving $2n$ variables.

Example 4-5

Consider Example 3-1 without the second constraint. It can be described so as to maximize

$$f(x_1, \dots, x_5) = \prod_{j=1}^{5}[1 - (1 - r_j)^{x_j}], \tag{4.5}$$

subject to

$$g_1(x_1, \dots, x_5) = \sum_{j=1}^{5} p_j x_j^2 \le 110,$$

$$g_2(x_1, \dots, x_5) = \sum_{j=1}^{5} w_j x_j \exp\left(\frac{x_j}{4}\right) \le 200,$$

x_j being a nonnegative integer.

Note that the above problem is a nonlinear integer programming problem with two nonlinear constraints. This example is used to demonstrate DP with Lagrange multipliers for a two-constraint problem.

4.2   The basic dynamic programming approach

The basic dynamic programming approach is generally used to solve optimization problems with, at most, one constraint and with or without bounds on the decision variables. This approach is now demonstrated for Examples 4-1, 4-2, and 4-4. While adopting the DP approach for Example 4-2, the lower limit on the variable x_j is improved as explained in Chapter 2, and a simple linear transformation of x_j is taken in order to reduce the computational effort in DP.

The DP approach for Example 4-1

Let $f_i(v_{i+1})$ denote the maximum expected net profit due to redundancy at stages $1, 2, \dots, i$, when v_{i+1} is the probability that the stages $i + 1, i + 2, \dots, 3$ work. Let $R_j(x_j) = 1 - (1 - r_j)^{x_j}$, and $v_4 = 1$. Then, for $i = 1, \dots, n$,

$$f_i(v_{i+1}) = \max_{x_1, x_2, \dots, x_i} \left\{ P' v_{i+1} \prod_{j=1}^{i} R_j(x_j) - \sum_{j=1}^{i} c_j x_j \right\}. \tag{4.6}$$

In our notation, $f_n(1.0)$ represents the maximum expected net profit. Equation (4.6) can be rewritten as

$$f_i(v_{i+1}) = \max_{x_i} \left\{ \max_{x_1, x_2, \dots, x_{i-1}} \left[P' v_{i+1} R_i(x_i) \prod_{j=1}^{i-1} R_j(x_j) - \sum_{j=1}^{i-1} c_j x_j \right] - c_i x_i \right\},$$

which can be expressed in a recursive form as

$$f_i(v_{i+1}) = \max_{x_i} \{ f_{i-1}(v_{i+1} R_i(x_i)) - c_i x_i \}, \tag{4.7}$$

for $i = 2, 3$. For $i = 3$, it is enough to consider only the value 1.0 for v_{i+1}. Let $x_i^*(v_{i+1})$ denote the value of x_i that maximizes $f_{i-1}[v_{i+1} R_i(x_i)] - c_i x_i$.

Equations (4.6) and (4.7) give

$$f_1(v_2) = \max_{x_1} \{ P' v_2 [1 - (1 - r_1)^{x_1}] - c_1 x_1 \},$$

$$f_2(v_3) = \max_{x_2} \{ f_1(v_3 [1 - (1 - r_2)^{x_2}]) - c_2 x_2 \}, \tag{4.8}$$

$$f_3(1.0) = \max_{x_3} \{ f_2(1 - (1 - r_3)^{x_3}) - c_3 x_3 \}.$$

Although recursive relations are available, it is not computationally feasible to evaluate $f_i(v)$ for all v in the continuous interval $[0, 1]$. Thus, $f_i(v)$ is evaluated on a grid of v for $i = 1$ and 2. Linear interpolation is done when $f_i(v)$ is needed, but not available, for a required value of v. Using the above equations, the values of $f_1(v)$ and $f_2(v)$ are obtained for $v = 1.0, 0.9, \dots, 0.2, 0.1$. The problem at stage 1 is to maximize

$$P' v_1 - c_1 x_1 = (10) v_2 [1 - (0.25)^{x_1}] - 1.0 x_1$$

over positive integer values of x_1 for $v_2 = 1.0, 0.9, \dots, 0.2, 0.1$. Any one-dimensional search technique can be used for this purpose. However, since x_1 usually takes a small integer value, a simple exhaustive search is carried out for each value of v_2. The computational results at stage 1 are given in Table 4.2.

Similarly the values of $f_2(v_3)$ and $x_2^*(v_3)$ are derived by maximizing

$$f_1(v_2) - c_2 x_2 = f_1(v_3 [1 - (0.5)^{x_2}]) - 1.0 x_2$$

over x_2 for $v_3 = 1.0, 0.9, \dots, 0.2, 0.1$ at stage 2. In the process of calculation, the value of $f_1(v)$ is obtained by interpolation when v is not a multiple of 0.1. For example, the value of $f_1(v)$ for $v = 0.88$ is determined by interpolation of $f_1(0.9)$ and $f_1(0.8)$ obtained in the stage 1 optimization. The computational results at stage 2 are also presented in Table 4.2.

Note that $f_3(1.0)$ is the maximum expected net profit. It is obtained by computing

$$f_2((1.0)[1 - (1 - r_3)^{x_3}]) - c_3 x_3 = f_2(1 - (0.667)^{x_3}) - 0.2 x_3$$

for several values of x_3 and finding its maximum. The optimal value of x_3 is $x_3^*(1.0) = 7$, and $f_3(1.0) = 1.3213$, which is the maximum expected net profit.

Table 4.2. Results of stage 1 of Example 4-1

v_2	$x_1^*(v_2)$	$f_1(v_2)$	v_3	$x_2^*(v_3)$	$f_2(v_3)$
1.0	2	7.3750	1.0	3	3.2031
0.9	2	6.4375	0.9	3	2.3828
0.8	2	5.5000	0.8	2	1.6250
0.7	2	4.5625	0.7	2	0.9687
0.6	2	3.6250	0.6	2	0.3750
0.5	1	2.7500	0.5	1	−0.1250
0.4	1	2.0000	0.4	1	−0.5000
0.3	1	1.2500	0.3	1	−0.8750
0.2	1	0.5000	0.2	1	−1.2500
0.1	1	−0.2500	0.1	1	−1.1250

The optimal solution is now obtained by the backtracking method. The optimal choice of x_3 is $x_3^* = x_3^*(1.0) = 7$. Correspondingly,

$$v_3 = v_4\left[1 - (1 - r_3)^{x_3^*}\right] = (1.0)\left[1 - (1 - 0.333)^7\right] = 0.94.$$

The optimal choice of x_2 is $x_2^* = x_2^*(0.94) = 3$ and, correspondingly,

$$v_2 = v_3\left[1 - (1 - r_2)^{x_2^*}\right] = (0.94)\left[1 - (1 - 0.5)^3\right] = 0.82.$$

The optimal choice of x_1 is $x_1^* = x_1^*(0.82) = 2$, and the corresponding value of v_1 is 0.77. Therefore, the optimal solution is $(x_1^*, x_2^*, x_3^*) = (2, 3, 7)$, and the maximum expected net profit is 1.3213. The system reliability for this optimal solution is 0.77.

The DP approach for Example 4-2

We first tighten the lower limit ℓ_j following the procedure described in Section 2.3: a feasible solution $(x_1, x_2, x_3, x_4) = (5, 3, 5, 3)$ is obtained. The corresponding system reliability is $f(x_1, x_2, x_3, x_4) = 0.918\,36$, which can be taken as a lower bound R_s^0 on the optimal system reliability. Then the reliability of each stage for optimal allocation $(x_1^*, x_2^*, x_3^*, x_4^*)$ must be at least 0.918 36, that is,

$$\left[1 - (1 - r_j)^{x_j^*}\right] \geq 0.918\,36,$$

that is,

$$x_j^* \geq \frac{\ln(1 - 0.918\,36)}{\ln(1 - r_j)}, \qquad \text{for } j = 1, 2, 3, 4. \tag{4.9}$$

Equation (4.9) gives the lower limits on the optimal values of x_1, x_2, x_3, and x_4 as 3, 2, 4, and 3, respectively. Now, the problem described in Example 4-2 is reduced so as to maximize

$$R_s = \prod_{j=1}^{4} R_j(x_j),$$

subject to

$$\sum_{j=1}^{4} c_j x_j \leq b,$$

$$\ell_j \leq x_j \leq u_j, \qquad \text{for } j = 1, \dots, 4,$$

x_j being a nonnegative integer, where $(\ell_1, \ell_2, \ell_3, \ell_4) = (3, 2, 4, 3)$. To reduce the computational effort further, we use the transformation $y_j = x_j - \ell_j$ and rewrite the above problem so as to maximize

$$R_s = \prod_{j=1}^{4} R'_j(y_j),$$

subject to

$$\sum_{j=1}^{4} c_j y_j \leq b',$$

$$y_j \leq u'_j, \qquad \text{for } j = 1, \dots, 4,$$

y_j being a nonnegative integer, where $R'_j(y_j) = [1 - (1 - r_j)^{y_j + \ell_j}]$, $b' = b - \sum_{j=1}^{4} c_j \ell_j$, and $u'_j = u_j - \ell_j$. We have $\sum_{j=1}^{4} c_j \ell_j = 22$, $b' = 8$, and $(u'_1, u'_2, u'_3, u'_4) = (3, 4, 2, 3)$. For $k = 1, \dots, 4$ and $d = 0, 1, \dots, b'$, let

$$f_k(d) = \max \left\{ \prod_{j=1}^{k} R'_j(y_j) : \sum_{j=1}^{k} c_j y_j \leq d, \ y_j \leq u'_j, \text{ and } y_j \text{ is a} \right.$$

$$\left. \text{nonnegative integer, for } j = 1, 2, \dots, k \right\}.$$

If the system consists of only the stages $1, 2, \dots, k$ in series and the total cost of components is bounded above by $d + \sum_{j=1}^{n} c_j \ell_j$, then $f_k(d)$ gives the optimal system reliability. We can write $f_k(d)$ in a recursive form as

$$f_k(d) = \max_{y_k \leq \min(\lfloor d/c_k \rfloor, u'_k)} R'_k(y_k) f_{k-1}(d - c_k y_k), \qquad (4.10)$$

for $k = 2, \dots, n$, where $\lfloor d/c_k \rfloor$ is the integer part of the ratio d/c_k. For $k = 1$, it is

$$f_1(d) = \max_{y_1 \leq \min(\lfloor d/c_1 \rfloor, u'_1)} R'_1(y_1).$$

Let $y_k^*(d)$ denote the value of y_k that maximizes $R'_k(y_k) f_{k-1}(d - c_k y_k)$ in (4.10). Since $R'_1(y)$ is an increasing function of y,

$$f_1(d) = \max_{y_1 \leq \min(\lfloor d/c_1 \rfloor, u'_1)} R'_1(y_1) = R'_1\left(\min\left(\left\lfloor \frac{d}{c_1} \right\rfloor, u'_1 \right) \right),$$

Table 4.3. Computational results of Example 4-2

d	$f_1(d)$	$y_1^*(d)$	$f_2(d)$	$y_2^*(d)$	$f_3(d)$	$y_3^*(d)$
0	0.936 00	0	0.877 50	0	0.822 66	0
1	0.936 00	0	0.877 50	0	0.850 08	1
2	0.974 40	1	0.913 50	0	0.863 79	2
3	0.974 40	1	0.921 37	1	0.884 95	1
4	0.989 76	2	0.927 90	0	0.899 23	2
5	0.989 76	2	0.959 17	1	0.906 98	2
6	0.995 90	3	0.959 17	1	0.929 20	1
7	0.995 90	3	0.974 29	1	0.944 19	2
8	0.995 90	3	0.974 29	1	0.944 19	2

and the optimal value of y_1 that gives $f_1(d)$ is $y_1^*(d) = \min(\lfloor d/c_1 \rfloor, u_1')$, and therefore the value of $f_1(d)$ can be directly computed for $d = 0, 1, \ldots, b'$. The recursive eq. (4.10) can be used to sequentially derive $f_2(d), f_3(d), \ldots, f_{n-1}(d)$ for $d = 0, 1, \ldots, b'$ and finally $f_n(b')$. Note that the derivation of $f_k(d)$ is now a single-variable problem, which can be solved using the values of $f_{k-1}(j)$, $j = 0, 1, \ldots, d$. At stage 1, we have $f_1(0) = R_1'(0) = 0.9360$ and $y_1^*(0) = 0$. Similarly, $f_1(d)$ and $y_1^*(d)$ are derived for $d = 1, 2, \ldots, 8$ and presented in Table 4.3. The values of $f_2(d)$ and $y_2^*(d)$ are derived from $f_1(j)$, $j = 0, 2, \ldots, 8$, as shown in the calculation below. For $d = 0$,

$$f_2(0) = \max_{y_2 \leq \min(\lfloor 0/3 \rfloor, 4)} R_2'(y_2) f_1(0 - 3y_2)$$

$$= \left[1 - (1 - 0.75)^{0+2}\right] f_1(0)$$

$$= (0.9375)(0.9360)$$

$$= 0.8775,$$

and $y_2^*(0) = 0$. Similarly, for $d = 6$,

$$f_2(6) = \max_{y_2 \leq \min(\lfloor 6/3 \rfloor, 4)} R_2'(y_2) f_1(6 - 3y_2)$$

$$= \max_{y_2 \in \{0,1,2\}} \left[1 - (1 - 0.75)^{y_2+2}\right] f_1(6 - 3y_2)$$

$$= \max\{(0.9375) f_1(6), (0.9844) f_1(3), (0.9961) f_1(0)\}$$

$$= \max\{(0.9375)(0.9959), (0.9844)(0.9744), (0.9961)(0.9360)\}$$

$$= 0.9592.$$

The value of y_2 that maximizes $R_2'(y_2) f_1(6 - 3y_2)$ over the choice of 0, 1, and 2 is $y_2^*(6) = 1$. The values of $f_2(d)$ and $y_2^*(d)$ are given in Table 4.3 for $d = 0, 1, 2, \ldots, 8$. The same procedure is repeated for deriving $f_3(d)$ and $y_3^*(d)$ for $d = 0, 1, 2, \ldots, 8$.

All these values are given in Table 4.3. Similarly this procedure gives $f_4(8) = 0.921\,67$ and $y_4^*(8) = 1$.

The optimal solution of the problem is derived using a backtracking method. The optimal choice of y_4 is $y_4^* = y_4^*(8) = 1$. The optimal choice of y_3 is

$$y_3^* = y_3^*(8 - c_4 y_4^*) = y_3^*(6) = 1.$$

The optimal choice of y_2 is

$$y_2^* = y_2^*(8 - c_3 y_3^* - c_4 y_4^*) = y_2^*(5) = 1.$$

The optimal choice of y_1 is

$$y_1^* = y_1^*(8 - c_2 y_2^* - c_3 y_3^* - c_4 y_4^*) = y_1^*(2) = 1.$$

Therefore, the optimal solution of Example 4-2 is

$$(y_1^*, y_2^*, y_3^*, y_4^*) = (1, 1, 1, 1),$$

and the corresponding system reliability is $f_4(8) = 0.921\,67$. The optimal allocation in Example 4-2 is

$$(x_1, x_2, x_3, x_4) = (\ell_1, \ell_2, \ell_3, \ell_4) + (y_1^*, y_2^*, y_3^*, y_4^*) = (4, 3, 5, 4).$$

The DP approach for Example 4-4
Let

$$f_k(d) = \max \left\{ \prod_{j=1}^{k} R_j'(x_j, a_j) : \sum_{j=1}^{k} g_j(x_j, a_j) \leq d, \, x_j \geq 0, \, a_j \geq 1, \text{ and } x_j \right.$$

$$\left. \text{and } a_j \text{ are integers, for } j = 1, 2, \ldots, k \right\}.$$

Then $f_k(d)$ can be written in recursive form as

$$f_k(d) = \max\{R_k'(x_k, a_k) f_{k-1}(d - g_k(x_k, a_k)): g_k(x_k, a_k) \leq d,$$

$$x_k \geq 0, \, a_k \geq 1, \text{ and } x_k \text{ and } a_k \text{ are integers}\}, \tag{4.11}$$

for $k = 2, \ldots, n$. Take $f_k(d) = 0$ for $d < 0$. For $k = 1$, it is

$$f_1(d) = \max\{R_1'(x_1, a_1): g_1(x_1, a_1) \leq d,$$

$$x_1 \geq 0, \, a_1 \geq 1, \text{ and } x_1 \text{ and } a_1 \text{ are integers}\}. \tag{4.12}$$

Now, we can solve Example 4-4 by the DP approach using eqs. (4.11) and (4.12).

4.3   The DP approach using Lagrange multipliers

Consider the two-constraint reliability optimization problem maximize

$$f(x_1, \ldots, x_n) = \prod_{j=1}^{n} R_j(x_j),$$

subject to

$$\sum_{j=1}^{n} g_{1j}(x_j) \leq b_1,$$

$$\sum_{j=1}^{n} g_{2j}(x_j) \leq b_2,$$

x_j being a nonnegative integer.

In order to reduce the search for an optimal solution, we can derive a lower bound ℓ_j on each x_j based on a lower bound on the maximum system reliability, as shown earlier in the basic DP approach, and add the constraints

$$x_j \geq \ell_j, \qquad \text{for } j = 1, \ldots, n,$$

to the above problem. The resulting problem can also be solved by the basic DP approach. However, as explained below, such an approach is not preferable from a computational point of view.

Let

$$f_k(d_1, d_2) = \max \left\{ \prod_{j=1}^{k} R_j(x_j): \sum_{j=1}^{k} g_{1j}(x_j) \leq d_1, \sum_{j=1}^{k} g_{2j}(x_j) \leq d_2, \text{ and} \right.$$

$$\left. x_j \geq \ell_j, x_j \text{ being an integer for } j = 1, 2, \ldots, k \right\}$$

for $k = 1, \ldots, n$, $d_1 = 0, 1, \ldots, b_1$, and $d_2 = 0, 1, \ldots, b_2$. Take $f_k(d_1, d_2) = 0$ when the corresponding system is infeasible. Then $f_k(d_1, d_2)$ can be written in recursive form as

$$f_k(d_1, d_2) = \max\{R_k(x_k) f_{k-1}(d_1 - g_{1k}(x_k), d_2 - g_{2k}(x_k)):$$

$$g_{1k}(x_k) \leq d_1, g_{2k}(x_k) \leq d_2, x_k \geq \ell_k, \text{ and}$$

$$x_k \text{ being an integer}\},$$

for $k = 2, \ldots, n$. For $k = 1$,

$$f_1(d_1, d_2) = \max\{R_1(x_1): g_{11}(x_1) \leq d_1, g_{21}(x_1) \leq d_2,$$

$$x_1 \geq \ell_1, \text{ and } x_1 \text{ is an integer}\}.$$

Optimal system reliability $f_n(b_1, b_2)$ is obtained by successively deriving $f_1(d_1, d_2)$, $f_2(d_1, d_2), \ldots, f_{n-1}(d_1, d_2)$; for all pairs (d_1, d_2) satisfying $d_1 \leq b_1$ and $d_2 \leq b_2$. This procedure obviously requires more computational effort and large memory capacity. Therefore, the basic DP approach is appropriate but less desirable for solving optimization problems with two constraints. In such a situation, the Lagrangian multipliers method reduces the problem to a set of single-constraint optimization problems which can be solved by the basic DP approach.

Problem 4.1

In order to solve Example 4-5 by the DP approach with Lagrange multipliers, consider the following optimization problem. Maximize

$$f(x_1, \ldots, x_n) = \exp\left\{ -\lambda\left[\sum_{j=1}^{n} g_{2j}(x_j) - b_2 \right] \right\} \prod_{j=1}^{n} R_j(x_j),$$

subject to

$$\sum_{j=1}^{n} g_{1j}(x_j) \leq b_1,$$

$$x_j \geq \ell_j, \qquad \text{for } j = 1, \ldots, n,$$

x_j being an integer, where λ is a penalty term.

Assume that the functions $g_{1j}(x_j)$ and $g_{2j}(x_j)$ are increasing in x_j. The Lagrange multipliers approach involves:

1. solving the above problem for a spectrum of λ values, and
2. selecting the value of λ for which $b_2 - \sum_{j=1}^{n} g_{2j}(x_j)$ is the least nonnegative.

Since b_2 is a constant, for any fixed λ, Problem 4.1 is equivalent to maximizing

$$R_s(\lambda) = \prod_{j=1}^{n} \{ R_j(x_j) \exp[-\lambda g_{2j}(x_j)] \}, \qquad (4.13)$$

subject to

$$\sum_{j=1}^{n} g_{1j}(x_j) \leq b_1,$$

$$x_j \geq \ell_j, \qquad \text{for } j = 1, \ldots, n,$$

x_j being an integer.

Let $G_k = \sum_{j=1}^{k} g_{1j}(\ell_j)$, $\overline{G}_k = b_1 - \sum_{j=k+1}^{n} g_{1j}(\ell_j)$, for $k = 1, \ldots, n$, and $\overline{G}_n = b_1$. Let

$$f_k(d) = \max\left\{ \prod_{j=1}^{k} \langle R_j(x_j) \exp[-\lambda g_{2j}(x_j)] \rangle : \sum_{j=1}^{k} g_{1j}(x_j) \leq d, \text{ and} \right.$$

$$\left. x_j \geq \ell_j, x_j \text{ being an integer, for } j = 1, 2, \ldots, k \right\},$$

for $G_k \leq d \leq \overline{G}_k$ and $k = 1, \ldots, n$. For $d < G_k$, no solution satisfies

$$\sum_{j=1}^{k} g_{1j}(x_j) \leq d, \text{ and } x_j \geq \ell_j, \qquad \text{for } j = 1, \ldots, k.$$

Similarly, for $d > \overline{G}_k$, there is no solution $(x_1, \ldots, x_n)$ that satisfies

$$\sum_{j=k+1}^{n} g_{1j}(x_j) \leq b_1 - m, \text{ and } x_j \geq \ell_j, \qquad \text{for } j = k+1, \ldots, n.$$

Therefore, $f_k(d)$ is not computed for $d < G_k$ and $d > \overline{G}_k$. We have the recursive relation

$$f_k(d) = \max\{R_k(x_k) \exp[-\lambda g_{2k}(x_k)] f_{k-1}[d - g_{1k}(x_k)]: g_{1k}(x_k) \leq d - G_{k-1},$$

$$\text{and } x_k \geq \ell_k, x_k \text{ being an integer}\}, \tag{4.14}$$

for $k = 2, \ldots, n$ and the equation

$$f_1(d) = \max\{R_1(x_1) \exp[-\lambda g_{21}(x_1)]: g_{11}(x_1) \leq d,$$

$$\text{and } x_1 \geq \ell_1, x_1 \text{ being an integer}\}. \tag{4.15}$$

Let $x_k^*(d)$ denote the value of x_k that gives $f_k(d)$ in the recursive eq. (4.14). Similarly, let $x_1^*(d)$ denote the value of x_1 that gives $f_1(d)$ in (4.15). Now, for any fixed λ, Problem 4.1 can be solved by the basic DP approach using eqs. (4.14) and (4.15).

Solution of Example 4-3 using Lagrange multipliers
In this example, $n = 5$, $b_1 = 110$, $b_2 = 200$, $g_{1j}(x_j) = p_j x_j^2$, and $g_{2j}(x_j) = w_j x_j \exp(x_j/4)$ for $j = 1, \ldots, 5$, where r_j, p_j, and w_j are as shown in Example 4-3. Consider a feasible solution $(x_1, \ldots, x_5) = (3, 3, 3, 2, 2)$, which is obtained by the simple procedure described in Chapter 2. The system reliability $R_s = 0.8125$ for this solution is obviously a lower bound on the optimal system reliability. As shown in the basic DP approach for Example 4-2, the lower bounds on variables are tightened as $(\ell_1, \ell_2, \ell_3, \ell_4, \ell_5) = (2, 1, 1, 2, 2)$. Now $g_{1j}(\ell_j)$, G_j and $\overline{G}_j$ are

j	$g_{1j}(\ell_j)$	G_j	$\overline{G}_j$
1	4	4	81
2	2	6	83
3	3	9	86
4	16	25	102
5	8	33	110

For several values of λ ranging from 0.0001 to 0.0018, Problem 4.1 is solved by deriving $f_k(d)$ for $G_k \leq d \leq \overline{G}_k$ for $k = 1, \ldots, n$. For each selected value of λ, the

Table 4.4. Optimal allocations for several values of λ

λ	g_1	g_2	System reliability	Optimal allocation
0.0002	108	241.3380	0.930 550	(3, 3, 3, 3, 3)
0.0004	93	216.9095	0.922 160	(3, 3, 2, 3, 3)
0.0006	93	216.9095	0.922 160	(3, 3, 2, 3, 3)
0.0008	83	192.4811	0.904 470	(3, 2, 2, 3, 3)
0.0010	83	192.4811	0.904 470	(3, 2, 2, 3, 3)
0.0011	83	192.4811	0.904 470	(3, 2, 2, 3, 3)
0.0012	83	192.4811	0.904 470	(3, 2, 2, 3, 3)
0.0013	83	192.4811	0.904 470	(3, 2, 2, 3, 3)
0.0014	83	192.4811	0.904 470	(3, 2, 2, 3, 3)
0.0015	83	192.4811	0.904 470	(3, 2, 2, 3, 3)
0.0016	78	171.1062	0.875 290	(2, 2, 2, 3, 3)
0.0017	78	171.1062	0.875 290	(2, 2, 2, 3, 3)
0.0018	68	143.6242	0.833 610	(2, 2, 2, 3, 2)
0.0020	68	143.6242	0.833 610	(2, 2, 2, 3, 2)

corresponding system reliability, optimal allocation and resource consumption values are given in Table 4.4.

Since $b_2 - \sum_{j=1}^{n} g_{2j}(x_j)$ is the least nonnegative value for $\lambda = 0.0015$, the latter is taken as the required optimal solution of Problem 4.1. For this choice of λ, the values of $f_k(d)$ and $x_k^*(d)$ for $k = 1, \ldots, n$ are given in Tables 4.5–4.7. The corresponding optimal values of $x_1, \ldots, x_5$ can be successively obtained from Tables 4.5–4.7 by the backtracking procedure, as given in the following calculations.

From Table 4.7, we have an optimal value of x_5 as $x_5^*(110) = 3$. Next, the optimal value of x_4 is

$$x_4^*(110 - p_5(3^2)) = x_4^*(92) = 3.$$

Similarly the optimal values of $x_2, \ldots, x_5$ are

$$x_3^*(110 - p_4(3^2) - p_5(3^2)) = x_3^*(56) = 2,$$
$$x_2^*(110 - p_3(2^2) - p_4(3^2) - p_5(3^2)) = x_2^*(44) = 2,$$
$$x_1^*(110 - p_2(2^2) - p_3(2^2) - p_4(3^2) - p_5(3^2)) = x_1^*(36) = 3.$$

Thus, the optimal solution of Problem 4.1 for $\lambda = 0.0015$ is

$$(x_1^*, x_2^*, x_3^*, x_4^*, x_5^*) = (3, 2, 2, 3, 3),$$

and the corresponding system reliability is 0.9045.

Table 4.5. DP solution for $\lambda = 0.0015$

d	$f_1(d)$	$x_1^*(d)$	$f_2(d)$	$x_2^*(d)$	$f_3(d)$	$x_3^*(d)$	$f_4(d)$	$x_4^*(d)$	$f_5(d)$	$x_5^*(d)$
0										
1										
2										
3										
4	0.927 33	2								
5	0.927 33	2								
6	0.927 33	2	0.776 18	1						
7	0.927 33	2	0.776 18	1						
8	0.927 33	2	0.776 18	1						
9	0.928 01	3	0.776 18	1	0.687 88	1				
10	0.928 01	3	0.776 18	1	0.687 88	1				
11	0.928 01	3	0.776 74	1	0.687 88	1				
12	0.928 01	3	0.871 30	2	0.687 88	1				
13	0.928 01	3	0.871 30	2	0.687 88	1				
14	0.928 01	3	0.871 30	2	0.688 38	1				
15	0.928 01	3	0.871 30	2	0.772 18	1				
16	0.928 01	3	0.871 30	2	0.772 18	1				
17	0.928 01	3	0.871 93	2	0.772 18	1				
18	0.928 01	3	0.871 93	2	0.772 18	1				
19	0.928 01	3	0.871 93	2	0.772 18	1				
20	0.928 01	3	0.871 93	2	0.772 74	1				
21	0.928 01	3	0.871 93	2	0.772 74	1				
22	0.928 01	3	0.871 93	2	0.772 74	1				
23	0.928 01	3	0.871 93	2	0.772 74	1				
24	0.928 01	3	0.871 93	2	0.829 12	2				
25	0.928 01	3	0.871 93	2	0.829 12	2	0.585 96	2		
26	0.928 01	3	0.871 93	2	0.829 12	2	0.585 96	2		
27	0.928 01	3	0.871 93	2	0.829 12	2	0.585 96	2		
28	0.928 01	3	0.871 93	2	0.829 12	2	0.585 96	2		
29	0.928 01	3	0.871 93	2	0.829 72	2	0.585 96	2		
30	0.928 01	3	0.871 93	2	0.829 72	2	0.586 39	2		
31	0.928 01	3	0.871 93	2	0.829 72	2	0.657 77	2		
32	0.928 01	3	0.871 93	2	0.829 72	2	0.657 77	2		
33	0.928 01	3	0.871 93	2	0.829 72	2	0.657 77	2	0.525 42	2
34	0.928 01	3	0.871 93	2	0.829 72	2	0.657 77	2	0.525 42	2
35	0.928 01	3	0.871 93	2	0.829 72	2	0.657 77	2	0.525 42	2
36	0.928 01	3	0.871 93	2	0.829 72	2	0.658 25	2	0.525 42	2
37	0.928 01	3	0.871 93	2	0.829 72	2	0.658 25	2	0.525 42	2
38	0.928 01	3	0.871 93	2	0.829 72	2	0.658 25	2	0.525 81	2
39	0.928 01	3	0.871 93	2	0.829 72	2	0.658 25	2	0.589 81	2
40	0.928 01	3	0.871 93	2	0.829 72	2	0.706 28	2	0.589 81	2

Table 4.6. DP solution for $\lambda = 0.0015$

d	$f_1(d)$	$x_1^*(d)$	$f_2(d)$	$x_2^*(d)$	$f_3(d)$	$x_3^*(d)$	$f_4(d)$	$x_4^*(d)$	$f_5(d)$	$x_5^*(d)$
41	0.928 01	3	0.871 93	2	0.829 72	2	0.706 28	2	0.589 81	2
42	0.928 01	3	0.871 93	2	0.829 72	2	0.706 28	2	0.589 81	2
43	0.928 01	3	0.871 93	2	0.829 72	2	0.706 28	2	0.589 81	2
44	0.928 01	3	0.871 93	2	0.829 72	2	0.706 28	2	0.590 24	2
45	0.928 01	3	0.871 93	2	0.829 72	2	0.706 79	2	0.590 24	2
46	0.928 01	3	0.871 93	2	0.829 72	2	0.706 79	2	0.590 24	2
47	0.928 01	3	0.871 93	2	0.829 72	2	0.706 79	2	0.590 24	2
48	0.928 01	3	0.871 93	2	0.829 72	2	0.706 79	2	0.633 31	2
49	0.928 01	3	0.871 93	2	0.829 72	2	0.706 79	2	0.633 31	2
50	0.928 01	3	0.871 93	2	0.829 72	2	0.706 79	2	0.633 31	2
51	0.928 01	3	0.871 93	2	0.829 72	2	0.706 79	2	0.633 31	2
52	0.928 01	3	0.871 93	2	0.829 72	2	0.706 79	2	0.633 31	2
53	0.928 01	3	0.871 93	2	0.829 72	2	0.706 79	2	0.633 77	2
54	0.928 01	3	0.871 93	2	0.829 72	2	0.706 79	2	0.633 77	2
55	0.928 01	3	0.871 93	2	0.829 72	2	0.706 79	2	0.633 77	2
56	0.928 01	3	0.871 93	2	0.829 72	2	0.706 79	2	0.633 77	2
57	0.928 01	3	0.871 93	2	0.829 72	2	0.706 79	2	0.633 77	2
58	0.928 01	3	0.871 93	2	0.829 72	2	0.706 79	2	0.638 12	3
59	0.928 01	3	0.871 93	2	0.829 72	2	0.706 79	2	0.638 12	3
60	0.928 01	3	0.871 93	2	0.829 72	2	0.749 48	3	0.638 12	3
61	0.928 01	3	0.871 93	2	0.829 72	2	0.749 48	3	0.638 12	3
62	0.928 01	3	0.871 93	2	0.829 72	2	0.749 48	3	0.638 12	3
63	0.928 01	3	0.871 93	2	0.829 72	2	0.749 48	3	0.638 58	3
64	0.928 01	3	0.871 93	2	0.829 72	2	0.749 48	3	0.638 58	3
65	0.928 01	3	0.871 93	2	0.829 72	2	0.750 03	3	0.638 58	3
66	0.928 01	3	0.871 93	2	0.829 72	2	0.750 03	3	0.638 58	3
67	0.928 01	3	0.871 93	2	0.829 72	2	0.750 03	3	0.638 58	3
68	0.928 01	3	0.871 93	2	0.829 72	2	0.750 03	3	0.672 05	2
69	0.928 01	3	0.871 93	2	0.829 72	2	0.750 03	3	0.672 05	2
70	0.928 01	3	0.871 93	2	0.829 72	2	0.750 03	3	0.672 05	2
71	0.928 01	3	0.871 93	2	0.829 72	2	0.750 03	3	0.672 05	2
72	0.928 01	3	0.871 93	2	0.829 72	2	0.750 03	3	0.672 05	2
73	0.928 01	3	0.871 93	2	0.829 72	2	0.750 03	3	0.672 54	2
74	0.928 01	3	0.871 93	2	0.829 72	2	0.750 03	3	0.672 54	2
75	0.928 01	3	0.871 93	2	0.829 72	2	0.750 03	3	0.672 54	2
76	0.928 01	3	0.871 93	2	0.829 72	2	0.750 03	3	0.672 54	2
77	0.928 01	3	0.871 93	2	0.829 72	2	0.750 03	3	0.672 54	2
78	0.928 01	3	0.871 93	2	0.829 72	2	0.750 03	3	0.677 15	3
79	0.928 01	3	0.871 93	2	0.829 72	2	0.750 03	3	0.677 15	3
80	0.928 01	3	0.871 93	2	0.829 72	2	0.750 03	3	0.677 15	3

Table 4.7. DP solution for $\lambda = 0.0015$

d	$f_1(d)$	$x_1^*(d)$	$f_2(d)$	$x_2^*(d)$	$f_3(d)$	$x_3^*(d)$	$f_4(d)$	$x_4^*(d)$	$f_5(d)$	$x_5^*(d)$
81	0.928 01	3	0.871 93	2	0.829 72	2	0.750 03	3	0.677 15	3
82			0.871 93	2	0.829 72	2	0.750 03	3	0.677 15	3
83			0.871 93	2	0.829 72	2	0.750 03	3	0.677 65	3
84					0.829 72	2	0.750 03	3	0.677 65	3
85					0.829 72	2	0.750 03	3	0.677 65	3
86					0.829 72	2	0.750 03	3	0.677 65	3
87							0.750 03	3	0.677 65	3
88							0.750 03	3	0.677 65	3
89							0.750 03	3	0.677 65	3
90							0.750 03	3	0.677 65	3
91							0.750 03	3	0.677 65	3
92							0.750 03	3	0.677 65	3
93							0.750 03	3	0.677 65	3
94							0.750 03	3	0.677 65	3
95							0.750 03	3	0.677 65	3
96							0.750 03	3	0.677 65	3
97							0.750 03	3	0.677 65	3
98							0.750 03	3	0.677 65	3
99							0.750 03	3	0.677 65	3
100							0.750 03	3	0.677 65	3
101							0.750 03	3	0.677 65	3
102							0.750 03	3	0.677 65	3
103									0.677 65	3
104									0.677 65	3
105									0.677 65	3
106									0.677 65	3
107									0.677 65	3
108									0.677 65	3
109									0.677 65	3
110									0.677 65	3

4.4   The DP approach using dominating allocations

When a redundancy allocation problem involves more constraints, even the Lagrangian approach involves a lot of computational effort. Due to its simple logic, Kettelle's [152] dynamic programming approach based on dominating allocations is quite useful in such a situation. Consider the following redundancy allocation problem with m constraints.

Maximize

$$f(x_1, \ldots, x_n) = \prod_{j=1}^{n} \left[1 - (1 - r_j)^{x_j} \right],$$

subject to

$$\sum_{j=1}^{n} g_{ij}(x_j) \leq b_i, \qquad \text{for } i = 1, \ldots, m,$$

$$x_j \geq \ell_j, \qquad \text{for } j = 1, \ldots, n,$$

x_j being an integer.

Let $(x_1, \ldots, x_k)$ denote allocation of x_j components to stage j, $j = 1, \ldots, k$. The partial vector $(x_1, \ldots, x_k)$ is called a *k-stage allocation*. A k-stage allocation $(x_1, \ldots, x_k)$ is feasible if

$$\sum_{j=1}^{k} g_{ij}(x_j) \leq b_i - \sum_{j=k+1}^{n} g_{ij}(\ell_j), \qquad \text{for } i = 1, \ldots, m,$$

and

$$x_j \geq \ell_j, \qquad \text{for } j = 1, \ldots, k.$$

Let $(x_1, \ldots, x_k)$ and $(y_1, \ldots, y_k)$ be two k-stage feasible allocations. Then $(x_1, \ldots, x_k)$ is said to be dominated by $(y_1, \ldots, y_k)$ if

$$\prod_{j=1}^{k} R_j(x_j) \leq \prod_{j=1}^{k} R_j(y_j)$$

and

$$\sum_{j=1}^{k} g_{ij}(x_j) \geq \sum_{j=1}^{k} g_{ij}(y_j), \qquad \text{for } i = 1, \ldots, m,$$

with one of the inequalities holding as a strict inequality.

When $y = (y_1, \ldots, y_k)$ dominates $x = (x_1, \ldots, x_k)$, we prefer y to x since y gives better reliability for stages $1, 2, \ldots, k$, while consuming less resources. Let H_k denote the set of all nondominated feasible k-stage allocations for $k = 1, \ldots, n$.

The DP approach using dominating allocations involves derivation of $H_1, \ldots, H_n$, recursively, and derivation of the optimal allocation in H_n by complete enumeration of H_n. The set of all x_1 integer values, which satisfy

$$x_1 \geq \ell_1,$$

$$g_{i1}(x_1) \leq b_i - \sum_{j=2}^{n} g_{ij}(\ell_j), \qquad \text{for } i = 1, \ldots, m,$$

is the set H_1. In order to derive H_{k+1} from H_k for $k = 1, \ldots, n-1$, first obtain the set of $(k+1)$-stage allocations $(x_1, \ldots, x_k, x_{k+1})$, which satisfy

$$(x_1, \ldots, x_k) \in H_k,$$

$$\sum_{j=1}^{k+1} g_{ij}(x_j) \leq b_i - \sum_{j=k+2}^{n} g_{ij}(\ell_j), \qquad \text{for } i = 1, \ldots, m,$$

$$x_{k+1} \geq \ell_{k+1}.$$

The elimination of all dominated $(k+1)$-stage allocations in the above set gives H_{k+1}.

Solution of Example 4-3 using dominating allocations

Consider Example 4-3 (Example 3-1) in which $n = 5$ and $m = 3$. Following the procedure for lower bounds on the optimal values of decision variables, the feasible allocation $(3, 3, 3, 2, 2)$ is obtained. The system reliability for this allocation is 0.8125, and the corresponding lower bounds are $(\ell_1, \ldots, \ell_5) = (2, 1, 1, 2, 2)$. Now the problem under consideration is equivalent to maximizing

$$R_s = \prod_{j=1}^{5} [1 - (1 - r_j)^{x_j}],$$

subject to

$$g_1 = \sum_{j=1}^{5} p_j x_j^2 \leq 110,$$

$$g_2 = \sum_{j=1}^{5} c_j \left[x_j + \exp\left(\frac{x_j}{4}\right) \right] \leq 175,$$

$$g_3 = \sum_{j=1}^{5} w_j x_j \exp\left(\frac{x_j}{4}\right) \leq 200,$$

$$x_j \geq \ell_j, \qquad \text{for } j = 1, \ldots, 5,$$

x_j being an integer.

Since the four terms $p_1 x_1^2$, $c_1[x_1 + \exp(x_1/4)]$, $w_1 x_1 \exp(x_1/4)$, and $1 - (1 - r_1)^{x_1}$ are strictly increasing in x_1, H_1 is the set of all values of x_1, which satisfy

$$p_1 x_1^2 \leq 110 - \sum_{j=2}^{5} p_j \ell_j^2,$$

$$c_1 \left[x_1 + \exp\left(\frac{x_1}{4}\right) \right] \leq 175 - \sum_{j=2}^{5} c_j \left[\ell_j + \exp\left(\frac{\ell_j}{4}\right) \right],$$

$$w_1 x_1 \exp\left(\frac{x_1}{4}\right) \leq 200 - \sum_{j=2}^{5} w_j \ell_j \exp\left(\frac{\ell_j}{4}\right),$$

$$x_1 \geq \ell_1.$$

Table 4.8. One-stage allocations

Allocation	g_1	g_2	g_3	Reliability
(2)	4.00	25.54	23.08	0.960 000
(3)	9.00	35.82	44.46	0.992 000
(4)	16.00	47.03	76.11	0.998 400
(5)	25.00	59.43	122.16	0.999 680

This means that H_1 is the set of all values of x_1 which satisfy

$$x_1^2 \leq 110 - [2(1)^2 + 3(1)^2 + 4(2)^2 + 2(2)^2] = 81,$$

$$7\left[x_1 + \exp\left(\frac{x_1}{4}\right)\right] \leq 175 - \left\langle 7\left[1 + \exp\left(\frac{1}{4}\right)\right] + 5\left[1 + \exp\left(\frac{1}{4}\right)\right]\right.$$

$$\left. + 9\left[2 + \exp\left(\frac{2}{4}\right)\right] + 4\left[2 + \exp\left(\frac{2}{4}\right)\right]\right\rangle = 100.16,$$

$$7x_1 \exp\left(\frac{x_1}{4}\right) \leq 200 - \left\langle 8\left[\exp\left(\frac{1}{4}\right)\right] + 8\left[\exp\left(\frac{1}{4}\right)\right]\right.$$

$$\left. + 6\left[2\exp\left(\frac{2}{4}\right)\right] + 9\left[2\exp\left(\frac{2}{4}\right)\right]\right\rangle = 129.99,$$

$$x_1 \geq 2.$$

Therefore, H_1 is the set of all values of x_1 which satisfy

$$x_1^2 \leq 81,$$

$$7\left[x_1 + \exp\left(\frac{x_1}{4}\right)\right] \leq 100.16,$$

$$7x_1 \exp\left(\frac{x_1}{4}\right) \leq 129.99,$$

$$x_1 \geq 2.$$

The first-stage allocations in H_1 are listed in Table 4.8, along with the corresponding resource consumptions $g_1(x_1) = p_1 x_1^2$, $g_2(x_1) = c_1[x_1 + \exp(x_1/4)]$, and $g_3(x_1) = w_1 x_1 \exp(x_1/4)$, and the value of $R_1(x_1)$. Now, H_2 is the set of all nondominated two-stage allocations (x_1, x_2), which satisfy

$x_1 \in H_1,$

$$p_1 x_1^2 + p_2 x_2^2 \leq 110 - \sum_{j=3}^{5} p_j \ell_j^2,$$

$$c_1 \left[x_1 + \exp\left(\frac{x_1}{4}\right) \right] + c_2 \left[x_2 + \exp\left(\frac{x_2}{4}\right) \right] \leq 175 - \sum_{j=3}^{5} c_j \left[\ell_j + \exp\left(\frac{\ell_j}{4}\right) \right],$$

$$w_1 x_1 \exp\left(\frac{x_1}{4}\right) + w_2 x_2 \exp\left(\frac{x_2}{4}\right) \leq 200 - \sum_{j=3}^{5} w_j \ell_j \exp\left(\frac{\ell_j}{4}\right),$$

$x_2 \geq \ell_2.$

By substituting the values of p_j, c_j, w_j, and ℓ_j, it can be seen that H_2 is the set of all (x_1, x_2), which satisfy

$$\left.\begin{array}{l} x_1 \in H_1, \\[4pt] x_1^2 + 2x_2^2 \leq 83, \\[4pt] 7\left[x_1 + \exp\left(\dfrac{x_1}{4}\right) \right] + 7\left[x_2 + \exp\left(\dfrac{x_2}{4}\right) \right] \leq 116.15, \\[8pt] 7x_1 \exp\left(\dfrac{x_1}{4}\right) + 8x_2 \exp\left(\dfrac{x_2}{4}\right) \leq 140.27, \\[8pt] x_2 \geq 1. \end{array}\right\} \tag{4.16}$$

For each (x_1) in H_1, find the values of x_2 that satisfy eq. (4.16). Note that if some value of x_2 violates (4.16) for a particular (x_1) in H_1, then all higher values of x_2 also violate the inequalities for the same (x_1).

Each two-stage allocation (x_1, x_2) for which $(x_1) \in H_1$ and eq. (4.16) hold, is given in Table 4.9 along with the corresponding resource consumptions $g_1(x_1) + g_1(x_2) = p_1 x_1^2 + p_2 x_2^2$, $g_2(x_1) + g_2(x_2) = c_1[x_1 + \exp(x_1/4)] + c_2[x_2 + \exp(x_2/4)]$, and $g_3(x_1) + g_3(x_2) = w_1 x_1 \exp(x_1/4) + w_2 x_2 \exp(x_2/4)$, and the reliability, $R_1(x_1)R_2(x_2)$, of the first two stages.

Note that all of the allocations with superscript d in Table 4.9 are dominated. H_2 is the set of all nondominated two-stage allocations in Table 4.9. The set of three-stage allocations derived from H_2 are given in Table 4.10. H_3 is the set of all allocations in Table 4.10 without the superscript d. Similarly, H_4 and H_5 are obtained, successively. The four-stage allocations derived from H_3 and the five-stage allocations derived from H_4 are given in Tables 4.11 and 4.12, respectively. The allocations without the superscript d in Tables 4.11 and 4.12 give the sets H_4 and H_5, respectively. The five-stage nondominated allocation $(3, 2, 2, 3, 3)$, which gives a maximum system reliability of 0.9044 in Table 4.12, is the required optimal allocation.

Table 4.9. Two-stage allocations

Allocation	g_1	g_2	g_3	Reliability
(2, 1)	6.00	41.53	33.35	0.816 000
(2, 2)	12.00	51.08	49.46	0.938 400
$(2, 3)^d$	22.00	61.36	73.89	0.956 760
$(2, 4)^d$	36.00	72.57	110.07	0.959 514
(3, 1)	11.00	51.81	54.73	0.843 200
(3, 2)	17.00	61.36	70.84	0.969 680
(3, 3)	27.00	71.64	95.27	0.988 652
$(3, 4)^d$	41.00	82.85	131.45	0.991 498
$(4, 1)^d$	18.00	63.02	86.38	0.848 640
(4, 2)	24.00	72.57	102.49	0.975 936
(4, 3)	34.00	82.85	126.92	0.995 030
$(5, 1)^d$	27.00	75.42	132.43	0.849 728

Table 4.10. Three-stage allocations

Allocation	g_1	g_2	g_3	Reliability
(2, 1, 1)	9.00	52.95	43.62	0.734 400
(2, 1, 2)	18.00	59.77	59.73	0.807 840
$(2, 1, 3)^d$	33.00	67.11	84.16	0.815 184
$(2, 1, 4)^d$	54.00	75.12	120.34	0.815 918
(2, 2, 1)	15.00	62.50	59.73	0.844 560
(2, 2, 2)	24.00	69.32	75.84	0.929 016
(2, 2, 3)	39.00	76.67	100.27	0.937 462
$(2, 2, 4)^d$	60.00	84.67	136.45	0.938 306
(3, 1, 1)	14.00	63.23	65.00	0.758 880
$(3, 1, 2)^d$	23.00	70.05	81.11	0.834 768
$(3, 1, 3)^d$	38.00	77.40	105.54	0.842 357
$(3, 1, 4)^d$	59.00	85.40	141.72	0.843 116
(3, 2, 1)	20.00	72.78	81.11	0.872 712
(3, 2, 2)	29.00	79.60	97.22	0.959 983
(3, 2, 3)	44.00	86.94	121.65	0.968 710
$(3, 3, 1)^d$	30.00	83.06	105.54	0.889 787
(3, 3, 2)	39.00	89.88	121.65	0.978 765
(3, 3, 3)	54.00	97.22	146.08	0.987 663
$(4, 2, 1)^d$	27.00	83.99	112.76	0.878 342
(4, 2, 2)	36.00	90.81	128.87	0.966 177
$(4, 3, 1)^d$	37.00	94.27	137.19	0.895 527

4.5   The DP approach for hierarchical series-parallel systems

As demonstrated previously, the DP approach is useful for maximizing series system reliability through redundancy allocation subject to separable constraints. This is true

Table 4.11. Four-stage allocations

Allocation	g_1	g_2	g_3	Reliability
(2, 1, 1, 2)	25.00	85.79	63.40	0.644 436
$(2, 1, 1, 3)^d$	45.00	99.00	81.73	0.702 913
$(2, 1, 1, 4)^d$	73.00	113.41	108.86	0.723 379
(2, 1, 2, 2)	34.00	92.61	79.51	0.708 880
$(2, 1, 2, 3)^d$	54.00	105.82	97.84	0.773 204
$(2, 1, 2, 4)^d$	82.00	120.23	124.97	0.795 717
(2, 2, 1, 2)	31.00	95.34	79.51	0.741 101
$(2, 2, 1, 3)^d$	51.00	108.55	97.84	0.808 349
$(2, 2, 1, 4)^d$	79.00	122.96	124.97	0.831 886
(2, 2, 2, 2)	40.00	102.16	95.62	0.815 212
(2, 2, 2, 3)	60.00	115.37	113.95	0.889 184
$(2, 2, 2, 4)^d$	88.00	129.78	141.08	0.915 075
(2, 2, 3, 2)	55.00	109.51	120.05	0.822 623
(2, 2, 3, 3)	75.00	122.72	138.38	0.897 268
(3, 1, 1, 2)	30.00	96.07	84.78	0.665 917
$(3, 1, 1, 3)^d$	50.00	109.28	103.11	0.726 343
$(3, 1, 1, 4)^d$	78.00	123.69	130.24	0.747 492
(3, 2, 1, 2)	36.00	105.62	100.89	0.765 805
$(3, 2, 1, 3)^d$	56.00	118.83	119.22	0.835 294
$(3, 2, 1, 4)^d$	84.00	133.24	146.35	0.859 616
(3, 2, 2, 2)	45.00	112.44	117.00	0.842 385
(3, 2, 2, 3)	65.00	125.65	135.33	0.918 824
(3, 2, 2, 4)	93.00	140.06	162.46	0.945 577
$(3, 2, 3, 2)^d$	60.00	119.78	141.43	0.850 043
(3, 2, 3, 3)	80.00	132.99	159.76	0.927 177
(3, 3, 2, 2)	55.00	122.72	141.43	0.858 866
(3, 3, 2, 3)	75.00	135.93	159.76	0.936 800
(3, 3, 3, 2)	70.00	130.06	165.86	0.866 674
(4, 2, 2, 2)	52.00	123.65	148.65	0.847 820
(4, 2, 2, 3)	72.00	136.86	166.98	0.924 752

because the objective function is a product of the functions of individual variables. Generally, it is not possible to adopt the DP approach when the objective function is inseparable. However, this approach can be adopted for optimal redundancy allocation in some systems, such as hierarchical series-parallel systems (HSP), which are described in Section 1.5.5. The DP approach has also been proposed by Hikita et al. [126] to solve reliability–redundancy allocation problems in HSP systems subject to a single constraint. We now explain and illustrate the DP approach for solving redundancy allocation problems in HSP systems.

Table 4.12. Nondominated five-stage allocations

Allocation	g_1	g_2	g_3	Reliability
(2, 1, 1, 2, 2)	33.00	100.38	93.08	0.604 159
(2, 1, 1, 2, 3)	43.00	106.26	120.56	0.634 367
(2, 1, 1, 2, 4)^d	57.00	112.66	161.26	0.641 919
(2, 1, 2, 2, 2)	42.00	107.20	109.19	0.664 575
(2, 1, 2, 2, 3)	52.00	113.08	136.67	0.697 804
(2, 1, 2, 2, 4)^d	66.00	119.48	177.37	0.706 111
(2, 2, 1, 2, 2)	39.00	109.93	109.19	0.694 782
(2, 2, 1, 2, 3)	49.00	115.81	136.67	0.729 521
(2, 2, 1, 2, 4)^d	63.00	122.21	177.37	0.738 206
(2, 2, 2, 2, 2)	48.00	116.75	125.30	0.764 261
(2, 2, 2, 2, 3)	58.00	122.63	152.78	0.802 474
(2, 2, 2, 2, 4)	72.00	129.03	193.48	0.812 028
(2, 2, 2, 3, 2)	68.00	129.96	143.63	0.833 610
(2, 2, 2, 3, 3)	78.00	135.84	171.11	0.875 291
(2, 2, 3, 2, 2)	63.00	124.10	149.73	0.771 209
(2, 2, 3, 2, 3)^d	73.00	129.98	177.21	0.809 770
(2, 2, 3, 3, 2)	83.00	137.31	168.06	0.841 189
(2, 2, 3, 3, 3)	93.00	143.19	195.54	0.883 248
(3, 1, 1, 2, 2)	38.00	110.66	114.46	0.624 297
(3, 1, 1, 2, 3)^d	48.00	116.54	141.94	0.655 512
(3, 1, 1, 2, 4)^d	62.00	122.94	182.64	0.663 316
(3, 2, 1, 2, 2)	44.00	120.21	130.57	0.717 942
(3, 2, 1, 2, 3)^d	54.00	126.09	158.05	0.753 839
(3, 2, 1, 2, 4)^d	68.00	132.49	198.75	0.762 814
(3, 2, 2, 2, 2)	53.00	127.03	146.68	0.789 736
(3, 2, 2, 2, 3)	63.00	132.91	174.16	0.829 223
(3, 2, 2, 3, 2)	73.00	140.24	165.01	0.861 398
(3, 2, 2, 3, 3)	83.00	146.12	192.49	0.904 467
(3, 2, 2, 4, 2)	101.00	154.65	192.14	0.886 478
(3, 2, 3, 3, 2)^d	88.00	147.58	189.44	0.869 228
(3, 3, 2, 2, 2)	63.00	137.31	171.11	0.805 187
(3, 3, 2, 2, 3)^d	73.00	143.19	198.59	0.845 446
(3, 3, 2, 3, 2)	83.00	150.52	189.44	0.878 250
(4, 2, 2, 2, 2)^d	60.00	138.24	178.33	0.794 831
(4, 2, 2, 3, 2)^d	80.00	151.45	196.66	0.866 955

Problem 4.2

Suppose an HSP system of n stages consists of k subsystems $S_1, \ldots, S_k$, in series or parallel. Let S_i denote the components in the subsystem. Let r_j denote the component reliability at stage j. Let x_j denote the number of components to be arranged in parallel to the existing component at stage j. Then, the reliability of stage j is $1 - (1 - r_j)^{x_j+1}$.

Consider the following single-constraint redundancy allocation problem.

Maximize

$$R_s = f(x_1, \ldots, x_n),$$

subject to

$$\sum_{j=1}^{n} c_j x_j \leq C,$$

x_j being a nonnegative integer, where $c_1, \ldots, c_n$ and C are positive integers.

Let $R(S_i; d_i)$ denote the maximum possible reliability of subsystem S_i, when amount d_i is provided for redundancy allocation in S_i. Let $N = \{1, 2, \ldots, n\}$ and $R(N; d)$ denote the maximum system reliability subject to the constraint $\sum_{j=1}^{n} c_j x_j \leq d$. For a given allocation $(d_1, \ldots, d_k)$ to k subsystems, the maximum system reliability is given by

$$\psi(d_1, \ldots, d_k) = \begin{cases} \prod_{i=1}^{k} R(S_i; d_i), & \text{when } S_1, \ldots, S_k \text{ are in series,} \\ 1 - \prod_{i=1}^{k}[1 - R(S_i; d_i)], & \text{when } S_1, \ldots, S_k \text{ are in parallel.} \end{cases}$$

Now, the problem under consideration is equivalent to maximizing

$$R_s = \psi(d_1, \ldots, d_k),$$

subject to

$$d_1 + \cdots + d_k \leq C,$$

d_i being a nonnegative integer.

If $R(S_i; d_i)$ is known for $i = 1, \ldots, k$ and $d_i = 0, 1, \ldots, C$, then the above problem can be solved by the DP approach. If $S_1, \ldots, S_k$ are connected in parallel, then $1 - \psi(d_1, \ldots, d_k)$ will be minimized by the DP approach. Since the system is HSP, each S_i is again a series or parallel arrangement of subsystems and $R(S_i; d)$ can be obtained by DP for $d = 0, 1, \ldots, C$ from optimal S_i subsystem reliability. The advantage of DP is that it simultaneously obtains $R(S_i; d_i)$ for $0, 1, \ldots, C$ values of d_i.

In the actual implementation of the DP approach, the optimal reliabilities of the smallest subsystems are derived first and used to obtain those of the larger subsystems. To derive the optimal redundancy allocation, a backtracking procedure is applied as shown in the example given below.

Example 4-6
Consider the five-stage HSP system described in Figure 1.8. Let the component reliabilities and costs at the five stages be as shown in Table 4.13. In addition to one basic unit in reserve in all five stages, $C = 20$ is made available for additional redundancy. The problem is to find a redundancy allocation that maximizes the system reliability, subject to the budget constraint.

Table 4.13. Reliabilities and costs for Example 4-6

j	1	2	3	4	5
r_j	0.60	0.75	0.50	0.70	0.80
c_j	2	3	1	2	3

Solution to Example 4-6

Let $S_1 = \{1, 2, 3\}$, $S_2 = \{4, 5\}$, and $\overline{S} = \{1, 2\}$. We need to determine $R(N; 20)$ and the corresponding redundancy allocation. We have

$$R(\overline{S}; d) = 1 - \min\{(1 - r_1)^{x_1+1}(1 - r_2)^{x_2+1} : c_1x_1 + c_2x_2 \le d,$$

$$\text{and } x_1, x_2 \in \{0, 1, 2, \dots\}\}, \tag{4.17}$$

$$R(S_1; d) = \max\{R(\overline{S}; d_1)[1 - (1 - r_3)^{x_3+1}] : d_1 + c_3x_3 \le d,$$

$$\text{and } d_1, x_3 \in \{0, 1, 2, \dots\}\}, \tag{4.18}$$

$$R(S_2; d) = \max\{[1 - (1 - r_4)^{x_4+1}][1 - (1 - r_5)^{x_5+1}] :$$

$$c_4x_4 + c_5x_5 \le d, \text{ and } x_4, x_5 \in \{0, 1, 2, \dots\}\}, \tag{4.19}$$

$$R(N; d) = 1 - \min\{[1 - R(S_1; d_1)][1 - R(S_2; d_2)] :$$

$$d_1 + d_2 \le d \text{ and } d_1, d_2 \in \{0, 1, 2, \dots\}\}. \tag{4.20}$$

The values of $R(\overline{S}; d)$ are first obtained from eq. (4.17) and used later in (4.18) to get the values of $R(S_1; d)$. The values of $R(S_2; d)$ are independently obtained from (4.19). Finally $R(N; d)$ is obtained from (4.20) using the $R(S_1; d)$ and $R(S_2; d)$ values. The $R(S_1; d)$, $R(S_2; d)$, and $R(N; d)$ values for $d = 0, 1, \dots, 20$ are given in Table 4.14. For the sake of brevity, the corresponding optimal allocations are not listed in the table.

The basic allocation with no additional redundancy at each stage is $(x_1, x_2, x_3, x_4, x_5) = (1, 1, 1, 1, 1)$. The optimal system reliability subject to budget constraint $\sum_{j=1}^{5} c_jx_j \le 20$ is $R(N; 20) = 0.99937$. It can be achieved by allocating $d_1 = 3$ resource units to subsystem $\{1, 2, 3\}$ and $d_2 = 17$ units to subsystem $\{4, 5\}$. Table 4.14 gives $R(S_1; 3) = 0.84375$ and $R(S_2; 17) = 0.99597$. All three resource units allocated to $\{1, 2, 3\}$ must be used for redundancy in component 3 in order to maximize subsystem reliability $\{1, 2, 3\}$. Such allocation gives redundancy levels $(x_1, x_2, x_3) = (1, 1, 4)$. Similarly, out of the 17 resource units allocated to $\{4, 5\}$, eight units are allocated for redundancy to component four, and nine units to component 5, in order to maximize the reliability of $\{4, 5\}$. This allocation gives redundancy levels $(x_4, x_5) = (4, 3)$. Therefore, the optimal redundancy allocation in the entire system is $(x_1, x_2, x_3, x_4, x_5) = (1, 1, 4, 5, 3)$.

Table 4.14. Optimal reliabilities of the system and subsystems of Example 4-6

d	$R(\overline{S}; d)$	$R(S_1; d)$	$R(S_2; d)$	$R(N; d)$
0	0.900 00	0.450 00	0.560 00	0.758 00
1	0.900 00	0.675 00	0.560 00	0.857 00
2	0.960 00	0.787 50	0.728 00	0.906 50
3	0.975 00	0.843 75	0.728 00	0.931 25
4	0.984 00	0.871 87	0.778 40	0.943 62
5	0.990 00	0.900 00	0.873 60	0.957 50
6	0.993 75	0.930 00	0.873 60	0.969 20
7	0.996 00	0.945 00	0.934 08	0.975 80
8	0.997 50	0.959 77	0.934 08	0.982 30
9	0.998 44	0.968 62	0.952 22	0.986 19
10	0.999 00	0.976 31	0.965 22	0.989 70
11	0.999 37	0.982 27	0.965 22	0.992 20
12	0.999 61	0.986 13	0.983 96	0.993 90
13	0.999 75	0.989 87	0.983 96	0.995 54
14	0.999 84	0.992 11	0.989 59	0.996 59
15	0.999 90	0.994 05	0.990 31	0.997 49
16	0.999 94	0.995 55	0.991 28	0.998 04
17	0.999 96	0.996 53	0.995 97	0.998 47
18	0.999 98	0.997 46	0.995 97	0.998 88
19	0.999 98	0.998 02	0.997 67	0.999 14
20	0.999 99	0.998 51	0.997 67	0.999 37

4.6   Discussion

Dynamic programming is a powerful solution methodology for solving various discrete optimization problems. It is a multi-stage decision-making approach in which an n-variable problem is solved by reducing it to a sequence of n single-variable problems and solving these successively using a recursive relationship. Implementation of the DP approach depends on the problem structure. Several variations of DP are available for solving discrete optimization problems, e.g. see the review article by Cooper [67]. The DP approach is mostly applicable to problems with separable objective functions and separable constraints. A redundancy allocation problem in a reliability system is a nonlinear integer programming problem which is solvable for some systems by the DP approach. The DP approach can be easily adopted for redundancy optimization if the system is a series or a hierarchical series-parallel system and there is a single constraint (separable). However, when there are two or more constraints or the system is more complex, some variation of this approach may be useful.

The advantage of DP is that it gives an exact optimal solution if the data used in the problem are integers. Otherwise, the data can be discretized on a grid, and DP can be used on the discretized problem. DP is very advantageous when the number of variables is large and the values on the right-hand side of the constraints are moderate. The main drawback of DP is the computational effort required to solve problems with more than one constraint. When multiple constraints are involved, other procedures, such as heuristics and those outlined in Section 2.3, can be incorporated to reduce the search space.

EXERCISES

4.1   What is Richard Bellman's principle of optimality? Explain how this principle is used in the application of dynamic programming?

4.2   In Table 4.2, how can we get $f_2(1.0) = 3.2031$ and $f_2(0.8) = 1.6250$?

4.3   Using Table 4.3, show the calculation of:
(a) $f_3(5)$ and x_3,
(b) $f_4(2)$ and x_4,
(c) $f_4(6)$ and x_4.

4.4   Illustrate and explain the advantage gained by tightening the bounds of the variables in the DP approach.

4.5   Consider a series system consisting of four stages with component reliabilities $(r_1, r_2, r_3, r_4) = (0.9, 0.75, 0.82, 0.95)$. Suppose the component costs at stages 1, 2, 3, and 4 are 50, 35, 42, and 60, respectively, and the profit from successful operation of the system is 1000. Find the cost-optimal redundancy allocation using the basic DP approach.

4.6   In Example 4-2, let a feasible solution be $(4, 4, 4, 3)$. If we apply the dynamic programming approach, find y_j, l'_j, u'_j, and b'.

4.7   In Example 4-5, let a feasible solution be $(2, 2, 3, 3, 3)$.
(a) Find l_j, $g_{1j}(l_j)$, G_j, and $\overline{G}_j$.
(b) Find $f_3(56)$ and x_3^* for $\lambda = 0.0015$.

4.8   Solve Example 4-3 without the third constraint by the Lagrange multiplier method.

4.9   Consider the following redundancy allocation problem. Maximize

$$R_s = \prod_{j=1}^{10} [1 - (1 - r_j)^{x_j}],$$

subject to

$$\sum_{j=1}^{10} c_j x_j \leq 70,$$

x_j being a nonnegative integer. The values of r_j and c_j are

j	1	2	3	4	5	6	7	8	9	10
r_j	0.6	0.75	0.5	0.7	0.8	0.65	0.7	0.6	0.9	0.55
c_j	2	3	1	2	3	2	3	2	4	1

Consider two subproblems, one with variables 1, 2, 3, 4, and 5, and the other with variables 6, 7, 8, 9, and 10. Solve both problems by the DP approach, and solve the original problem using the computational results of the subproblems.

4.10 Solve Example 4-3 using the dominating allocation approach, taking additional constraints $x_1 \leq 2$ and $x_4 \leq 2$.

4.11 Consider the problem of maximizing

$$R_s = \prod_{j=1}^{5}[1 - (1 - r_j)^{x_j}],$$

subject to

$$g_1 = \sum_{j=1}^{5} w_j x_j \exp\left(\frac{x_j}{4}\right) \leq 180,$$

$$g_2 = \sum_{j=1}^{5} p_j x_j^2 \leq 100, \qquad j = 1, 2, \ldots, 5,$$

x_j being a nonnegative integer.

Let $\mathbf{w} = (6, 6, 9, 7, 7)$, $\mathbf{c} = (5, 6, 6, 9, 7)$, and $\mathbf{r} = (0.8, 0.85, 0.90, 0.65, 0.75)$. Find the first-stage allocations derived from H_1, and the two-stage allocations derived from H_2.

4.12 Can we apply the dynamic programming approach to a complex system? Discuss.

4.13 Consider a two-component system. Maximize

$$R_s = r_{1j}r_{2j}, \qquad j = 0, 1, 2, 3,$$

subject to $g_1 = c_{ij}x_i \leq 10$, $i = 1, 2$, $j = 0, 1, 2, 3$. Find the optimal redundancy allocations for the coefficients given below.

j	r_{1j}	r_{2j}	j	c_{1j}	c_{2j}
0	0.4	0.3	0	0	0
1	0.6	0.5	1	2	3
2	0.7	0.8	2	4	5
3	0.9	0.9	3	6	7

4.14 Consider the three-component HSP problem listed below.   Let **r** = $(0.7, 0.8, 0.75)$, **c** $= (2, 3, 2)$, and $C = 10$. Find the recursive functions and a redundancy allocation that maximizes the system reliability subject to the budget constraint.

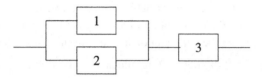

4.15 Consider the reliability system shown below:

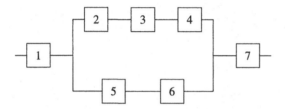

The reliability and cost of components at seven stages are

j	1	2	3	4	5	6	7
r_j	0.9	0.85	0.9	0.75	0.85	0.9	0.95
c_j	3	1	1	2	1	1	4

Find a redundancy allocation that maximizes system reliability without increasing the cost beyond 15.

4.16 Consider a machine that can be in one of two states, good or bad. Suppose that the machine produces an item at the end of each period. The item produced is either good or bad depending on whether the machine is in a good or bad state, respectively. Assume that once the machine is in a bad state it remains in that state until it is replaced. If the machine is in a good state at the beginning of a certain period, then with probability q it will be in a bad state at the end of the period. Once an item is produced, we may inspect the item at a cost C, or not inspect it. If an inspected item is found to be bad, the machine is replaced with a machine in a good state at a cost C'. The cost for producing a bad item is $C'' > 0$. Write a DP algorithm for obtaining an optimal inspection policy assuming that the initial machine is in a good state and has a horizon of n periods. Solve the problem for $q = 0.2$, $C = 1$, $C' = 3$, $C'' = 2$, and $n = 8$. (The optimal policy is to inspect at the end of the third period and not inspect in any other period.) (Bertsekas [35].)

5 Redundancy allocation by discrete optimization methods

5.1 Introduction

The majority of reliability optimization problems are nonlinear integer programming problems for which there are no efficient (polynomially bounded) solution procedures. Discrete optimization methods, such as dynamic programming, branch-and-bound techniques, enumeration methods, etc., are available for solving such problems, although their computational complexity is very high. Cooper [67] presented a classification and discussion of algorithms for nonlinear integer programming problems. This classification was based on the mathematical form of the problems. Algorithms for problems with separable objective functions and separable constraints are useful for solving redundancy allocation problems in series systems. Many combine dynamic programming and branch-and-bound strategies, e.g. see Morin and Marsten [238], Marsten and Morin [210], Aust [18], Denardo and Fox [72], and Alekseev and Volodos [10]. Cooper [67] provided an excellent discussion on algorithms developed for specially structured problems. Abadie [1], [2], and Abadie et al. [3] provided several branch-and-bound strategies for minimizing a convex objective function over a convex constraint set. A branch-and-bound approach in which the branching variables are selected in lexicographic order is proposed in Prasad and Kuo [270]. Glankwahmdee et al. [108] studied the application of continuous variable search techniques to unconstrained discrete optimization problems. Gupta and Ravindran [117] conducted a survey of algorithms for nonlinear integer programming problems. Later, Gupta and Ravindran [118] presented several branch-and-bound strategies for convex nonlinear integer programming problems.

Nonlinear integer optimization problems can be transformed into nonlinear 0–1 optimization problems. Several methods are available in the literature for solving such transformed problems, e.g. see Hansen [121] and Hansen et al. [122]. The partial enumeration method of Lawler and Bell [186] has been adopted by Misra [221], [226] to solve redundancy allocation problems in reliability systems. As shown in Section 5.2, nonlinear integer optimization problems with separable functions can also be transformed into 0–1 linear programming problems. For solution methodologies for integer and 0–1 linear programming, refer to Nemhauser and Wolsey [253]. Enumerative methods play an important role in discrete optimization. They are widely used for both linear and nonlinear 0–1 programming problems. Methods such as the partial search of Lawler and Bell [186], the implicit enumeration of Geoffrion [106], and lexicographic search, belong to the class of enumeration methods.

Problem 5.1

Several redundancy allocation problems in reliability systems can be mathematically formulated so as to maximize

$$R_s = f(x_1, \ldots, x_n),$$

subject to

$$g_i(x_1, \ldots, x_n) \le b_i, \qquad \text{for } i = 1, \ldots, m,$$
$$\ell_j \le x_j \le u_j, \qquad \text{for } j = 1, \ldots, n,$$

x_j being an integer.

Problem 5.2

When the functions f and g_i, $1 \le i \le m$, are separable, the problem can be transformed into the following form. Maximize

$$R_s = \sum_{j=1}^{n} f_j(x_j),$$

subject to

$$\sum_{j=1}^{n} g_{ij}(x_j) \le b_i, \qquad \text{for } i = 1, \ldots, m,$$
$$\ell_j \le x_j \le u_j, \qquad \text{for } j = 1, \ldots, n,$$

x_j being an integer.

Problem 5.2 can be formulated as a 0–1 linear programming (ZOLP) problem using the bounds on the decision variables. A ZOLP problem can be solved by several methods, e.g. by the Gomory cutting plan method, the branch-and-bound method, etc. Several interesting applications of this approach to reliability optimization can be found, e.g. Tillman [301], Hyun [143], and Gen et al. [104] used implicit enumeration methods to solve reliability optimization problems. Tillman adopted the Gomory method to solve the ZOLP formulation of a constrained reliability problem with several failure modes, while Hyun (Gen) used Geoffrion's implicit enumeration method to solve the same problem.

For optimal redundancy allocation by discrete optimization methods, refer to Ghare and Taylor [107], Tillman and Littschwager [311], McLeavey [212], [213], Banerjee et al. [22], Misra [221], [226], Luus [204], Hwang et al. [132], McLeavey and McLeavey [214], Mizukami [234], Proschan and Bray [274], and Nakagawa et al. [248].

Misra [228] has proposed a search algorithm for obtaining an exact optimal solution to Problem 5.1 when $f(x_1, x_2, \ldots, x_n)$ is nondecreasing in each x_j, and $g_i(x)$ is separable and nondecreasing in each x_j for all i. This method does not always give an optimal solution. For example consider the problem of maximizing

$$f(\mathbf{x}) = (1 - 0.4^{x_1})(1 - 0.3^{x_2})(1 - 0.1^{x_3})(1 - 0.1^{x_4}),$$

subject to

$$g_1(\mathbf{x}) = 10x_1 + 20x_2 + 5x_3 + 100x_4 \leq 181,$$

$$g_2(\mathbf{x}) = 10x_1 + 20x_2 + 100x_3 + 5x_4 \leq 181,$$

$$(1, 1, 1, 1) \leq (x_1, x_2, x_3, x_4) \leq (5, 3, 1, 1),$$

x_j being an integer.

Misra's search algorithm presents the solution $(5, 1, 1, 1)$, with an optimal system reliability of 0.5612, whereas the actual optimal solution is $(3, 2, 1, 1)$, with a corresponding system reliability of 0.6899.

Section 5.2 contains the ZOLP formulation for redundancy allocation problems. Section 5.3 describes two branch-and-bound methods for redundancy allocation: one for ZOLP formulations and another for a broad class of nonlinear integer programming problems. Section 5.4 describes the partial enumeration of Lawler and Bell, whereas Section 5.5 contains a lexicographic method for solving redundancy allocation problems.

5.2 The 0–1 linear programming formulation

Consider Problem 5.2. Problem 5.1 can be formulated as a 0–1 linear programming problem. Let

$$m_j = u_j - \ell_j, \qquad \text{for } j = 1, \ldots, n,$$

and for a feasible solution $\mathbf{x} = (x_1, \ldots, x_n)$ define,

$$y_{jk} = \begin{cases} 1, & \text{for } k = 1, \ldots, x_j - \ell_j, \\ 0, & \text{for } k = x_j - \ell_j + 1, \ldots, m_j, \end{cases}$$

for $j = 1, \ldots, n$.

Note that there is a one–one correspondence between the solution $\mathbf{x}$ and vector $\mathbf{y} = (y_{11}, \ldots, y_{1m_1}, y_{21}, \ldots, y_{2m_2}, \ldots, y_{n1}, \ldots, y_{nm_n})$. For example, let $(\ell_1, u_1) = (2, 7)$. Then $m_1 = 7 - 2 = 5$. For $x_1 = 4$, the corresponding values of $y_{11}, y_{12}, \ldots, y_{15}$ are $1, 1, 0, 0, 0$, respectively. Similarly,

$$(y_{11}, y_{12}, \ldots, y_{15}) = (0, 0, 0, 0, 0), \qquad \text{for } x_1 = 2,$$

and

$$(y_{11}, y_{12}, \ldots, y_{15}) = (1, 1, 1, 1, 1), \qquad \text{for } x_1 = 7.$$

Now, we can write the objective function of Problem 5.2 as

$$\sum_{j=1}^{n} f_j(x_j) = \sum_{j=1}^{n} \left(\sum_{k=1}^{x_j - \ell_j} [f_j(\ell_j + k) - f_j(\ell_j + k - 1)] \right) + \sum_{j=1}^{n} f_j(\ell_j)$$

$$= \sum_{j=1}^{n} \sum_{k=1}^{m_j} f_{jk} y_{jk} + \sum_{j=1}^{n} f_j(\ell_j),$$

where

$$f_{jk} = f_j(\ell_j + k) - f_j(\ell_j + k - 1).$$

Similarly, for $i = 1, \ldots, m$, we can write

$$\sum_{j=1}^{n} g_{ij}(x_j) = \sum_{j=1}^{n} \left(\sum_{k=1}^{x_j - \ell_j} [g_{ij}(\ell_j + k) - g_{ij}(\ell_j + k - 1)] \right) + \sum_{j=1}^{n} g_{ij}(\ell_j)$$

$$= \sum_{j=1}^{n} \sum_{k=1}^{m_j} g_{ijk} y_{jk} + \sum_{j=1}^{n} g_{ij}(\ell_j),$$

where

$$g_{ijk} = g_{ij}(\ell_j + k) - g_{ij}(\ell_j + k - 1).$$

Problem 5.3

Now, Problem 5.2 is equivalent to maximizing

$$R_s = \sum_{j=1}^{n} \sum_{k=1}^{m_j} f_{jk} y_{jk} + \sum_{j=1}^{n} f_j(\ell_j),$$

subject to

$$\sum_{j=1}^{n} \sum_{k=1}^{m_j} g_{ijk} y_{jk} \leq b_i', \qquad \text{for } i = 1, \ldots, m,$$

$$y_{jk} - y_{j(k-1)} \leq 0, \qquad \text{for } k = 2, \ldots, m_j \text{ and } j = 1, \ldots, n,$$

and $y_{jk} = 0$ or 1 for $k = 1, \ldots, m_j$, and $j = 1, \ldots, n$, where

$$b_i' = b_i - \sum_{j=1}^{n} g_{ij}(\ell_j), \qquad \text{for } i = 1, \ldots, m.$$

Note that Problem 5.3 is a ZOLP. Several methods for solving it, such as the Gomory cutting plane method, the branch-and-bound method, etc., are available in the literature.

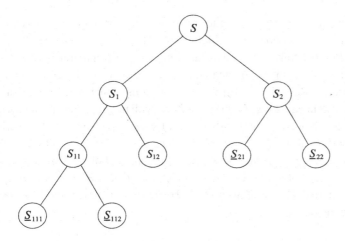

Figure 5.1. Schematic description of the branch-and-bound method

5.3   Branch-and-bound methods

The branch-and-bound method is one of the most useful techniques for solving discrete optimization problems. One of the earliest papers that contributed to the development of branch-and-bound methods is by Land and Doig [185].

Conceptually, the method starts with a set of all feasible solutions, and partitions the set into several subsets as it proceeds. Each subset is associated with a node, and an upper bound on the value of the objective function (if the objective is to maximize) is computed for each subset. The upper bound is usually derived by solving a continuous version of the problem with additional bound restrictions on the variables. Sometimes the upper bound is derived using some other criterion. The computational efficiency of a branch-and-bound method depends on the sharpness of the bound and the effort involved in computing the bound. If the upper bound of any subset is less than, or equal to, the objective value of a known feasible solution, the subset is ignored for further partition. The method stops when the upper bound of each subset to be partitioned is less than, or equal to, the objective value of the best known feasible solution. A schematic description of a branch-and-bound method is given in Figure 5.1.

Initially, the set S of all feasible solutions of a problem is partitioned into two subsets S_1 and S_2. The node S_1 is in turn partitioned into two subsets S_{11} and S_{12}. Similarly, node S_2 is partitioned into two subsets S_{21} and S_{22}. When a subset is ignored for further partitioning based on the upper bound, the corresponding node is fathomed. In the branching tree of Figure 5.1, the fathomed nodes are underlined. Each subset in the partition is represented by a node. The partitioning of a subset into two or more subsets is done on the basis of either the partition of a set of possible variable values or some other criterion.

Nakagawa et al. [248] developed a good branch-and-bound method for discrete optimization problems in general reliability systems. The development of branch-

and-bound methods for redundancy allocation problems in reliability systems can be found in Ghare and Taylor [107], McLeavey [212], [213], Banerjee et al. [22], Misra [221], [226], and Nakagawa and Nakashima [247]. Several branching strategies are available for implementing a branch-and-bound method for discrete optimization problems. Gupta and Ravindran [118] described various branch-and-bound strategies for convex nonlinear integer programming problems with minimization as the objective. They presented two branch-and-bound heuristics for these problems, and numerically compared the performance of 27 branching strategies: the heurience differed by (i) the method of selecting a branching variable, (ii) the method of selecting a branching node, and (iii) the usage of heuristics. These strategies can be used to maximize system reliability through redundancy allocation when the system reliability is a concave function of the component reliabilities.

5.3.1   Redundancy allocation in a series system

We now describe the branch-and-bound method for solving a redundancy allocation problem in series systems.

Problem 5.4

Consider the redundancy allocation problem so as to maximize

$$\ln R_s = \sum_{j=1}^{n} \ln\left[1 - (1 - r_j)^{x_j + 1}\right],$$

subject to

$$\sum_{j=1}^{n} a_{ij} x_j \le b_i, \qquad \text{for } i = 1, \ldots, m,$$

$$0 \le x_j \le u_j, \qquad \text{for } j = 1, \ldots, n,$$

x_j being an integer.

Assume that a_{ij} has a nonnegative value. This assumption is valid in almost all real-life problems. Let

$$f_{jk} = \ln\left[1 - (1 - r_j)^{k+1}\right] - \ln\left[1 - (1 - r_j)^k\right], \qquad \text{for } k = 1, \ldots, u_j,$$

$$f_{j0} = \ln[1 - (1 - r_j)],$$

for $j = 1, \ldots, n$. Let $g_{ijk} = a_{ij}$, for $k = 1, \ldots, u_j$ for all (i, j). Note that f_{jk} decreases as k increases.

Problem 5.5

As explained in Section 5.2, Problem 5.4 is equivalent to maximizing

$$\ln R_s = \sum_{j=1}^{n} \sum_{k=1}^{u_j} f_{jk} y_{jk} + \sum_{j=1}^{n} f_{j0},$$

subject to

$$\sum_{j=1}^{n}\sum_{k=1}^{u_j} a_{ij} y_{jk} \le b_i, \qquad \text{for } i = 1, \ldots, m,$$

$$y_{jk} - y_{j(k-1)} \le 0, \qquad \text{for } k = 2, \ldots, u_j \text{ and } j = 1, \ldots, n, \tag{5.1}$$

and $y_{jk} = 0$ or 1, for $k = 1, \ldots, u_j$, and $j = 1, \ldots, n$.

It can be easily verified that the optimal objective value in Problem 5.4 is the optimal objective value in Problem 5.5 plus $\sum_{j=1}^{n} f_{j0}$.

Derivation of an upper bound

We now describe a procedure to derive an upper bound on the objective function $\sum_{j=1}^{n}\sum_{k=1}^{u_j} f_{jk} y_{jk}$ for a given partial solution. Let

$$\gamma_{ijk} = \frac{f_{jk}}{a_{ij}}, \qquad \text{for all } i, j, \text{ and } k,$$

and

$$\gamma_{ir}^* = \max\{\gamma_{ijk} : r \le j \le n, 1 \le k \le u_j\}.$$

Consider a set S' of feasible solutions for which

$$y_{jk} = \bar{y}_{jk}, \qquad \text{for } k = 1, \ldots, u_j, \text{ and } j = 1, \ldots, r-1,$$

$$y_{rk} = \bar{y}_{rk}, \qquad \text{for } k = 1, \ldots, s \ (<u_r),$$

where $\bar{y}_{jk}$ and $\bar{y}_{rk}$ are fixed 0–1 values. Then for any solution, say $\mathbf{y} = (\bar{y}_{11}, \ldots, \bar{y}_{rs}, y_{r(s+1)}, \ldots, y_{nu_n})$, in S' the objective value is

$$Z = \ln R_s = g + \sum_{k=s+1}^{u_r} f_{rk} y_{rk} + \sum_{j=r+1}^{n}\sum_{k=1}^{u_j} f_{jk} y_{jk}, \tag{5.2}$$

where

$$g = \sum_{j=1}^{r-1}\sum_{k=1}^{u_j} f_{jk} \bar{y}_{jk} + \sum_{k=1}^{s} f_{rk} \bar{y}_{rk}.$$

For any i, $1 \le i \le m$, (5.2) can be written as

$$\ln R_s = g + \sum_{k=s+1}^{u_r} \gamma_{irk} \, a_{ir} \, y_{rk} + \sum_{j=r+1}^{n}\sum_{k=1}^{u_j} \gamma_{ijk} \, a_{ij} \, y_{jk} + \sum_{j=1}^{n} f_{j0}$$

$$\le g + \gamma_{ir}^* \left(\sum_{k=s+1}^{u_r} a_{ir} \, y_{rk} + \sum_{j=r+1}^{n}\sum_{k=1}^{u_j} a_{ij} \, y_{jk} \right)$$

$$\le g + \gamma_{ir}^* \left(b_i - \sum_{j=1}^{r-1}\sum_{k=1}^{u_j} a_{ij} \bar{y}_{jk} - \sum_{k=1}^{s} a_{ir} \bar{y}_{rk} \right).$$

The above inequalities hold since γ_{ijk} and a_{ij} are nonnegative, and the solution $\mathbf{y}$ is feasible. Thus, we have

$$\ln R_s \leq \min_{1 \leq i \leq m} (g + \gamma_{ir}^* h_i), \tag{5.3}$$

where

$$h_i = b_i - \sum_{j=1}^{r-1} \sum_{k=1}^{u_j} a_{ij} \bar{y}_{jk} - \sum_{k=1}^{s} a_{ir} \bar{y}_{rk}, \qquad \text{for } i = 1, \dots, m.$$

Equation (5.3) provides an upper bound on the objective value for all feasible solutions in set S'.

In the branch-and-bound implementation for Problem 5.5, we start with two initial nodes, which jointly represent all feasible solutions and branch off to generate more nodes. A node is denoted by a partial sequence of variables. The node representing the set S' of feasible solutions in which $y_{jk} = \bar{y}_{jk}$ for $j = 1, \dots, r-1$, and $k = 1, \dots, u_j$ and $y_{rk} = \bar{y}_{rk}$ for $k = 1, \dots, s$, is denoted by the sequence $(\bar{y}_{11}, \dots, \bar{y}_{rs})$. The two initial nodes are (1) and $(0, 0, \dots, 0)_{1 \times u_1}$. Whenever a node, say $(\bar{y}_{11}, \dots, \bar{y}_{rs})$, is generated, the corresponding upper bound on the objective value, as given in eq. (5.3), is computed. The node with the largest upper bound is selected for branching. If a node $(\bar{y}_{11}, \dots, \bar{y}_{rs})$ is selected for branching, two nodes are generated from it by taking values 0 and 1 for the variable following y_{rs}. If the upper bound of any node is less than or equal to the objective value of a known solution, the node is fathomed. When a node corresponds to a single feasible solution, the corresponding objective value is directly computed and the best known objective value is updated (if necessary). If a new node is generated from $(\bar{y}_{11}, \dots, \bar{y}_{rs})$ with $s < u_r$ by taking $y_{r(s+1)} = 0$, then the node is replaced by $(\bar{y}_{11}, \dots, \bar{y}_{rs}, \bar{y}_{r(s+1)}, \dots, \bar{y}_{ru_r})$, where $\bar{y}_{r\ell} = 0$ for $s + 1 \leq \ell \leq u_r$. This is to ensure the validity of eq. (5.1).

The branch-and-bound algorithm

- Step 0: Take a feasible solution $\mathbf{y}^* = (y_{11}, y_{12}, \dots, y_{nu_n})$ and compute its objective value Z^*. Let $H = \phi$, and compute the upper bound for nodes (1) and $(0, 0, \dots, 0)_{1 \times u_1}$. Add them to set H if the bounds exceed Z^*. If both bounds exceed Z^*, then $H = \{(1), (0, 0, \dots, 0)\}$. Go to step 1.

- Step 1: If $H = \phi$, or the upper bound of each node in H does not exceed Z^*, go to step 4. Otherwise, select the node, say $\bar{\sigma} = (\bar{y}_{11}, \dots, \bar{y}_{jk})$, with the largest upper bound. By taking values 0 and 1 for the variable immediately following y_{jk} in the order $(y_{11}, \dots, y_{nu_n})$, generate the following two nodes from node $\bar{\sigma}$:

$$\sigma' = (\bar{y}_{11}, \dots, \bar{y}_{jk}, 1) \quad \text{and} \quad \sigma'' = (\bar{y}_{11}, \dots, \bar{y}_{jk}, 0).$$

Let $H = H - \{\bar{\sigma}\}$. If $j = n$ and $k = u_n - 1$, that is, if either of σ' and σ'' corresponds to a single solution, go to step 2. If σ' corresponds to a nonempty set of feasible solutions and its upper bound exceeds Z^*, let $H = H \cup \{\sigma'\}$. Consider the node $\hat{\sigma} = (\bar{y}_{11}, \ldots, \bar{y}_{jk}, \bar{y}_{j(k+1)}, \ldots, \bar{y}_{ju_j})$, where $\bar{y}_{j\ell} = 0$ for $k + 1 \leq \ell \leq u_j$. If $j = n$, go to step 3. Otherwise, compute the upper bound for $\hat{\sigma}$. If this bound exceeds Z^*, let $H = H \cup \{\hat{\sigma}\}$, and repeat step 1.

- Step 2: If the solution $\sigma' = (\bar{y}_{11}, \ldots, \bar{y}_{n(u_n-1)}, 1)$ is feasible, compute the objective value Z for σ'. If Z exceeds Z^*, let $\mathbf{y}^* = \sigma'$ and $Z^* = Z$. Repeat the same for σ'', and go to step 1.

- Step 3: Compute the objective value Z for the solution $\hat{\sigma}$. If $Z^* < Z$, let $Z^* = Z$ and $\mathbf{y}^* = \hat{\sigma}$. Go to step 1.

- Step 4: Take $\mathbf{y}^*$ and Z^* as the optimal solution and the optimal objective value, respectively, and stop.

Example 5-1

Consider the problem of maximizing

$$\ln R_s = \sum_{j=1}^{3} \ln\left[1 - (1 - r_j)^{x_j+1}\right],$$

subject to

$5x_1 + 4x_2 + 9x_3 \leq 34,$

$7x_1 + 7x_2 + 8x_3 \leq 37,$

$0 \leq x_j \leq 3, \qquad \text{for } j = 1, 2, 3,$

x_j being an integer, where $(r_1, r_2, r_3) = (0.65, 0.90, 0.75)$.

This problem is equivalent to the ZOLP problem of maximizing

$$Z = \sum_{j=1}^{3} \sum_{k=1}^{3} f_{jk} y_{jk},$$

subject to

$$5 \sum_{k=1}^{3} y_{1k} + 4 \sum_{k=1}^{3} y_{2k} + 9 \sum_{k=1}^{3} y_{3k} \leq 34,$$

$$7 \sum_{k=1}^{3} y_{1k} + 7 \sum_{k=1}^{3} y_{2k} + 8 \sum_{k=1}^{3} y_{3k} \leq 37,$$

and

$$y_{jk} - y_{j(k-1)} \leq 0,$$

for $k = 2$ and 3, and for $j = 1, 2, 3$; $y_{jk} = 0$ or 1 for $j = 1, 2, 3$ and $k = 1, 2, 3$, where f_{jk} values are given below.

j	k		
	1	2	3
1	0.3001	0.0869	0.0287
2	0.0953	0.0090	0.0009
3	0.2231	0.0488	0.0118

Note that if Z is the objective value for a feasible solution $\mathbf{y}$ in the above ZOLP problem, $\exp(Z + \sum_{j=1}^{3} f_{j0})$ is the objective value for the corresponding solution $\mathbf{x}$ in the original problem.

The values of γ_{ir}^{*} are given below.

i	r		
	1	2	3
1	0.060	0.025	0.025
2	0.043	0.028	0.028

The initial feasible solution considered is

$$\mathbf{y}^{*} = (1, 0, 0, 1, 0, 0, 1, 0, 0).$$

The corresponding value of $\sum_{j=1}^{3} \sum_{k=1}^{3} f_{jk} y_{jk}$ is 0.6185, and the value of $\sum f_{j0} = -0.8238$. The nodes obtained by the branch-and-bound method are listed in Table 5.1, along with their upper bounds. If a node A is generated from node B, then B is called the parent node of A. For each node (except the first two), the parent node is also given in Table 5.1. The optimal solution is

$$\mathbf{y}^{*} = (1, 1, 0, 1, 0, 0, 1, 1, 0),$$

and the corresponding objective value is $Z^{*} = 0.7542$. The corresponding solution of the original problem is $x^{*} = (2, 1, 2)$, and the corresponding system reliability is 0.9328.

5.3.2   Redundancy allocation in a complex system

Nakagawa et al. [248] developed a branch-and-bound method (NNH) for solving nonlinear integer programming formulations of reliability optimization problems. The method is applicable for maximization as well as minimization of a monotonic function of integer variables subject to constraints involving monotonic functions. Nakagawa et al.'s [248] branch-and-bound method is described below, with minor modification for a maximization problem. As explained at the end of this section, a minimization problem can easily be transformed into a maximization problem.

Problem 5.6

Consider the general problem of maximizing

$$R_{s} = f(x_{1}, \dots, x_{n}),$$

Table 5.1. Nodes generated in branch-and-bound application for Example 5-1

Node	Node	Parent node	Upper bound	Residual resources
1	(1)		1.59	29, 30
2	(0)		1.59	34, 37
3	(1, 1)	1	1.37	24, 23
4	(1, 0, 0)	1	1.59	29, 30
5	(0, 0, 0, 1)	2	0.84	30, 30
6	(0, 0, 0, 0, 0, 0)	2	0.84	34, 37
7	(1, 0, 0, 1)	4	1.02	25, 23
8	(1, 0, 0, 0, 0, 0)	4	1.02	29, 30
9	(1, 1, 1)	3	1.10	19, 16
10	(1, 1, 0)	3	1.37	24, 23
11	(1, 1, 0, 1)	10	0.93	20, 16
12	(1, 1, 0, 0, 0, 0)	10	0.98	24, 23
13	(1, 1, 1, 1)	9	0.76	15, 9
14	(1, 1, 1, 0, 0, 0)	9	0.86	19, 16
15	(1, 0, 0, 0, 0, 0, 1)	8	1.02	20, 22
16	(1, 0, 0, 0, 0, 0, 1, 1)	15	0.85	11, 14
17	(1, 0, 0, 1, 1)	7	0.85	21, 16
18	(1, 0, 0, 1, 0, 0)	7	1.02	25, 23
19	(1, 0, 0, 1, 0, 0, 1)	18	1.02	16, 15
20	(1, 0, 0, 1, 0, 0, 1, 1)	19	0.84	7, 7
21	(1, 1, 0, 0, 0, 0, 1)	12	0.98	15, 15
22	(1, 1, 0, 0, 0, 0, 1, 1)	21	0.81	6, 7
23	(1, 1, 0, 1, 1)	11	0.74	16, 9
24	(1, 1, 0, 1, 0, 0)	11	0.93	20, 16
25	(1, 1, 0, 1, 0, 0, 1)	24	0.93	11, 8
26	(1, 1, 0, 1, 0, 0, 1, 1)	25	0.75	2, 0
27	(1, 1, 1, 0, 0, 0, 1)	14	0.86	10, 8
28	(1, 0, 0, 1, 1, 0)	17	0.85	21, 16
29	(1, 0, 0, 1, 1, 0, 1)	28	0.85	12, 8
30	(0, 0, 0, 0, 0, 1)	6	0.84	25, 29
31	(0, 0, 0, 1, 1)	5	0.75	26, 23
32	(0, 0, 0, 1, 0, 0)	5	0.84	30, 30
33	(0, 0, 0, 1, 0, 0, 1)	32	0.84	21, 22
34	(1, 1, 1, 1, 0, 0)	13	0.76	15, 9
35	(1, 1, 1, 1, 0, 0, 1)	34	0.76	6, 1

subject to

$$g_i(x_1, \ldots, x_n) \le b_i, \qquad \text{for } i = 1, \ldots, m,$$

$$\ell_j \le x_j \le u_j, \qquad \text{for } j = 1, \ldots, n,$$

x_j being an integer, where the functions f and g_i, $1 \le i \le s$, are nondecreasing in each x_j, and the functions g_i, $s + 1 \le i \le m$, are nonincreasing in each x_j.

Tightening of bounds on variables

Using the monotonic nature of g_i, the bounds on the variables can be tightened as follows. Find

$$\bar{x}_j = \max\{x_j: x_j \le u_j \text{ and } g_i(\ell_1, \dots, \ell_{j-1}, x_j, \ell_{j+1}, \dots, \ell_n) \le b_i, 1 \le i \le s\}, \quad (5.4)$$

for $j = 1, \dots, n$, and subsequently find

$$\underline{x}_j = \min\{x_j: x_j \ge \ell_j \text{ and } g_i(\bar{x}_1, \dots, \bar{x}_{j-1}, x_j, \bar{x}_{j+1}, \dots, \bar{x}_n) \le b_i, s + 1 \le i \le m\},$$

$$(5.5)$$

for $j = 1, \dots, n$. Note that $\ell_j \le \underline{x}_j$ and $\bar{x}_j \le u_j$ for $j = 1, \dots, n$, and for any feasible solution $(x_1, \dots, x_n)$ we have $\underline{x}_j \le x_j \le \bar{x}_j$ for $j = 1, \dots, n$.

Problem 5.7

Rewrite Problem 5.6 so as to maximize

$$R_s = f(x_1, \dots, x_n),$$

subject to

$$g_i(x_1, \dots, x_n) \le b_i, \qquad \text{for } i = 1, \dots, m, \qquad (5.6)$$

$$\underline{x}_j \le x_j \le \bar{x}_j, \qquad \text{for } j = 1, \dots, n, \qquad (5.7)$$

x_j being an integer.

The bounds on the variables can be further tightened using a feasible solution and a suitable upper bound on the objective function, as follows. Suppose $\mathbf{x} = (x_1, \dots, x_n)$ is a known feasible solution, and let $f^* = f(\mathbf{x})$ be the corresponding objective value. Let $W_j(\beta)$ be an upper bound on

$$\max\{f(\mathbf{x}): \mathbf{x} \text{ satisfies eqs. (5.6) and (5.7), and } x_j = \beta\}.$$

For $j = 1, \dots, n$, let

$$\bar{x}_j^c = \max\{\beta: \underline{x}_j \le \beta \le \bar{x}_j \text{ and } W_j(\beta) > f^*\}, \qquad (5.8)$$

$$\underline{x}_j^c = \min\{\beta: \underline{x}_j \le \beta \le \bar{x}_j \text{ and } W_j(\beta) > f^*\}. \qquad (5.9)$$

For any feasible solution with x_j less than $\underline{x}_j^c$, or greater than $\bar{x}_j^c$, the objective value cannot be more than f^*. Therefore, Problem 5.7 is equivalent to the problem of maximizing

$$R_s = f(x_1, \dots, x_n),$$

subject to

$$g_i(x_1, \ldots, x_n) \le b_i, \qquad \text{for } i = 1, \ldots, m,$$

$$\underline{x}_j^c \le x_j \le \overline{x}_j^c, \qquad \text{for } j = 1, \ldots, n,$$

x_j being an integer.

The upper bound $W_j(\beta)$ can be derived in several ways, as explained below. However, we need to make a tradeoff between sharpness of the bound and the computational effort required. In the branch-and-bound method of Nakagawa et al. [248], for branching purposes the variables are selected in increasing order of the ranges $\overline{x}_j^c - \underline{x}_j^c$. This order is expected to reduce the overall computational effort.

Upper bound on objective function

In their method, Nakagawa and Nakashima [247] compute and use an upper bound on the subproblem optimal objective.

Problem 5.8

Maximize

$$R(\alpha_1, \ldots, \alpha_{k-1}) = f(\alpha_1, \ldots, \alpha_{k-1}, x_k, \ldots, x_n),$$

subject to

$$g_i(\alpha_1, \ldots, \alpha_{k-1}, x_k, \ldots, x_n) \le b_i, \qquad \text{for } i = 1, \ldots, m,$$

$$\underline{x}_j^c \le x_j \le \overline{x}_j^c, \qquad \text{for } j = k, \ldots, n,$$

x_j being an integer, where $\alpha_1, \ldots, \alpha_{k-1}$ are some fixed integer values of $x_1, \ldots, x_{k-1}$, respectively.

An upper bound on the objective value of Problem 5.8 can be obtained by solving its relaxed version (without integer restrictions on the variables). There are several ways to derive the bound. For example, we can use either

1. $f(\alpha_1, \ldots, \alpha_{k-1}, \overline{x}_k^c, \ldots, \overline{x}_n^c)$, or

2. $\min\limits_{1 \le i \le m} \max\{f(\alpha_1, \ldots, \alpha_{k-1}, x_k, \ldots, x_n): g_i(\alpha_1, \ldots, \alpha_{k-1}, x_k, \ldots, x_n) \le b_i$ and x_j is real$\}$.

When the functions f and g_i, $1 \le i \le m$, are separable, an upper bound can be derived by solving the linear programming problem: maximize

$$R_s = \sum_{j=k}^{n} \sum_{t=0}^{\Delta_j} y_{jt} f_j(\underline{x}_j^c + t) + \sum_{j=1}^{k-1} f_j(\alpha_j),$$

subject to

$$\sum_{j=k}^{n}\sum_{t=0}^{\Delta_j} y_{jt}g_{ij}(\underline{x}_j^c + t) \le b_i - \sum_{j=1}^{k-1} g_{ij}(\alpha_j),$$

$$\sum_{t=0}^{\Delta_j} y_{jt} = 1, \qquad \text{for } j = k, \dots, n,$$

$$y_{jt} \ge 0,$$

where $\Delta_j = \overline{x}_j^c - \underline{x}_j^c$. Similarly, we can derive the upper bound $W_j(\beta)$ in Problem 5.7.

Tightening of bounds on variables of a subproblem

For Problem 5.8, the subproblem, the bounds on the first variable x_k, can be tightened as follows. Obtain an upper bound $W_k'(\alpha_1, \dots, \alpha_{k-1}, \beta)$ on the objective value of Problem 5.8, with x_k fixed at β for $\beta = \underline{x}_k^c, \dots, \overline{x}_k^c$. Now, compute the bounds

$$\overline{x}_k' = \max\{\beta : \underline{x}_k^c \le \beta \le \overline{x}_k^c, g_i(\alpha_1, \dots, \alpha_{k-1}, \beta, \underline{x}_{k+1}^c, \dots, \underline{x}_n^c) \le b_i, 1 \le i \le s$$

$$\text{and } W_k'(\alpha_1, \dots, \alpha_{k-1}, \beta) > f^*\}, \tag{5.10}$$

$$\underline{x}_k' = \min\{\beta : \underline{x}_k^c \le \beta \le \overline{x}_k^c, g_i(\alpha_1, \dots, \alpha_{k-1}, \beta, \overline{x}_{k+1}^c, \dots, \overline{x}_n^c) \le b_i, s+1 \le i \le n$$

$$\text{and } W_k'(\alpha_1, \dots, \alpha_{k-1}, \beta) > f^*\}, \tag{5.11}$$

for the variable x_k. It may be noted that the value of x_k in the optimal solution of Problem 5.8 lies between $\underline{x}_k'$ and $\overline{x}_k'$. These bounds are used in the branch-and-bound method of Nakagawa et al. [248].

NNH algorithm

- Step 1: Select an appropriate method to derive the upper bounds $W(\beta)$ and $W_k'(\alpha_1, \dots, \alpha_{k-1}, \beta)$. Find an initial feasible solution $\mathbf{x}^0$. Let $\mathbf{x}^* = \mathbf{x}^0$, and compute $f^* = f(\mathbf{x}^*)$.

- Step 2: First, calculate $\overline{x}_j$ for $j = 1, \dots, n$, using eq. (5.4), and later $\underline{x}_j$ for $j = 1, \dots, n$, using eq. (5.5). Next, calculate $\overline{x}_j^c$ and $\underline{x}_j^c$ for $j = 1, \dots, n$, using eqs. (5.8) and (5.9).

- Step 3: Renumber the variables such that the ranges $\overline{x}_j^c - \underline{x}_j^c$ are in increasing order. Obtain $\overline{x}_1'$ and $\underline{x}_1'$ using eqs. (5.10) and (5.11), and let $x_1 = \overline{x}_1'$ and $k = 1$.

- Step 4: Let $k = k + 1$. If $k = n$, go to step 6.

- Step 5: Calculate $\underline{x}_k'$ and $\overline{x}_k'$ using eqs. (5.10) and (5.11), with $(\alpha_1, \dots, \alpha_{k-1}) = (x_1, \dots, x_{k-1})$. If they exist, go to step 4. Otherwise, go to step 8.

- Step 6: Calculate $\overline{x}_n'$ using eq. (5.10). If it does not exist, go to step 8. Otherwise, let $x_n = \overline{x}_n'$.

- Step 7: If $f(x_1, \dots, x_n) > f^*$, update x^* as $(x_1, \dots, x_n)$ and let $f^* = f(x_1, \dots, x_n)$.

- Step 8: Find $\hat{k} = \max\{r: x_r > x'_r, 1 \le r \le k-1\}$. If $\hat{k}$ exists, let $x_{\hat{k}} = x_{\hat{k}} - 1$, and $k = \hat{k}$, and go to step 4. If $\hat{k}$ does not exist, then restore the original numbering of the variables to get the optimal solution corresponding to x^*, and stop.

If the functions f and g_i, $1 \le i \le m$, are of the form

$$f(x_1, \dots, x_n) = \prod_{j=1}^{n} f_j(x_j),$$

$$g_i(x_1, \dots, x_n) = \sum_{j=1}^{n} g_{ij}(x_j),$$

then we can use the recursive relations

$$f^{k+1} = f^k[f_{k+1}(x_{k+1})],$$
$$g_i^{k+1} = g_i^k + g_{i,k+1}(x_{k+1}), \qquad \text{for } 1 \le i \le m$$

to reduce the computations. The values of f_i^0 and g_i^0, $1 \le i \le m$, are to be taken as one and zero, respectively.

If the objective in Problem 5.1 is to minimize $f(\mathbf{x})$, then by taking $\mathbf{y} = \mathbf{u} - \mathbf{x}$, where $\mathbf{u} = (u_1, \dots, u_n)$, Problem 5.1 can be transformed into the problem of maximizing $-f(\mathbf{u} - \mathbf{y})$ subject to $g_i(\mathbf{u} - \mathbf{y}) \le b_i$, $1 \le i \le m$, y_j being an integer. This problem can be solved directly by Nakagawa and Nakashima's [247] branch-and-bound method.

5.4 A partial enumeration method

Lawler and Bell [186] developed a partial enumeration method for solving 0–1 optimization problems of the form: minimize

$$z = f(y_1, \dots, y_n),$$

subject to

$$g_{i1}(y_1, \dots, y_n) - g_{i2}(y_1, \dots, y_n) \ge 0, \qquad \text{for } i = 1, \dots, m,$$

$$y_j = 0 \text{ or } 1, \qquad \text{for } j = 1, \dots, n,$$

where the functions f, g_{i1}, and g_{i2} are monotonically nondecreasing in each of the variables. Define

$$d(\mathbf{y}) = y_1(2^{n-1}) + y_2(2^{n-2}) + \dots + y_{n-1}(2^1) + y_n(2^0).$$

Then $d(0, 0, \dots, 0) = 0$, $d(1, 1, \dots, 1) = 2^n - 1$, and there is a one–one correspondence between 0–1 vectors of n elements and the integers $0, 1, 2, 3, \dots, 2^n-1$. Arrange all 2^n zero–one vectors of n elements in increasing order of $d(\mathbf{y})$, and denote this order by H. The Lawler and Bell method performs a partial search in the order H.

Let $\mathbf{a} = (a_1, \ldots, a_n)$ and $\mathbf{b} = (b_1, \ldots, b_n)$ be two distinct vectors in the order H. If $a_j \geq b_j$ for $j = 1, \ldots, n$, then $\mathbf{a}$ is said to be larger than $\mathbf{b}$ (denoted by $\mathbf{a} \geq \mathbf{b}$). Let $\mathbf{y}$ be an arbitrary 0–1 vector and let $\mathbf{y}^*$ be the vector that follows $\mathbf{y}$ in the order H, such that

1. all vectors between $\mathbf{y}$ and $\mathbf{y}^*$ in the order H are larger than $\mathbf{y}$, and

2. $\mathbf{y}^*$ is not larger than $\mathbf{y}$.

For example, let $n = 5$ and consider vector $\mathbf{y} = (1, 0, 1, 0, 0)$. Note that the partial order of vectors

$(1, 0, 1, 0, 0)$,

$(1, 0, 1, 0, 1)$,

$(1, 0, 1, 1, 0)$,

$(1, 0, 1, 1, 1)$,

$(1, 1, 0, 0, 0)$,

is part of H. The second, third, and fourth vectors in the above order are larger than $\mathbf{y}$ whereas the fifth vector $(1, 1, 0, 0, 0)$ is not larger than $\mathbf{y}$. Thus $\mathbf{y}^*$ in this case is $(1, 1, 0, 0, 0)$. For any vector $\mathbf{y}$, let $\mathbf{y}_+$ denote the vector immediately following $\mathbf{y}$ in the order H, and let $\mathbf{y}_-$ denote the vector immediately preceding $\mathbf{y}$ in the order H. The following algorithm is a detailed description of the Lawler and Bell method.

Lawler and Bell algorithm

- Step 0: Let $\mathbf{y}^0 = (0, \ldots, 0)_{1 \times n}$, $f^0 = \infty$, and $\mathbf{y} = (0, \ldots, 0, 1)_{1 \times n}$.

- Step 1: Find $\mathbf{y}^*$ corresponding to $\mathbf{y}$. If $\mathbf{y}^*$ does not exist for vector $\mathbf{y}$, go to step 4. If $f(\mathbf{y}) < f^0$, go to step 2. Otherwise, let $\mathbf{y} = \mathbf{y}^*$, and repeat step 1.

- Step 2: If $g_{i1}(\mathbf{y}^*_-) - g_{i2}(\mathbf{y}) < 0$, for some i, $1 \leq i \leq m$, let $\mathbf{y} = \mathbf{y}^*$ and go to step 1. Otherwise go to step 3.

- Step 3: If $g_{i1}(\mathbf{y}) - g_{i2}(\mathbf{y}) \geq 0$, for $i = 1, \ldots, m$, let $f^0 = f(\mathbf{y})$ and $\mathbf{y}^0 = \mathbf{y}$. If $\mathbf{y}_+$ does not exist, go to step 4. Otherwise, let $\mathbf{y} = \mathbf{y}_+$, and go to step 1.

- Step 4: Stop. If $\mathbf{y}^0 = (0, \ldots, 0)$ and is infeasible, then the problem does not have any feasible solution (under the assumption that no feasible solution has $f(\mathbf{y}) = \infty$). Otherwise, take $\mathbf{y}^0$ as the optimal solution and $f(\mathbf{y}^0)$ as the optimal objective value.

The majority of reliability optimization problems can be formulated as required by the Lawler and Bell method. However, such a formulation sometimes requires intuition and ingenuity. In order to solve Example 5-1 by the Lawler and Bell method, we can formulate the problem as follows.

Since x_j is restricted to taking an integer value in the set $\{0, 1, 2, 3\}$, it can be represented by

$$x_j = (1 - z_{j1}) + 2(1 - z_{j2}), \qquad \text{for } j = 1, 2, \text{ and } 3, \tag{5.12}$$

where $y_{jk} = 0$ or 1, for $j = 1, 2$, and 3, and $k = 1$ and 2. There is a one–one correspondence between the value of x_j and those of y_{j1} and y_{j2}. Now define

$$- \ln R_s = f(y_{ij}) = - \sum_{j=1}^{3} \ln\left[1 - (1 - r_j)^{1+(1-y_{j1})+2(1-y_{j2})}\right].$$

Using eq. (5.12), it can easily be verified that the problem in Example 5-1 is equivalent to minimizing

$$- \ln R_s = f(y_{11}, y_{12}, y_{21}, y_{22}, y_{31}, y_{32}),$$

subject to

$$5y_{11} + 10y_{12} + 4y_{21} + 8y_{22} + 9y_{31} + 18y_{32} - 20 \geq 0,$$
$$7y_{11} + 14y_{12} + 7y_{21} + 14y_{22} + 8y_{31} + 16y_{32} - 29 \geq 0,$$

and $y_{jk} = 0$ or 1, for $j = 1, 2$, and 3, and $k = 1$ and 2.

This problem can be directly solved by the Lawler and Bell method. It was adopted by Misra [221], [226] to solve redundancy allocation problems in reliability systems. Although the method can be used to solve the vast majority of discrete optimization problems, its scope is limited to small-scale problems due to its high computational complexity.

5.5　The lexicographic method

To solve a wide class of discrete reliability optimization problems, Prasad and Kuo [270] have developed an implicit enumeration method based on a lexicographic search and a bound on the objective value. In all real-life reliability optimization problems, the decision variables are bounded. The method carries out a search among all solutions (feasible and infeasible), satisfying the bounds in lexicographic order. It also uses an upper bound on the objective function and constraints for reducing the search for an optimal solution.

Problem 5.9
A general reliability optimization problem involving redundancy options and component choices can be formulated so as to maximize

$$R_s = f(x_1, \ldots, x_n),$$

subject to

$$\sum_{j=1}^{n} g_{ij}(x_j) \leq b_i, \quad \text{for } i = 1, \ldots, m,$$

$$\ell_j \leq x_j \leq u_j, \quad \text{for } j = 1, \ldots, n,$$

x_j being an integer.

Lexicographic order

Let $\mathbf{x} = (x_1, x_2, \ldots, x_n)$ and $\mathbf{y} = (y_1, y_2, \ldots, y_n)$ be two vectors in R^n. If $x_1 < y_1$ or $x_j = y_j$ for $j = 1, \ldots, k$ and $x_{k+1} < y_{k+1}$ for some $k \geq 1$, then vector $\mathbf{x}$ is said to be *lexicographically* smaller than vector $\mathbf{y}$. We write $\mathbf{x} \overset{L}{<} \mathbf{y}$ when $\mathbf{x}$ is lexicographically smaller than vector $\mathbf{y}$. If $\mathbf{x}$ and $\mathbf{y}$ are distinct vectors in R^n, then one of them must be lexicographically smaller than the other.

The lexicographic order is transitive, that is, if $\mathbf{x} \overset{L}{<} \mathbf{y}$ and $\mathbf{y} \overset{L}{<} \mathbf{z}$, then $\mathbf{x} \overset{L}{<} \mathbf{z}$. If m vectors $\mathbf{x}^{(1)}, \mathbf{x}^{(2)}, \ldots, \mathbf{x}^{(m)}$ satisfy $\mathbf{x}^{(1)} \overset{L}{<} \mathbf{x}^{(2)} \overset{L}{<} \cdots \overset{L}{<} \mathbf{x}^{(m)}$, then the order $(\mathbf{x}^{(1)}, \mathbf{x}^{(2)}, \ldots, \mathbf{x}^{(m)})$ is called the *lexicographic order*. Let

$$S = \{(x_1, x_2, \ldots, x_n): \ell_j \leq x_j \leq u_j, \text{ for } 1 \leq j \leq n, \text{ and } x_j \text{ being an integer}\}.$$

We can obtain the optimal solution of Problem 5.9 by carrying out a search in set S following the lexicographic order of all vectors in S.

Upper bound

For $j = 1, \ldots, n$, define a vector of increments

$$
\begin{bmatrix}
\Delta_{1j}(d) \\
\Delta_{2j}(d) \\
\vdots \\
\Delta_{mj}(d)
\end{bmatrix}
=
\begin{bmatrix}
g_{1j}(d) - g_{1j}(\ell_j) \\
g_{2j}(d) - g_{2j}(\ell_j) \\
\vdots \\
g_{mj}(d) - g_{mj}(\ell_j)
\end{bmatrix},
\qquad \text{for } \ell_j < d \leq u_j.
$$

Note that for $\ell_j < d \leq u_j$, $[\Delta_{1j}(d), \Delta_{2j}(d), \ldots, \Delta_{mj}(d)]^T$ is the vector of increments in the resource consumptions due to the increase in x_j from ℓ_j to d. If $g_{ij}(x)$ is a linear function for $i = 1, 2, \ldots, m$ for any fixed j, say, $g_{ij}(x) = a_{ij} x$, then

$$
\begin{bmatrix}
\Delta_{1j}(d) \\
\Delta_{2j}(d) \\
\vdots \\
\Delta_{mj}(d)
\end{bmatrix}
= (d - \ell_j)
\begin{bmatrix}
a_{1j} \\
a_{2j} \\
\vdots \\
a_{mj}
\end{bmatrix}.
$$

Consider a feasible solution $\mathbf{d} = (d_1, d_2, \ldots, d_k, \ell_{k+1}, \ldots, \ell_n)$. Now we derive an upper bound on the maximum of $f(d_1, \ldots, d_k, x_{k+1}, \ldots, x_n)$ over all feasible solutions $(d_1, \ldots, d_k, x_{k+1}, \ldots, x_n)$. The values of slack variables for the solution $\mathbf{d}$ are

$$s_i = b_i - \sum_{j=1}^{k} \Delta_{ij}(d_j) - \sum_{j=k+1}^{n} \Delta_{ij}(\ell_j)$$

$$= b_i' - \sum_{j=1}^{k} \Delta_{ij}(d_j),$$

where

$$b_i' = b_i - \sum_{j=1}^{n} \Delta_{ij}(\ell_j), \qquad \text{for } i = 1, 2, \ldots, m.$$

We have $s_i \geq 0$ for $i = 1, 2, \ldots, m$, since solution d is feasible. For each $j, k + 1 \leq j \leq n$, find the maximum x_j, say x'_j, in the set $\{\ell_j, \ldots, u_j\}$ such that

$$
\begin{bmatrix}
\Delta_{1j}(x'_j) \\
\Delta_{2j}(x'_j) \\
\vdots \\
\Delta_{mj}(x'_j)
\end{bmatrix}
\leq
\begin{bmatrix}
s_1 \\
s_2 \\
\vdots \\
s_m
\end{bmatrix}.
$$

Since the system reliability $f(x_1, \ldots, x_n)$ is nondecreasing in each x_j, $f(d_1, \ldots, d_k, x'_{k+1}, \ldots, x'_n)$ is the upper bound on the objective value for all feasible solutions, with $x_j = d_j$ for $j = 1, 2, \ldots, k$. For any particular j, if $g_{ij}(x)$ is linear, that is, $g_{ij}(x) = a_{ij}x$, for $1 \leq i \leq m$, then

$$
x'_j = \min\left[u_j, \ell_j + \min\left(\left\lfloor \frac{s_i}{a_{ij}} \right\rfloor : 1 \leq i \leq m, a_{ij} > 0\right)\right].
$$

The algorithm given below searches for an optimal solution in the lexicographic order using the upper bound for discarding several solutions during the search. Whenever an upper bound is computed, it is compared with the current best feasible solution. If the upper bound $f(d_1, \ldots, d_k, x'_{k+1}, \ldots, x'_n)$ is less than or equal to $f(y)$ for some feasible solution y, then all solutions with $x_j = d_j$, for $j = 1, 2, \ldots, k$, are eliminated in the search. In order to reduce the computational effort further, a good feasible solution can be taken as the initial best solution. The procedure described in Section 2.3 gives a good feasible solution. This solution is obtained from $x = (\ell_1, \ldots, \ell_n)$ by increasing each variable one after another by the value 1 until no such increment is possible for any variable. In fact, this is used as the initial solution in several heuristics proposed in the literature.

The Prasad–Kuo algorithm

- Step 0: Obtain a feasible solution x^* following the simple procedure of Proschan and Bray [274]. Let $f^* = f(x^*)$ and $(x_1, \ldots, x_{n-1}) = (\ell_1, \ldots, \ell_{n-1})$, and go to step 4.

- Step 1: Let $k = n - 1$.

- Step 2: If $x_k = u_k$, go to step 3. Otherwise let $x_k = x_k + 1$ and $x_j = \ell_j$, for $j = k + 1, \ldots, n - 1$ (if $k = n - 1$, let $x_{n-1} = x_{n-1} + 1$). If the solution $(x_1, \ldots, x_k, \ell_{k+1}, \ldots, \ell_n)$ is infeasible, go to step 3. If this solution is feasible but the bound $f(x_1, \ldots, x_k, x'_{k+1}, \ldots, x'_n)$ is less than or equal to f^*, repeat step 2. If the solution is feasible and the bound $f(x_1, \ldots, x_k, x'_{k+1}, \ldots, x'_n)$ is greater than or equal to f^*, go to step 4.

- Step 3: Let $k = k - 1$. If $k = 0$, take x^* as the optimal solution and stop. Otherwise, go to step 2.

- Step 4: Find the largest integer r such that $(x_1, \ldots, x_{n-1}, r)$ satisfies the constraints (5.4) and (5.13). If $f^* < f(x_1, \ldots, x_{n-1}, r)$, update f^* and x^* as $f^* = f(x_1, \ldots, x_{n-1}, r)$ and $x^* = (x_1, \ldots, x_{n-1}, r)$, respectively, and go to step 1.

If $g_{in}(x) = a_{in}x$ for $1 \leq i \leq m$, then the integer r in step 4 is given by

$$r = \min\left[u_n, \ell_n + \min\left(\left\lfloor\frac{s_i}{a_{in}}\right\rfloor : 1 \leq i \leq m, a_{in} > 0\right)\right]$$

where s_i is a slack corresponding to the feasible solution $(x_1, \ldots, x_{n-1}, \ell_n)$. If $g_{ij}(x)$ is linear for $i = 1, 2, \ldots, m$, for at least one j, renumber the variables such that all functions $g_{in}(x)$ are linear, which simplifies the computations in step 4.

Example 5-2

We now solve the reliability optimization problem which was first presented by Nakagawa and Nakashima [247]. This problem is more complex than a simple series-parallel redundancy problem.

Consider a system composed of three stages operating in series. The system reliability can be increased by choosing the most reliable component out of four candidates at stage 1, adding redundant components in parallel at stage 2, and using a 2-out-of-n: G configuration at stage 3. The problem is to maximize

$$f(x_1, x_2, x_3) = R_1(x_1)R_2(x_2)R_3(x_3),$$

subject to

$$4\exp\left[\frac{0.02}{1 - R_1(x_1)}\right] + 5x_2 + 2(x_3 + 1) \leq 45,$$

$$5\exp\left(\frac{x_1}{8}\right) + 3\left[x_2 + \exp\left(\frac{x_2}{4}\right)\right] + 5\left[x_3 + 1 + \exp\left(\frac{x_3}{4}\right)\right] \leq 75,$$

$$10 + 8x_2\exp\left(\frac{x_2}{4}\right) + 6x_3\exp\left(\frac{x_3}{4}\right) \leq 240,$$

and

$$(1, 1, 1) \leq (x_1, x_2, x_3) \leq (4, 5, 5),$$

x_j being an integer, where

$R_1(x_1) = 0.88, 0.92, 0.98,$ and 0.99 for $x_1 = 1, 2, 3, 4$, respectively,

$R_2(x_2) = 1 - (1 - 0.81)^{x_2}$,

$$R_3(x_3) = \sum_{k=2}^{x_3+1}\binom{x_3 + 1}{k}(0.77)^k(1 - 0.77)^{x_3+1-k}.$$

The values of $\triangle_{1j}(d)$ for $d = 2, 3,$ and 4 are:

j	d		
	2	3	4
1	0.4107	6.1477	24.8308
2	0.7544	1.6092	2.5779
3	0.0000	0.0000	0.0000

Table 5.2. Feasible solutions of Example 5-2 evaluated in step 4

No.	Solution	System reliability
1	(1, 1, 5)	0.7106
2	(1, 2, 5)	0.8456
3	(1, 3, 5)	0.8712
4	(1, 4, 5)	0.8761
5	(2, 1, 5)	0.7429
6	(2, 2, 5)	0.8840
7	(2, 3, 5)	0.9108
8	(2, 4, 5)	0.9159
9	(3, 1, 5)	0.7913
10	(3, 2, 5)	0.9417
11	(3, 3, 5)	0.9702
12	(3, 4, 5)	0.9757

The values of $\triangle_{2j}(d)$ for $d = 2, 3, 4$, and 5 are:

j	d			
	2	3	4	5
1	5.0000	10.0000	15.0000	20.0000
2	4.0941	8.4989	13.3028	18.6190
3	16.1073	40.5358	76.7128	129.3415

The values of $\triangle_{3j}(d)$ for $d = 2, 3, 4$, and 5 are:

j	d			
	2	3	4	5
1	2.0000	4.0000	6.0000	8.0000
2	6.8235	14.1649	22.1713	31.0316
3	12.0805	30.4018	57.5346	97.0061

Using the above algorithm, the lexicographic search is carried out over a set of 100 solutions. The solution (1, 1, 1) is taken as the initial feasible solution. The system reliability corresponding to (1, 1, 1) is 0.5489.

System reliability is evaluated for 12 feasible solutions in step 4. These solutions are provided in Table 5.2. The upper bounds for partial sequences (1, 5), (2, 5), and (3, 5) eliminate five solutions each in the search, whereas the partial sequence (4) eliminates as many as 25 solutions. The optimal solution derived by the method is $\mathbf{x}^* = (3, 4, 5)$, and the corresponding system reliability is 0.9757.

5.6   Discussion

Most of the optimization problems in reliability systems are formulated as nonlinear integer programming models. The majority of such problems deal with redundancy allocation for system-reliability maximization or cost minimization. In the previous chapters, we have primarily discussed systems in which each stage has a 1-out-of-n: G configuration. A stage may sometimes have a k-out-of-n: G configuration, with $k > 1$. As discussed for Model V in Chapter 2, there could be several choices for a component at a stage with or without provision for redundancy. In such cases, the reliability function becomes more complex. However, we may still be able to formulate the optimization problem as a nonlinear integer programming model. From a computational point of view, these models are highly complex. Several researchers have adopted appropriate discrete optimization methods to solve a variety of discrete reliability optimization problems. The most prominent of these are implicit enumeration methods, integer programming methods, partial search methods, branch-and-bound methods, and lexicographic search procedures. References to the application of discrete methods to reliability optimization problems are listed in Table 2.3. The NNH branch-and-bound method and the partial enumeration method of Lawler and Bell are applicable to a general class of discrete optimization problems in reliability systems. However, they do not exploit special structures, such as separability, to reduce the computations. The lexicographic procedure of Prasad and Kuo takes advantage of the separability of constraints.

In this chapter, we have formulated a discrete reliability optimization problem with separable objective functions and separable constraints as a 0–1 linear programming problem for which there are several excellent solution procedures available in the literature. This formulation is useful in series and parallel systems. We have also described two branch-and-bound methods: a partial enumeration method and a lexicographic search procedure. It is not easy to compare, in a general context, the performance of discrete methods for system-reliability optimization. Their performance depends on the structure of the problem, the number of variables, and the number of constraints. There is no unique method that can be considered to be the best for all discrete reliability optimization problems. A singular advantage of these methods is that they give an exact optimal solution. However, they also involve a large amount of computational effort. Development of a good discrete method for such problems remains a challenge in reliability design.

EXERCISES

5.1   Give a real example of the 0–1 knapsack problem. Is this problem NP hard?

5.2   Let the feasible region be

$$S_1 = \{x \in B^3 : 9x_1 + 3x_2 + 5x_3 \leq 10\},$$

or

$$S_2 = \{x \in B^3 : x_1 + x_2 \leq 1, x_1 + x_3 \leq 1\}.$$

Which formulation is better? Why?

5.3   Given the two knapsack-type constraints below,

$$4x_1 + 5x_2 + 3x_3 \leq 7 \quad \text{and}$$

$$3x_1 - 2x_2 \leq 2,$$

$$x \in B^3,$$

find the values of x by using minimal cover. (Hint: If $x_i + x_j = 1$ and $x_i \leq x_j$, then $x_i = 0$ and $x_j = 1$.)

5.4   Briefly explain the cutting plane method. (Hint: extreme points.)

5.5   Let one of the constraints in the final simplex tableau be $x_1 + 2.25x_2 - 3.25x_3 = 2.67$. Generate Gomory fractional cut, and show a necessary condition for satisfying integrality.

5.6   Discuss the advantages and disadvantages of transforming a reliability optimization problem into a 0–1 LP.

5.7   Determine the reliability structures for which the redundancy allocation problem can be transformed into a 0–1 LP.

5.8   Give three methods for transforming an optimization integer variable problem into 0–1 LP. (The methods must be distinct with respect to the 0–1 representation of bounded integer variables.)

5.9   Explain both depth-first and breadth-first search in node selection of the branch-and-bound algorithm. Discuss the advantages and disadvantages of both types of search. What are the stopping and fathoming criteria?

5.10   Explain how the tightening of bounds on the variables reduces the computational effort of the methods discussed in this chapter.

5.11   Develop a simple branch-and-bound method to solve Example 5-1 without transforming the problem into 0–1 LP. (Use, if necessary, a trivial upper bound on the objective function of the subproblems.)

5.12   In Table 5.1, show the calculations of:

   (a) the upper bound for node 26, and
   (b) the upper bound for node 29.

5.13   Solve the problem described in Example 5-2 using the NNH algorithm.

5.14 Discuss the advantages and disadvantages of the Lawler and Bell method for nonlinear integer programming problems.

5.15 In Lawler and Bell's algorithm, let $\mathbf{y} = (0, 1, 1, 0, 0, 0)$. Find y^*, y_+, and y_-^*.

5.16 Determine which features of the Prasad and Kuo method reduce the computational effort.

5.17 Is it advantageous to apply the Prasad and Kuo method to a redundancy allocation problem after transforming the problem into 0–1 LP? Explain your reasons.

5.18 Let $\mathbf{x} = (x_1, x_2, x_3)$, where $1 \leq x_j \leq 3$, for all $j = 1, 2, 3$. Show the lexicographic order of all 27 vectors.

5.19 In Example 5-2, let $\mathbf{b} = (30, 50, 100)$. Find the optimal solution and draw a table for the feasible solutions evaluated in step 4 similar to those given in Table 5.2.

5.20 Compare the computational performance of all the methods in this chapter using Example 5-1.

6 Reliability optimization by nonlinear programming

6.1 Introduction

The vast majority of reliability optimization problems are nonlinear ones involving integer and/or continuous variables. One of the approaches for dealing with a nonlinear optimization problem in integer variables is to solve its continuous version and appropriately round off the variable optimal values. There are three primary approaches for solving a nonlinear programming (NLP) problem. The first approach involves iterative feasible direction methods; while in every iteration, we move from one feasible point to another in the direction in which the objective value decreases (for minimization objective). The approach is quite useful for problems involving linear constraints. The second approach is based on Lagrange multipliers; these methods are termed Lagrangian. A simple Lagrange multiplier method is used when the problem has the following characteristics:

1. no inequalities appear in the constraints,

2. no nonnegativity or discreteness restrictions are imposed on the variables,

3. the number of constraints is less than the number of variables, and

4. the objective and constraint functions possess second-order partial derivatives.

This method can easily be implemented when the system involves a single equality constraint. The method can be generalized to solve problems involving inequality constraints and nonnegative variables. The general version uses Kuhn–Tucker (K–T) conditions, which are the necessary conditions for local optimality of a feasible solution. These conditions are also sufficient for a global minimum when the objective is to minimize a convex function on a convex set. They can be also used to maximize a concave function on a convex set. The necessary and sufficient conditions for optimality in NLP problems are developed using Taylor's series expansion. Many authors have derived necessary and sufficient conditions for various cases of NLP problems utilizing K–T conditions. As we shall see later in this chapter, it is possible to solve a constrained NLP problem by solving a sequence of unconstrained NLP problems. There are several methods, such as Newton, gradient, pattern search, etc., for solving unconstrained NLP problems. When the constraints are linear equalities, variable-reduction algorithms can also be used for solving NLP problems. Some algorithms can transform an NLP problem into an unconstrained NLP problem involving fewer variables. The convex simplex method and the reduced gradient method of Wolfe [323] are quite useful for solving such problems. For a good description of these methods, refer to Reklaitis et al. [276] and McCormick [211].

In the third approach, the solution of a constrained optimization problem is obtained by solving a sequence of unconstrained optimization problems, whose objective functions include a penalty for violating the constraints. Solutions to the unconstrained problems approach an optimal solution of the original constrained problem. The penalization methods belong to this class. These methods are widely used due to their simplicity, ability to handle nonlinear constraints, and dependence only on unconstrained minimization techniques. Their main disadvantages are that the convergence may be slow, and the unconstrained problems may be ill-conditioned when the penalty is large.

There are two types of methods for solving an unconstrained optimization problem: (1) derivative-type methods, such as the steepest ascent method, the Newton method, etc; and (2) search methods, such as the direct search of Hooke and Jeeves [130], the conjugate direction method of Powell [264], and the flexible polygon search of Nelder and Mead [251], etc. Derivative-type methods require evaluation of partial derivatives of the objective function in every iteration, whereas search methods do not. Although the derivative-type methods converge faster than search methods, they involve a lot of preparatory work prior to implementation when the number of variables is large. Due to currently available high power computing facilities, search methods may be more convenient than derivative-type methods for large-scale problems.

There are several NLP applications for reliability optimization problems. Hwang et al. [133] used sequential unconstrained minimization techniques (SUMT) to optimize system reliability. Hwang et al. [136] proposed an augmented Lagrangian method and a generalized reduced gradient method for system-reliability optimization. Kuo et al. [178] developed a branch-and-bound method to solve reliability optimization nonlinear integer programming formulations. In their method, the bound at each node is derived by a Lagrange multiplier method involving K–T conditions. This approach can also be followed for mixed integer reliability optimization programming formulations. For systems with a general large-scale structure, Li and Haimes [195] adopted a three-level decomposition approach for optimization of reliability.

Section 6.2 describes Lagrangian methods based on necessary conditions for local optimality whereas Section 6.3 presents four types of penalization methods which belong to the class of SUMT. The following two examples are used to illustrate NLP methods for system-reliability optimization. Throughout this chapter, we denote the vector of component reliabilities by $\mathbf{r} = (r_1, \ldots, r_n)$ and the vector of redundancy levels by $\mathbf{x} = (x_1, \ldots, x_n)$.

Example 6-1

Consider a reliability system as shown in Figure 6.1. Let r_i represent the reliability of the ith component. The problem is to select optimal values of r_i so as to maximize the

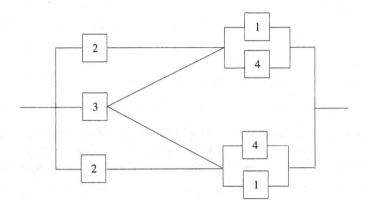

Figure 6.1. A schematic diagram of a complex system

system reliability

$$R_\text{s} = 1 - r_3[(1 - r_1)(1 - r_4)]^2 - (1 - r_3)[1 - r_2 + r_2(1 - r_1)(1 - r_4)]^2, \qquad (6.1)$$

subject to a nonlinear separable constraint and bounds on r_i. More specifically, the problem is to maximize

$$R_\text{s},$$

subject to

$$r_1^{0.6} + r_2^{0.6} + r_3^{0.6} + 1.5r_4^{0.6} \le 4.0,$$

and

$$0.65 \le r_i \le 1.0, \qquad \text{for } i = 1, 2, 3, 4.$$

Example 6-2
Maximize

$$R_\text{s} = \prod_{j=1}^{4} \left[1 - (1 - r_j)^{x_j+1} \right],$$

subject to

$$\sum_{j=1}^{4} c_j x_j \le 44.6,$$

$$\sum_{j=1}^{4} w_j x_j \le 96,$$

x_j being a nonnegative integer for $j = 1, 2, 3, 4$; where the values of r_j, c_j, and w_j are as listed below.

j	r_j	c_j	w_j
1	0.80	1.2	5
2	0.70	2.3	4
3	0.75	3.4	8
4	0.85	4.5	7

Note that the above problem is a redundancy allocation problem in a series system involving linear constraints. The continuous version of this problem can be solved by NLP methods, where the resulting solutions can be rounded off to yield integer solutions.

6.2   The Lagrangian method

We now describe the generalized Lagrangian method based on the necessary conditions for local optimality. Throughout this chapter, $f(\cdot)$ denotes a function to be minimized.

Problem 6.1

Consider the problem of minimizing

$$z = f(r_1, \ldots, r_n),$$

subject to

$$g_i(r_1, \ldots, r_n) \geq 0, \qquad \text{for } i = 1, \ldots, m,$$
$$r_j \geq 0, \qquad \text{for } j = 1, \ldots, n.$$

It is assumed that the objective and constraint functions have continuous first-order derivatives. Define the Lagrangian function

$$L(\mathbf{r}, \boldsymbol{\lambda}) = f(\mathbf{r}) - \sum_{i=1}^{m} \lambda_i g_i(\mathbf{r}),$$

where $\lambda_1, \ldots, \lambda_m$ are Lagrange multipliers. We then have

$$\nabla_r L(\mathbf{r}, \boldsymbol{\lambda}) = \nabla f(\mathbf{r}) - \sum_{i=1}^{m} \lambda_i \nabla g_i(\mathbf{r}),$$

where

$$\nabla_r L(\mathbf{r}, \boldsymbol{\lambda}) = \begin{bmatrix} \partial L(\mathbf{r}, \boldsymbol{\lambda})/\partial r_1 \\ \partial L(\mathbf{r}, \boldsymbol{\lambda})/\partial r_2 \\ \vdots \\ \partial L(\mathbf{r}, \boldsymbol{\lambda})/\partial r_n \end{bmatrix}, \qquad \nabla f(\mathbf{r}) = \begin{bmatrix} \partial f(\mathbf{r})/\partial r_1 \\ \partial f(\mathbf{r})/\partial r_2 \\ \vdots \\ \partial f(\mathbf{r})/\partial r_n \end{bmatrix},$$

and

$$\nabla g_i(\mathbf{r}) = \begin{bmatrix} \partial g_i(\mathbf{r})/\partial r_1 \\ \partial g_i(\mathbf{r})/\partial r_2 \\ \vdots \\ \partial g_i(\mathbf{r})/\partial r_n \end{bmatrix}.$$

The necessary conditions for a solution $\mathbf{r}^*$ to be the local optimal for Problem 6.1 are

$$g_i(\mathbf{r}^*) \geq 0, \qquad \text{for } i = 1, \ldots, m, \tag{6.2}$$

$$r_j^* \geq 0, \qquad \text{for } j = 1, \ldots, n, \tag{6.3}$$

and there exists a vector $\boldsymbol{\lambda}^* = (\lambda_1^*, \ldots, \lambda_m^*)^T$ of Lagrange multipliers such that

$$\nabla f(\mathbf{r}^*) - \sum_{i=1}^{m} \lambda_i^* \nabla g_i(\mathbf{r}^*) \geq 0, \tag{6.4}$$

$$\left[\nabla f(\mathbf{r}^*) - \sum_{i=1}^{m} \lambda_i^* \nabla g_i(\mathbf{r}^*) \right]^T \mathbf{r}^* = 0, \tag{6.5}$$

$$\lambda_i^* g_i(\mathbf{r}^*) = 0, \qquad \text{for } i = 1, \ldots, m, \tag{6.6}$$

$$\lambda_i^* \geq 0, \qquad \text{for } i = 1, \ldots, m. \tag{6.7}$$

Equations (6.4)–(6.7) are the Kuhn–Tucker (K–T) conditions, and point $(\mathbf{r}^*, \boldsymbol{\lambda}^*)$ is the K–T point. An additional necessary condition involving the second-order derivatives of $L(\mathbf{r}, \boldsymbol{\lambda})$ is ignored, in order to keep the discussion simple.

One of the standard approaches to solving an NLP problem is to find a K–T point $(\mathbf{r}^*, \boldsymbol{\lambda}^*)$ and take $\mathbf{r}^*$ as the optimal solution of Problem 6.1. If the function $f(\mathbf{r})$ is convex and the feasible region is a convex set, then $\mathbf{r}^*$ is also the global minimum. Standard computer packages for solving nonlinear systems can be used for solving K–T conditions.

Solution of Example 6-1 by the Lagrangian method
Problem 6.2
Example 6-1 can be written so as to minimize

$$f(\mathbf{r}) = r_3[(1 - r_1)(1 - r_4)]^2 + (1 - r_3)[1 - r_2 + r_2(1 - r_1)(1 - r_4)]^2,$$

subject to

$$g_i(\mathbf{r}) \geq 0, \qquad \text{for } i = 1, \ldots, 9,$$

where

$$g_1(\mathbf{r}) = 4.0 - (r_1^{0.6} + r_2^{0.6} + r_3^{0.6} + 1.5r_4^{0.6}),$$

$$g_{1+i}(\mathbf{r}) = 1.0 - r_i,$$

$$g_{5+i}(\mathbf{r}) = r_i - 0.65,$$

for $i = 1, 2, 3, 4$.

Since there is no explicit nonnegativity assumption on the variables, the inequality (6.4) (among K–T conditions) is replaced by

$$\nabla f(\mathbf{r}^*) - \sum_{i=1}^{m} \lambda_i^* \nabla g_i(\mathbf{r}^*) = 0.$$

We have

$$[\partial f(\mathbf{r})/\partial r_1] = -2r_3(1 - r_1)(1 - r_4)^2$$
$$- 2(1 - r_3)[1 - r_2 + r_2(1 - r_1)(1 - r_4)](r_2 - r_2r_4),$$

$$[\partial f(\mathbf{r})/\partial r_2] = 2(1 - r_3)[1 - r_2 + r_2(1 - r_1)(1 - r_4)](r_1r_4 - r_1 - r_4),$$

$$[\partial f(\mathbf{r})/\partial r_3] = [(1 - r_1)(1 - r_4)]^2 - [1 - r_2 + r_2(1 - r_1)(1 - r_4)]^2,$$

$$[\partial f(\mathbf{r})/\partial r_4] = -2r_3(1 - r_1)^2(1 - r_4)$$
$$- 2(1 - r_3)[1 - r_2 + r_2(1 - r_1)(1 - r_4)](r_2 - r_2r_1).$$

The K–T conditions are

$$4.0 - r_1^{0.6} - r_2^{0.6} - r_3^{0.6} - 1.5r_4^{0.6} \geq 0,$$

$$1 - r_i \geq 0, \qquad \text{for } i = 1, 2, 3, 4,$$

$$r_i - 0.65 \geq 0, \qquad \text{for } i = 1, 2, 3, 4,$$

$$[\partial f(\mathbf{r})/\partial r_1] + \lambda_1(0.6)r_1^{-0.4} + \lambda_2 - \lambda_6 = 0,$$

$$[\partial f(\mathbf{r})/\partial r_2] + \lambda_1(0.6)r_2^{-0.4} + \lambda_3 - \lambda_7 = 0,$$

$$[\partial f(\mathbf{r})/\partial r_3] + \lambda_1(0.6)r_3^{-0.4} + \lambda_4 - \lambda_8 = 0,$$

$$[\partial f(\mathbf{r})/\partial r_4] + \lambda_1(0.9)r_4^{-0.4} + \lambda_5 - \lambda_9 = 0,$$

$$\lambda_1(4.0 - r_1^{0.6} - r_2^{0.6} - r_3^{0.6} - 1.5r_4^{0.6}) = 0,$$

$$\lambda_{1+i}(1 - r_i) = 0, \qquad \text{for } i = 1, 2, 3, 4,$$

$$\lambda_{5+i}(r_i - 0.65) = 0, \qquad \text{for } i = 1, 2, 3, 4,$$

and $\lambda_i \geq 0$, for $i = 1, 2, 3, 4$.

The K–T point satisfying the above conditions is

$$\mathbf{r}^* = (0.9958, 0.9991, 0.7122, 0.6771), \text{ and } \lambda_i^* = 0 \text{ for } i = 1, 2, 3, 4.$$

This solution is obtained using the Newton method. The corresponding objective value $R(\mathbf{r}^*)$ in Example 6-1 is 0.999 997.

Solution of Example 6-2 by the Lagrangian method
Problem 6.3
The continuous version of the problem Example 6-2 can be written so as to minimize

$$f(\mathbf{x}) = -\prod_{j=1}^{4} \left[1 - (1 - r_j)^{x_j+1} \right],$$

subject to

$$44.6 - \sum_{j=1}^{4} c_j x_j \geq 0,$$

$$96 - \sum_{j=1}^{4} w_j x_j \geq 0,$$

$$x_j \geq 0, \qquad \text{for } j = 1, 2, 3, 4.$$

Let

$$Q_j(x_j) = -\frac{(1 - r_j)^{x_j+1}}{1 - (1 - r_j)^{x_j+1}}, \qquad \text{for } j = 1, 2, 3, 4.$$

Then the partial derivatives of $f(\mathbf{x})$ and $g_i(\mathbf{x})$ are

$$[\partial f(\mathbf{x})/\partial x_j] = f(\mathbf{x})Q_j(x_j)\ln(1 - r_j), \qquad \text{for } j = 1, 2, 3, 4,$$
$$\nabla g_1(\mathbf{x}) = -(c_1, c_2, c_3, c_4)^T,$$

and

$$\nabla g_2(\mathbf{x}) = -(w_1, w_2, w_3, w_4)^T.$$

The necessary conditions for local optimality are

$$44.6 - \sum_{j=1}^{4} c_j x_j \geq 0,$$

$$96.0 - \sum_{j=1}^{4} w_j x_j \geq 0,$$

$$f(\mathbf{x})Q_j(x_j)\ln(1 - r_j) + \lambda_1 c_j + \lambda_2 w_j \geq 0, \qquad \text{for } j = 1, 2, 3, 4,$$

$$[f(\mathbf{x})Q_j(x_j)\ln(1 - r_j) + \lambda_1 c_j + \lambda_2 w_j]x_j = 0, \qquad \text{for } j = 1, 2, 3, 4,$$

$$\lambda_1 \left(44.6 - \sum_{j=1}^{4} c_j x_j\right) = 0,$$

$$\lambda_2 \left(96 - \sum_{j=1}^{4} w_j x_j\right) = 0,$$

and

$$\lambda_1 \geq 0, \quad \lambda_2 \geq 0 \quad \text{and } x_j \geq 0, \qquad \text{for } j = 1, 2, 3, 4.$$

The above nonlinear system of nonnegative variables can be solved using numerical methods such as the Newton method. The solution of the above system is

$$(x_1, x_2, x_3, x_4) = (4.2604, 5.1442, 3.6124, 3.3842),$$

$$(\lambda_1, \lambda_2) = (0.000\,000\,3, 0.000\,000\,3).$$

The integer solution $\mathbf{x}^* = (4, 5, 4, 3)$, obtained by rounding off the above without violating the constraints, is taken as the optimal solution of Example 6-2. The corresponding objective value, that is, the system reliability, is 0.9975.

6.3   Penalization methods

The main feature of these methods is that they allow us to solve a constrained minimization problem by solving a sequence of unconstrained minimization problems. The objective function of each unconstrained problem is that of the constrained problem plus a penalty term involving the constraint functions and a positive parameter. The value of the parameter determines the degree of penalty for constraint violation. The solution of an unconstrained problem in the sequence is used as a starting point for solving the next unconstrained problem by an iterative method. The sequence of unconstrained problems is generated by selecting a monotonic sequence of values for the parameter.

In this section, we shall describe four types of penalization methods: (1) barrier, (2) penalty, (3) mixed penalty function, and (4) augmented Lagrangian. In the barrier method, the objective function of each unconstrained problem involves a penalty for a feasible solution which is close to the boundary of the feasible region. The minimizer of each unconstrained problem is a feasible solution of the original constrained problem, and the sequence of such minimizers converges to a constrained optimal solution. These methods belong to the class of *interior point methods*. In penalty methods, the objective function involves a penalty for violating the constraints. The minimizers of unconstrained problems are usually infeasible solutions of the constrained problem. However, their sequence converges to the required optimal solution. These methods are called *exterior point methods*. In mixed methods, the objective function of each unconstrained problem involves both barrier and penalty terms. In augmented Lagrangian methods, the objective function involves Lagrangian terms, as described in Section 6.2 in addition to penalty terms. For a good description of penalization methods, refer to Bertsekas [35] and McCormick [211].

The Hooke–Jeeves (H–J) pattern search method [130] is used for solving each unconstrained minimization problem encountered in the numerical examples of this section, and is described in Appendix 2.

6.3.1   The barrier method

Problem 6.4
Consider the nonlinear optimization problem of minimizing

$$z = f(r_1, \ldots, r_n),$$

subject to

$$g_i(r_1, \ldots, r_n) \geq 0, \qquad \text{for } i = 1, \ldots, m.$$

Note that this problem is the same as Problem 6.1 without nonnegative restrictions on r_j. Let

$$S = \{\mathbf{r} : g_i(\mathbf{r}) \geq 0, \qquad \text{for } i = 1, \ldots, m\},$$

and assume that the interior of the feasible region S is nonempty and f assumes a minimum in the interior of S.

Problem 6.5
Now consider an unconstrained minimization problem:

$$\phi_\tau(\mathbf{r}) = f(\mathbf{r}) + \tau \sum_{i=1}^{m} \frac{1}{g_i(\mathbf{r})},$$

where τ is a positive parameter.

The parameter τ is called a *barrier parameter*. For any fixed τ, the value of $\phi_\tau(\mathbf{r})$ tends to infinity as feasible $\mathbf{r}$ approaches the boundary of the feasible region. In other words, $\phi_\tau(\mathbf{r}) \to \infty$ as $g_i(\mathbf{r}) \to 0$ for at least one i. Therefore, if an unconstrained minimization technique is used to solve Problem 6.5, starting with an interior point in the feasible set S, the minimizer of this problem will also be an interior point in S.

If Problem 6.5 is solved for a monotonic sequence $\{\tau_k\}$ of τ values, with $\tau_k \to 0$ as $k \to \infty$, the sequence of resulting solutions $\{\mathbf{r}^k\}$ converges to an optimal solution of Problem 6.4. The term $\tau \sum_{i=1}^m [g_i(\mathbf{r})]^{-1}$ can be replaced by $-\tau \sum_{i=1}^m \ln g_i(\mathbf{r})$ in Problem 6.5. The function $f(\mathbf{r}) + \tau \sum_{i=1}^m [g_i(\mathbf{r})]^{-1}$ is called an *inverse barrier function*, whereas $f(\mathbf{r}) - \tau \sum_{i=1}^m \ln g_i(\mathbf{r})$ is called a *logarithmic barrier function*. The barrier method for Problem 6.4 can be described by the following steps.

1. Select a small positive value ϵ and an initial value $\tau_1 > 0$. Solve the unconstrained Problem 6.5 by an appropriate unconstrained minimization method, starting with a point in the feasible region S. Suppose $\mathbf{r}^1$ minimizes $\phi_{\tau_1}(\mathbf{r})$. Let $k = 2$.

2. Select τ_k such that $0 < \tau_k < \tau_{k-1}$ and find the minimizer $\mathbf{r}^k$, of $\phi_{\tau_k}(\mathbf{r})$, by an unconstrained minimization method, starting with the feasible point $\mathbf{r}^{k-1}$. Go to step 3.

3. If $\left\| \mathbf{r}^k - \mathbf{r}^{k-1} \right\| < \epsilon$, stop and take $\mathbf{r}^k$ as the required solution of Problem 6.4.

The barrier method is applicable when the constraints are inequalities. Its main disadvantage is that the unconstrained Problem 6.5 becomes ill-conditioned and increasingly difficult to solve as the parameter value decreases to zero.

Solution of Example 6-1 by the barrier method

Consider Problem 6.2, which is equivalent to Example 6-1. In the barrier method, the objective function of the corresponding unconstrained minimization problem is

$$\phi_\tau(\mathbf{r}) = r_3[(1 - r_1)(1 - r_4)]^2 + (1 - r_3)[1 - r_2 + r_2(1 - r_1)(1 - r_4)]^2$$
$$+ \tau \left(4.0 - r_1^{0.6} - r_2^{0.6} - r_3^{0.6} - 1.5 r_4^{0.6} \right)^{-1}$$
$$+ \tau \sum_{i=1}^4 \left[(1 - r_i)^{-1} + (r_i - 0.65)^{-1} \right].$$

To solve this problem, we select $\epsilon = 0.001$ and $\tau_k = (0.1)^k$, for $k \geq 1$.

The solutions are obtained by the H–J method. The initial solution considered in the H–J method is $(0.75, 0.75, 0.75, 0.75)$. The optimal solution of each unconstrained problem is taken as the initial solution of the succeeding problem. The procedure is terminated at $k = 8$, as the solution $\mathbf{r}^* = (0.9905, 0.9963, 0.6778, 0.6840)$ is optimal for both the seventh and eighth unconstrained problems. Thus, $\mathbf{r}^*$ is as an optimal solution of Problem 6.2 and, consequently, an optimal solution of

Table 6.1. Optimal solutions of unconstrained problems

k	τ_k	r_1^k	r_2^k	r_3^k	r_4^k	$\phi_{\tau_k}(\mathbf{r}^k)$
1	10^{-1}	0.7831	0.7837	0.7806	0.7648	5.705 84
2	10^{-2}	0.7831	0.7869	0.7865	0.7648	0.585 11
3	10^{-3}	0.7863	0.8446	0.7865	0.7589	0.070 13
4	10^{-4}	0.8783	0.9376	0.7375	0.7150	0.011 75
5	10^{-5}	0.9661	0.9727	0.7065	0.6840	0.002 08
6	10^{-6}	0.9864	0.9922	0.6819	0.6840	0.000 36
7	10^{-7}	0.9905	0.9963	0.6778	0.6840	0.000 07
8	10^{-8}	0.9905	0.9963	0.6778	0.6840	0.000 03

Example 6-1 also. The system reliability corresponding to these component reliabilities is 0.999 98. Optimal solutions of unconstrained problems are summarized in
Table 6.1.

Solution of Example 6-2 by the barrier method

Consider Problem 6.3 which is a continuous version of Example 6-2. The objective
function of the unconstrained minimization problems associated with Problem 6.3 is

$$\phi_\tau(\mathbf{x}) = -\prod_{j=1}^{4}[1 - (1 - r_j)^{x_j+1}]$$

$$+ \tau\left[\left(44.6 - \sum_{j=1}^{4} c_j x_j\right)^{-1} + \left(96 - \sum_{j=1}^{4} w_j x_j\right)^{-1} + \sum_{j=1}^{4}\frac{1}{x_j}\right].$$

The last term $\sum 1/x_j$ is included due to nonnegative restriction of the variables in the
original problem. The sequence $\{\tau_k\}$ is taken as $\tau_k = (1/2)^{k-1}$ for $k \geq 1$. The H–J
method is used to solve each unconstrained minimization problem, and is modified
to avoid exploration of solutions infeasible for Example 6-2. The feasible solution
$(2, 2, 2, 2)$ is taken as the initial solution for the first unconstrained problem and the
optimal solution of each unconstrained problem is taken as the initial solution of the
succeeding problem. The value of ϵ is selected as 0.01 and the procedure is terminated
at $k = 14$ since $\|\mathbf{x}^{14} - \mathbf{x}^{13}\| < 0.01$. The results of all 14 unconstrained problems
are given in Table 6.2. The solution of Problem 6.3 obtained by the barrier method,
is $(4.120, 5.357, 4.138, 2.716)$. It is rounded off to the integer solution $(4, 5, 4, 3)$, for
which the objective value of Example 6-2 is 0.9975.

Table 6.2. Optimal solutions of Example 6-2 by the barrier method

k	τ_k	x_1^k	x_2^k	x_3^k	x_4^k	$\phi_{\tau_k}(x^k)$
1	1.0	4.355	3.840	3.069	2.880	0.3810
2	0.5	4.355	3.881	3.069	2.880	−0.3060
3	0.25	4.355	3.965	3.069	2.880	−0.6496
4	0.125	4.304	4.211	3.233	2.716	−0.8219
5	0.062	4.009	4.285	3.364	2.716	−0.9081
6	0.031	4.140	4.429	3.495	2.716	−0.9520
7	0.015	4.015	4.618	3.626	2.716	−0.9745
8	0.007	4.015	4.762	3.626	2.716	−0.9861
9	0.0035	3.874	4.877	3.880	2.716	−0.9914
10	0.0017	4.069	5.082	3.882	2.716	−0.9942
11	0.0008	3.864	5.226	4.138	2.716	−0.9958
12	0.0004	4.120	5.226	4.138	2.716	−0.9966
13	0.0002	4.120	5.357	4.138	2.716	−0.9971
14	0.0001	4.120	5.357	4.138	2.716	−0.9973

6.3.2 The penalty method

Problem 6.6
Consider the optimization problem of minimizing

$$z = f(r_1, \dots, r_n),$$

subject to

$$g_i(r_1, \dots, r_n) \geq 0, \quad \text{for } i = 1, \dots, m,$$
$$h_i(r_1, \dots, r_n) = 0, \quad \text{for } i = 1, \dots, v.$$

The penalty method is very similar to the barrier method. In this method, the constrained Problem 6.6 is solved by a sequence of unconstrained minimization problems in which the objective function involves $f(\mathbf{r})$ and a penalty for violating the constraints.

Problem 6.7
Consider the unconstrained problem of minimizing

$$\psi_\tau(\mathbf{r}) = f(\mathbf{r}) + \tau \left\{ \sum_{i=1}^{m} (\min[0, g_i(\mathbf{r})])^2 + \sum_{i=1}^{v} h_i^2(\mathbf{r}) \right\},$$

where τ is a positive parameter.

In the penalty method, $\psi_\tau(\mathbf{r})$ is minimized iteratively for a strictly increasing sequence of τ penalty parameter values. Solution of Problem 6.6 is reached through a sequence of unconstrained problem minimizers. The minimizers are usually infeasible solutions of Problem 6.6. As τ increases to infinity, the objective function of the unconstrained Problem 6.7 involves an increasing penalty for violating the constraints, thereby forcing the minimizers toward the feasible region. If the minimizer $\psi_\tau(\mathbf{r})$ is a feasible solution of Problem 6.6 for any particular $\tau > 0$, then it is also the minimizer for Problem 6.6, too. The penalty method is more appropriate when all of the constraints are equalities. In the penalty method, when the value of the penalty parameter τ is large, the unconstrained optimization problem may become ill-conditioned and difficult to solve. Newton-type unconstrained minimization methods become ineffective due to ill-conditioning of the problem. The effect of ill-conditioning can be eliminated if a starting point for the unconstrained minimization problem is close to the minimizer of the problem. While solving a sequence of unconstrained minimization problems with the penalty method, the minimizer of a problem can be used as the starting point of the next problem. Such a procedure is effective in dealing with ill-conditioning when the successive minimizers x^{k-1} and x^k are close. This can happen only when τ_k is close to τ_{k-1}. It means that when the successive values in the sequence $\{\tau_k\}$ are close, the ill-conditioning can be overcome. However, the convergence is very slow in such cases. It is suggested in the literature that the sequence $\{\tau_k\}$, satisfying $\tau_k = \beta \tau_{k-1}, \beta \in [4, 10]$, can be used in the penalty method to determine the initial value τ_1 by trial and error.

The penalty term in the unconstrained minimization problem can be defined in several ways. Extensive work on the penalty method can be found in Fiacco and McCormick [93].

Solution of Example 6-1 by the penalty method
In this example, the objective of an unconstrained problem is to minimize

$$\psi_{\tau_k}(\mathbf{r}) = r_3\left[(1 - r_1)(1 - r_4)\right]^2 - (1 - r_3)\left[1 - r_2 + r_2(1 - r_1)(1 - r_4)\right]^2$$
$$+ \tau_k \max\left(0, r_1^{0.6} + r_2^{0.6} + r_3^{0.6} + 1.5r_4^{0.6} - 4.0\right)$$
$$+ \tau_k \sum_{i=1}^{4}[\max(0, r_i - 1) + \max(0, 0.65 - r_i)].$$

For Problem 6.2, it is noticed that this objective is found to be more effective than the one described above as part of the penalty method. The $\{\tau_k\}$ sequence is taken as $\tau_k = 4^{k-1}$ for $k \geq 1$. With $\tau = \tau_1$, the unconstrained minimization problem is solved by the H–J method starting with the initial solution $(0.75, 0.75, 0.75, 0.75)$. Its optimal solution is $\mathbf{r}^* = (0.9987, 0.9987, 0.7094, 0.6753)$, with the corresponding objective value $\psi_{\tau_1}(\mathbf{r}^*) = 0.999\,999$. Since this solution is feasible for Example 6-1, it is also optimal for Example 6-1. The corresponding objective value in the original problem is $R_s(\mathbf{r}^*) = 0.999\,999$.

Solution of Example 6-2 by the penalty method

The objective function of the unconstrained minimization problems associated with Problem 6.3 is

$$\psi_{\tau_k}(\mathbf{x}) = -\prod_{j=1}^{4}[1 - (1 - r_j)^{x_j+1}]$$

$$+ \tau_k \left\langle \left[\min\left(0, 44.6 - \sum_{j=1}^{4} c_j x_j\right)\right]^2 + \left[\min\left(0, 96 - \sum_{j=1}^{4} w_j x_j\right)\right]^2 \right\rangle.$$

The sequence $\{\tau_k\}$ is taken as $\tau_k = 4^{k-1}$ for $k \geq 1$. Here also, the H–J search method is used to solve each unconstrained minimization problem. This method is modified to avoid exploration of solutions which violate the nonnegative restriction on the variables. The feasible solution $(3, 3, 3, 3)$ is taken as the initial solution for the first unconstrained problem.

The solution yielded by the H–J method is $(4.12, 5.44, 4.0, 3.0)$, which is feasible for Example 6-2. Therefore, the procedure is terminated and the solution is taken as the optimal one for the original problem. This solution is rounded off to obtain the integer feasible solution $(4, 5, 4, 3)$, for which the objective value in Example 6-2 is 0.9975.

6.3.3   The mixed penalty function method

This method is also a sequential unconstrained minimization method in which the penalty function includes both barrier and penalty parameters. Consider the NLP Problem 6.6 with two types of constraints. In this case, the objective function of the kth unconstrained minimization problem is

$$P(\mathbf{r}, \tau_k) = f(\mathbf{r}) + \tau_k \sum_{i=1}^{m} \frac{1}{g_i(\mathbf{r})} + (\tau_k)^{-1/2} \sum_{i=1}^{v} h_i^2(\mathbf{r}).$$

The sequence $\{\tau_k\}$ of parameter values strictly decreases to zero. One of the choices for the sequence $\{\tau_k\}$ recommended by Fiacco and McCormick [93] is $\tau_1 = 1$ and $\tau_k = \tau_{k-1}/c$, where $c > 1$ (usually $c = 4$).

6.3.4   The penalization method with Lagrange multipliers

This method is a penalization method in which the objective function of each unconstrained minimization problem also involves Lagrange multipliers. The ill-conditioning of an unconstrained problem resulting from barrier and penalty methods can be overcome by this method, and the requirement that τ tends to zero or infinity is avoided. The method was first introduced, independently, by Hestenes [125] and Powell [265]. A good description of this type of method can be found in Bertsekas [35].

Problem 6.8

Consider the constrained NLP problem of minimizing

$$z = f(r_1, \ldots, r_n),$$

subject to

$$g_i(r_1, \ldots, r_n) \leq 0, \qquad \text{for } i = 1, \ldots, m,$$
$$h_i(r_1, \ldots, r_n) = 0, \qquad \text{for } i = 1, \ldots, v.$$

By introducing m slack variables, this can be transformed into the following NLP problem. Minimize

$$z = f(r_1, \ldots, r_n),$$

subject to

$$g_i(r_1, \ldots, r_n) + y_i^2 = 0, \qquad \text{for } i = 1, \ldots, m,$$
$$h_i(r_1, \ldots, r_n) = 0, \qquad \text{for } i = 1, \ldots, v.$$

Now, define the augmented Lagrangian function

$$L_\tau(\mathbf{r}, \mathbf{y}, \boldsymbol{\lambda}) = f(\mathbf{r}) + \sum_{i=1}^{m} \lambda_i \left[g_i(\mathbf{r}) + y_i^2 \right] + \sum_{i=1}^{v} \lambda_{m+i} h_i(\mathbf{r})$$

$$+ \frac{\tau}{2} \left\{ \sum_{i=1}^{m} [g_i(\mathbf{r}) + y_i^2]^2 + \sum_{i=1}^{v} [h_i(\mathbf{r})]^2 \right\}.$$

For any given vectors $\boldsymbol{\lambda}$ and $\mathbf{r}$, $L_\tau(\mathbf{r}, \mathbf{y}, \boldsymbol{\lambda})$ is minimized over $y_1, \ldots, y_m$, when

$$y_i^2 = \max \left\{ -\frac{\lambda_i}{\tau}, g_i(\mathbf{r}) \right\} - g_i(\mathbf{r}), \qquad \text{for } i = 1, \ldots, m.$$

This is true since the term

$$\sum_{i=1}^{m} \left\{ \lambda_i [g_i(\mathbf{r}) + z_i] + \frac{\tau}{2} [g_i(\mathbf{r}) + z_i]^2 \right\},$$

for fixed $\boldsymbol{\lambda}$ and $\mathbf{r}$, is the sum of separable functions of variable $z_1, \ldots, z_m$, and the summand is minimized over nonnegative $z_1, \ldots, z_m$ by $z_i = \max\{-[\lambda_i/\tau + g_i(\mathbf{r})], 0\}$ for $i = 1, \ldots, m$. Let

$$\bar{L}_\tau(\mathbf{r}, \boldsymbol{\lambda}) = \min_{y_1, \ldots, y_m} L_\tau(\mathbf{r}, \mathbf{y}, \boldsymbol{\lambda}).$$

Then, for fixed vectors $\boldsymbol{\lambda}$ and $\mathbf{r}$ we have

$$\bar{L}_\tau(\mathbf{r}, \boldsymbol{\lambda}) = f(\mathbf{r}) + \sum_{i=1}^{m} \left\{ \lambda_i \max \left[-\frac{\lambda_i}{\tau}, g_i(\mathbf{r}) \right] + \frac{\tau}{2} \left\langle \max \left[-\frac{\lambda_i}{\tau}, g_i(\mathbf{r}) \right] \right\rangle^2 \right\}$$

$$+ \sum_{i=1}^{v} \lambda_{m+i} h_i(\mathbf{r}) + \frac{\tau}{2} \sum_{i=1}^{v} [h_i(\mathbf{r})]^2.$$

Performing simple algebraic operations on the second term on the right-hand side, we can write

$$\overline{L}_\tau(\mathbf{r}, \boldsymbol{\lambda}) = f(\mathbf{r}) + \frac{1}{2\tau} \sum_{i=1}^{m} \left\{ \langle \max[\lambda_i + \tau g_i(\mathbf{r}), 0] \rangle^2 - \lambda_i^2 \right\}$$

$$+ \sum_{i=1}^{v} \lambda_{m+i} h_i(\mathbf{r}) + \frac{\tau}{2} \sum_{i=1}^{v} [h_i(\mathbf{r})]^2.$$

In the augmented Lagrangian method for Problem 6.8, the unconstrained problem of minimizing $\overline{L}_{\tau_k}(\mathbf{r}, \boldsymbol{\lambda}^k)$, subject to $\mathbf{r} \in R^n$, is solved sequentially for a strictly increasing sequence $\{\tau_k\}$ of positive τ values and for the sequence $\{\boldsymbol{\lambda}^k\}$ of multiplier vectors, satisfying

$$\lambda_i^{k+1} = \lambda_i^k + \tau_k \max\left\{ \frac{-\lambda_i^k}{\tau_k}, g_i(\mathbf{r}^k) \right\}, \qquad \text{for } i = 1, \ldots, m,$$

that is,

$$\lambda_i^{k+1} = \max\{0, \lambda_i^k + \tau_k g_i(\mathbf{r}^k)\}, \qquad \text{for } i = 1, \ldots, m,$$

and

$$\lambda_{m+i}^{k+1} = \lambda_{m+i}^k + \tau_k h_i(\mathbf{r}^k), \qquad \text{for } i = 1, \ldots, v,$$

where $\mathbf{r}^k$ is the minimizer of $\overline{L}_{\tau_k}(\mathbf{r}, \boldsymbol{\lambda}^k)$. The above unconstrained problems can be solved by the H–J method, which does not require any assumption on continuity and differentiability of the functions involved.

In the augmented Lagrangian method, the rate of convergence to an optimal or local optimal point of the original problem is quite high. There is no need to increase the parameter value to infinity.

Solution of Example 6-1 by the augmented Lagrangian method
Example 6-1 can be written so as to minimize

$$f(\mathbf{r}) = r_3[(1 - r_1)(1 - r_4)]^2 + (1 - r_3)[1 - r_2 + r_2(1 - r_1)(1 - r_4)]^2,$$

subject to

$$g_i(\mathbf{r}) \leq 0, \qquad \text{for } i = 1, \ldots, 9,$$

where

$$g_1(\mathbf{r}) = \left(r_1^{0.6} + r_2^{0.6} + r_3^{0.6} + 1.5 r_4^{0.6}\right) - 4.0,$$

$$g_{1+i}(\mathbf{r}) = r_i - 1.0,$$

$$g_{5+i}(\mathbf{r}) = 0.65 - r_i,$$

for $i = 1, 2, 3, 4.$

Table 6.3. Optimal solutions of Example 6-1 by the augmented Lagrangian method

k	τ_k	$(\lambda_1, \ldots, \lambda_9)$	r_1^k	r_2^k	r_3^k	r_4^k	$\bar{L}_{\tau_k}(\mathbf{r}, \boldsymbol{\lambda}^k)$
1	1	$(1, 1, 1, 1, 1, 1, 1, 1, 1)$	0.7100	0.7500	0.6870	0.5836	$-1.489\,29$
2	4	$(0.540, 0.710, 0.750, 0.687, 0.584,$ $0.940, 0.900, 0.963, 1.067)$	0.8242	0.8236	0.8141	0.8029	$-0.655\,65$
3	16	$(0.457, 0.007, 0.044, 0, 0,$ $0.243, 0.206, 0.307, 0.455)$	0.8886	0.8880	0.8785	0.6741	$-0.015\,34$
4	64	$(0.008, 0, 0, 0, 0, 0, 0, 0, 0.07)$	0.9746	0.9887	0.7443	0.6741	$0.000\,11$
5	256	$(0, 0, 0, 0, 0, 0, 0, 0, 0)$	0.9746	0.9887	0.7443	0.6741	$0.000\,15$

The objective function of the kth unconstrained problem is

$$\bar{L}_{\tau_k}(\mathbf{r}, \boldsymbol{\lambda}^k) = f(\mathbf{r}) + \frac{1}{2\tau_k}\Bigg[\Big[\big\langle \max[0, \lambda_1^k + \tau_k g_1(\mathbf{r})]\big\rangle^2 - (\lambda_1^k)^2\Big]$$

$$+ \sum_{j=1}^4 \Big\{\big\langle \max[0, \lambda_{1+j}^k + \tau_k(r_j - 1)]\big\rangle^2 - (\lambda_{1+j}^k)^2\Big\}$$

$$+ \sum_{j=1}^4 \Big\{\big\langle \max[0, \lambda_{5+j}^k + \tau_k(0.65 - r_j)]\big\rangle^2 - (\lambda_{5+j}^k)^2\Big\}\Big]\Bigg].$$

Let $(r_1^k, r_2^k, r_3^k, r_4^k)$ denote the optimal solution of the kth unconstrained problem. Let us take

$(\lambda_1^1, \ldots, \lambda_9^1) = (1, \ldots, 1),$

$\tau_k = 4^{k-1},$ for $k > 1,$

$\lambda_1^{k+1} = \max[0, \lambda_1^k + \tau_k g_1(\mathbf{r}^k)],$

$\lambda_{1+j}^{k+1} = \max[0, \lambda_{1+j}^k + \tau_k(r_j^k - 1)],$ for $j = 1, 2, 3, 4,$

$\lambda_{5+j}^{k+1} = \max[0, \lambda_{5+j}^k + \tau_k(0.65 - r_j^k)],$ for $j = 1, 2, 3, 4.$

As explained in the barrier and the penalty methods, the H–J method is used for solving unconstrained minimization problems. In the H–J method, the initial solution considered for the first unconstrained minimization problem is $(0.7, 0.7, 0.7, 0.7)$. The optimal solution of each unconstrained problem is used as the initial solution for the next problem in the sequence. The procedure is terminated after solving the fifth unconstrained problem since the optimal solutions obtained for the fourth and fifth problems are the same. The solutions of the three unconstrained problems are given in Table 6.3.

The optimal solution of Example 6-1 is

$$\mathbf{r}^* = (0.9746, 0.9887, 0.7443, 0.6741),$$

which is obtained from both the third and fourth unconstrained problems.   The corresponding system reliability is $R_s(\mathbf{r}^*) = 0.999\,85$.

Solution of Example 6-2 by the augmented Lagrangian method

Problem 6.3 (a continuous version of Example 6-2) can be rewritten so as to minimize

$$f(\mathbf{x}) = -\prod_{j=1}^{4}[1 - (1 - r_j)^{x_j+1}],$$

subject to

$$\sum_{j=1}^{4} c_j x_j - 44.6 \leq 0,$$

$$\sum_{j=1}^{4} w_j x_j - 96.0 \leq 0,$$

$$-x_i \leq 0, \qquad \text{for } i = 1, 2, 3, 4.$$

The objective function of the kth unconstrained problem is

$$\overline{L}_{\tau_k}(\mathbf{x}, \boldsymbol{\lambda}^k) = f(\mathbf{x}) + \frac{1}{2\tau_k}\left\{\left\langle \max\left[0, \lambda_1^k + \tau_k\left(\sum_{j=1}^{4} c_j x_j - 44.6\right)\right]\right\rangle^2 - (\lambda_1^k)^2\right.$$

$$+ \left\langle \max\left[0, \lambda_2^k + \tau_k\left(\sum_{j=1}^{4} w_j x_j - 96.0\right)\right]\right\rangle^2 - (\lambda_2^k)^2$$

$$+ \left. \sum_{j=1}^{4}\left\langle[\max(0, \lambda_{2+j}^k - \tau_k x_j)]^2 - (\lambda_{2+j}^k)^2\right\rangle\right\}.$$

Let us take $(\lambda_1^1, \ldots, \lambda_6^1) = (1, 1, 1, 1, 1, 1)$, $\tau_1 = 1$, $\tau_k = 4^{k-1}$, for $k > 1$, and

$$\lambda_1^{k+1} = \max\left[0, \lambda_1^k + \tau_k\left(\sum_{j=1}^{4} c_j x_j^k - 44.6\right)\right],$$

$$\lambda_2^{k+1} = \max\left[0, \lambda_2^k + \tau_k\left(\sum_{j=1}^{4} w_j x_j^k - 96.0\right)\right],$$

$$\lambda_{2+j}^{k+1} = \max(0, \lambda_{2+j}^k - \tau_k x_j^k), \qquad \text{for } j = 1, 2, 3, 4,$$

where $(x_1^k, x_2^k, x_3^k, x_4^k)$ is the optimal solution of the kth unconstrained problem. As explained in the barrier and penalty methods, the H–J method is used for solving unconstrained minimization problems. The procedure is terminated after solving the third unconstrained problem, since the optimal solutions obtained for the second and third problems are the same. The solutions of the three unconstrained problems are given in Table 6.4.

Table 6.4. Optimal solutions of Example 6-2 by the augmented Lagrangian method

k	τ_k	$(\lambda_1, \dots, \lambda_6)$	x_1^k	x_2^k	x_3^k	x_4^k	$\overline{L}_{\tau_k}(\mathbf{x}, \boldsymbol{\lambda}^k)$
1	1	$(1, 1, 1, 1, 1, 1)$	4.164	5.000	4.000	3.000	$-1.997\,545$
2	4	$(0, 0, 0, 0, 0, 0)$	4.306	5.344	4.000	3.000	$-0.997\,841$
3	16	$(0, 0, 0, 0, 0, 0)$	4.306	5.344	4.000	3.000	$-0.997\,841$

The optimal solution of the constrained Problem 6.3 derived by the augmented Lagrangian method is $(4.306, 5.344, 4.000, 3.000)$. It is rounded off to the integer solution $(4, 5, 4, 3)$ for which the objective value in Example 6-2 is 0.9975.

6.4   Discussion

Several NLP methods have been used to solve reliability optimization problems. Most of these problems are discrete, mixed integer and nonlinear optimization problems. When component reliabilities are to be selected on a continuous scale for optimization, the problem is of NLP type, and these methods can be used directly to solve it. When the problem is to select component reliabilities (on a continuous scale) and redundancy levels of components for optimization, then the problem is of mixed integer programming type. When component redundancies are to be determined, the problem is of discrete optimization type. All of the reliability optimization problems that can be described as mathematical programming models can be solved heuristically, or exactly, by using NLP methods and appropriate rounding off procedures for integer variables.

In an approximation approach, NLP methods are used to solve continuous mixed integer and discrete optimization problems, and the resulting solutions are rounded off to get approximate optimal solutions. NLP methods are used by Kuo et al. [178] in the development of branch-and-bound methods for discrete and mixed integer reliability optimization problems. The most prominent methods for solving general NLP problems are: gradient, Lagrangian, and penalization. Hwang et al. [136] used reduced gradient methods for solving some reliability optimization problems. In the Lagrangian method for a problem with general constraints, Kuhn–Tucker conditions are solved to obtain a local optimal solution. Hwang et al. [136] applied Lagrangian methods for reliability optimization. Penalization methods are also quite useful for solving NLP problems. These methods transform a constrained NLP problem into a sequence of unconstrained NLP problems and solve them iteratively by an appropriate unconstrained optimization method. Such methods can be used even when the functions do not satisfy assumptions on continuity and differentiability. In such cases, a direct search method such as the Hooke–Jeeves method, Powell's conjugate method, etc., can be used. Barrier and penalty methods, and other sequential unconstrained minimization methods, belong to this class. Hwang et al. [133] adopted sequential unconstrained minimization techniques developed by Fiacco and McCormick [92] to solve reliability optimization problems. Excellent software packages are available to solve NLP problems. A major disadvantage of the NLP approach for solving discrete optimization problems in reliability systems is that rounding off does not necessarily give an optimal solution.

EXERCISES

6.1 What are the approaches commonly used to solve NLP problems? Briefly explain each of them.

6.2 Show the Kuhn–Tucker (K–T) conditions for a general NLP problem. Explain each of the K–T conditions. Suppose the objective function and constraints are linear. Is the global optimal solution the point which satisfies the K–T conditions? What if the objective function is not linear?

6.3 Show the K–T conditions in Example 3-2.

6.4 In Chapter 6, both H–J and Newton methods are used to solve NLP problems. When is the H–J method used, and when is the Newton method used? Discuss other methods to solve NLP problems.

6.5 Derive the K–T conditions for the problem: maximize

$$f = r_1r_2 + r_3r_4 + r_1r_5r_4 + r_2r_3r_5 - r_1r_2r_3r_4 - r_1r_2r_3r_5 - r_1r_2r_4r_5 - r_1r_3r_4r_5$$
$$- r_2r_3r_4r_5 + 2r_1r_2r_3r_4r_5,$$

subject to

$$5e^{r_1} + 3e^{r_2} + 6e^{r_3} + 4e^{r_4} + 9e^{r_5} \leq 70,$$
$$0.9 \leq r_i \leq 1.0, \qquad i = 1, 2, \ldots, 5.$$

6.6 Describe a reliability optimization problem for which an exact optimal solution can be derived by K–T conditions.

6.7 Explain the motivation of penalization methods. Describe the advantages and disadvantages of each penalization method.

6.8 Solve Exercise 6.5 by the barrier method.

6.9 Solve Problem 6.2 by the barrier method, taking $\tau_k = (0.4)^k$, $k \geq 1$; and compare the computational results with those of Table 6.1.

6.10 Apply the mixed penalty method to Problem 6.2 and compare the results with those of Table 6.1.

6.11 Find the inverse barrier function for Example 3-2.

6.12 Show the unconstrained nonlinear function of the penalty method for Example 3-3.

6.13 Find $\overline{Lr}(r, \lambda_k)$ and λ_i^{k+1} of the augmented Lagrangian method for Example 4-3.

6.14 In Example 4-3, let the third constraint be $\sum_{j=1}^{n} w_j x_j \exp(x_j/4) = 200$. Find the mixed penalty function.

6.15 Solve the relaxed version of Example 5-1 (by ignoring the integer restriction on x_i) by any NLP method described in Chapter 6 and round off the solution in a feasible manner. Compare the resulting solution with the exact optimal one.

6.16 What is the disadvantage of the NLP approach for solving discrete optimization problems in reliability systems.

- In the paper by Tillman et al. [303], both the reliability function of a complex system and the cost function are derived: solve the following exercises by referring to this paper.

6.17 Verify the reliability function and express it in the correct format. Set up the optimization problem so as to maximize the reliability function subject to the cost constraint.

6.18 Solve the optimization problem in Exercise 6.17 by the Lagrangian method.

6.19 Solve the optimization problem in Exercise 6.17 by the Hooke–Jeeves algorithm (see Appendix 2).

6.20 Predict the global optimal solution (by any method you choose) to Exercise 6.17.

6.21 Compare the solutions of the Lagrangian method (Exercise 6.18) and the Hooke–Jeeves algorithm (Exercise 6.19) in terms of:
(a) deviation from the global optimal solution,
(b) number of iterations,
(c) computation time, and
(d) your working experience.

7 Metaheuristic algorithms for optimization in reliability systems

7.1 Introduction

As described in Chapters 3–6, an abundance of methods have been used to solve various reliability optimization problems. However, methods that include the maximum principle, once applied to reliability optimization problems, are not emphasized in this book because they are difficult to understand and often produce bad solutions. These methods were generated from research ideas and most were used for only a short time following publication.

In recent years, metaheuristics have been selected and successfully applied to handle a number of reliability optimization problems. These heuristics, founded more on artificial reasoning than classical mathematics-based optimization, include genetic algorithms, simulated annealing, and tabu search. Genetic algorithms seek to imitate the biological phenomenon of evolutionary production through the parent–children relationship; simulated annealing is based on a physical process in metallurgy; and the philosophy of tabu search is to derive and exploit a collection of intelligent problem-solving principles. In addition to these three primary methods, in Chapter 7 we briefly discuss several other techniques for reliability optimization.

Simulated annealing and tabu search methods can be used for a wide range of complex discrete optimization problems, and both have been implemented in various design optimization problems. The simulated annealing method is based on Markov (probabilistic) transitions among the solutions, whereas tabu search depends significantly on past information at any stage of its progress. A common feature of both methods is the goal of each to provide a near-optimal solution. Although the majority of reliability design problems are of a discrete nature, not many applications of the two methods are available in the literature on reliability optimization. We therefore introduce simulated annealing and tabu search methods in this book as they have great potential, in our view, for application in reliability design. We illustrate the methods using examples in reliability optimization.

7.2 Genetic algorithms

A genetic algorithm is a probabilistic method for solving an optimization problem. It solves a problem by imitating the natural evolution process in which populations undergo continuous change through cross breeding, mutation, and natural selection. A genetic algorithm can easily be implemented on a computer for a wide spectrum of problems. The method is particularly useful for solving complex optimization problems which are cumbersome for direct mathematical treatment. Genetic algorithms can be very effective in dealing with large-scale, real-life discrete, and continuous optimization problems without making unrealistic assumptions and approximations. Keeping the imitation of natural evolution as the foundation, genetic algorithms can be appropriately designed and modified to exploit special features of the problem to be solved. Therefore, genetic algorithms constitute an approach or solution methodology for a broad spectrum of problems.

During the last decade significant progress has been made in genetic algorithm methodology. Holland [129] has made pioneering contributions to the development of genetic algorithms. Several, including Davis [69], Goldberg [111], Kinnear [156], Koza [166], [167], Leipins and Hilliard [192], Michalewicz [217], Mitchell [233], Gen and Cheng [99], and Chambers [45] provide good general descriptions of genetic algorithms. Coit et al. [64] and a few others have contributed significantly to reliability optimization by applying the genetic algorithmic approach.

Almost all classical optimization methods are uni-directional in the sense that they move from one solution to another until they reach a termination criterion. In contrast, a genetic algorithm carries out a multi-directional search by generating and selecting multiple solutions at every stage. A genetic algorithm starts with a set of solutions (feasible or infeasible) and alters the set iteratively, by a random procedure, generating new solutions and discarding some current solutions in every iteration. It stops according to a termination criterion, which is usually a number of iterations without improvement or an upper limit on the number of iterations. Since the genetic algorithm imitates the process of natural evolution, the following vocabulary of natural genetics is used to describe it:

Chromosome	a coded version of a solution as a string or vector
Population	a set of chromosomes
Population size	the number of chromosomes in the population
Chromosome fitness	objective function value of the solution associated with a chromosome
Gene	an indivisible part of a chromosome
Locus	the position of a gene in a chromosome
Parents	chromosomes involved in a cross breeding operation
Mutation	random change in a gene
Offspring	newly generated chromosomes

Chromosomes are also called *strings* or *individuals*: these three words are used synonymously throughout this chapter. The fitness of a chromosome $\mathbf{v}$ is denoted by $eval(\mathbf{v})$ and the population size by s. The most important aspects of a genetic algorithm are:

1. representation of a general solution as a chromosome, that is, coding of a solution as a vector of discrete elements;

2. evaluation of chromosome fitness;

3. generation of new chromosomes from the available ones; and

4. selection of chromosomes to maintain a population of fixed size.

Two operations, that is, crossover and mutation, are performed to generate new chromosomes. These are called genetic operations. In a crossover operation, a pair of chromosomes exchange their genetic material (parts of the solutions) to yield one or more offspring. In a mutation operation, some genes of a chromosome undergo random changes to produce a new chromosome. The following example illustrates the genetic algorithm terminology and the two operations.

Example 7-1
Suppose there are six objects $1, 2, \ldots, 6$, with object i having weight w_i and value p_i for $i = 1, 2, \ldots, 6$. The problem is how to select a subset of these six objects such that the total value of the selected objects is a maximum, with their total weight not exceeding a specified weight W. This problem is well-known in optimization literature as the knapsack problem.

Let $x_i = 1$ if object i is selected and 0 otherwise. Then the 0–1 vector $(x_1, x_2, \ldots, x_6)$ represents a solution of the problem. The solution is feasible only if $\sum_{i=1}^{6} w_i x_i \leq W$. In this example, a vector of six 0–1 elements can be taken as a chromosome. For instance, $(1\,0\,0\,1\,0\,1)$ is a chromosome representing the selection of objects 1, 4, and 6. The set

$$\{(1\,0\,1\,0\,0\,1), (0\,1\,0\,0\,1\,1), (1\,1\,1\,0\,1\,1), (1\,0\,1\,0\,1\,0), (0\,0\,1\,1\,1\,0)\}$$

can be taken as a population of size $s = 5$. The gene at locus (position) 4 of chromosome $(1\,0\,1\,1\,0\,1)$ is the element 1. The fitness of chromosome $\mathbf{v} = (v_1, \ldots, v_6)$ is

$$eval(\mathbf{v}) = \sum_{i=1}^{6} p_i v_i.$$

For example, the fitness of chromosome $(1\,1\,0\,0\,0\,1)$ is $p_1 + p_2 + p_6$.

A simple crossover operation (for generation of new chromosomes) is described as follows. Consider the pair of chromosomes $(1\,0\,1\,0\,0\,1)$ and $(0\,1\,1\,1\,0\,1)$. Cut each chromosome between positions 3 and 4 to form two segments, and interchange the segments on the right-hand side of the chromosomes:

```
101001        101|001        101101
       ⟶            ↑↓   ⟶
011101        011|101        011001
```

The operation yields two new chromosomes $(1\,0\,1\,1\,0\,1)$ and $(0\,1\,1\,0\,0\,1)$. This is a crossover operation in which the parents are the chromosomes $(1\,0\,1\,0\,0\,1)$ and $(0\,1\,1\,1\,0\,1)$, whereas the offspring are $(1\,0\,1\,1\,0\,1)$ and $(0\,1\,1\,0\,0\,1)$.

A mutation operation on chromosome $(1\,1\,0\,0\,1\,1)$ may yield $(1\,1\,0\,0\,0\,1)$: this can happen when gene 1 in position 5 is altered to gene 0 due to the underlying random mechanism. It is possible that genes at two or more positions are altered in a mutation operation on a chromosome.

The major steps in a typical genetic algorithm are:

1. chromosome representation,

2. random generation of a fixed number of chromosomes to form an initial population,

3. evaluation of fitness for each chromosome,

4. selection of parents for the crossover operation,

5. generation of offspring by the crossover operation,

6. mutation operation on chromosomes, and

7. selection of s chromosomes for the population of the next generation according to the fitness values.

Chromosome representation

Chromosome representation is a crucial aspect of any genetic algorithm. There may be several ways of chromosome representation for any given problem. In the initial work of Holland [129], a chromosome is defined as a fixed-length string of elements 0 and 1 for any specific problem. It is not easy to provide such a representation in all situations. In some cases, the 0–1 string representation may also increase the chromosome length drastically. For this reason, several researchers have adopted problem-specific chromosome representation which facilitates design of effective crossover and mutation operations. Consider the problem of maximizing

$$z = \sum_{i=1}^{n} p_i x_i,$$

subject to

$$\sum_{i=1}^{n} w_i x_i \leq W,$$

$$\ell_i \leq x_i \leq u_i, \qquad \text{for } i = 1, \ldots, n,$$

x_i being an integer.

For example, to solve this problem by the genetic algorithm approach, one can take the solution vector itself as the chromosome representation which is amenable to both of the genetic operations. It is better to have a representation that gives a one–one correspondence between chromosomes and solutions of the problem.

Fitness evaluation

Whenever a chromosome is generated, the objective function is evaluated for the corresponding solution and the value is taken as the chromosome fitness. Analytical methods are generally used for fitness evaluation. If the evaluation is very complex, or impossible by direct computation, any appropriate method can be used for this purpose. For example, Coit and Smith [60] have used a neural network approach to evaluate the system reliability in their genetic algorithm for finding optimal component redundancy. If the problem involves any constraints on the solutions, the fitness value may also include some penalty (a negative value for maximization problems) for the infeasibility of solutions. Fitness is used to cull the chromosomes while forming the next generation.

Initial population

A fixed number of chromosomes are randomly generated to form an initial population. A gene at any particular locus of a chromosome belongs to a set of values. In the random generation of a chromosome, a gene at any locus is randomly generated from the corresponding set. For instance, in Example 7-1 the gene at any locus belongs to the set $\{0, 1\}$. While generating a chromosome randomly, one of the elements 0 and 1 is randomly selected as the gene at every locus. Creation of chromosomes in a random fashion for the initial population, is different from the mutation operation in which the genes of existing chromosomes undergo random changes to yield new chromosomes.

Crossover operation

Chromosomes are randomly selected for crossover operation. Several procedures are adopted in the literature for selecting parents for crossover. In some selection procedures, a predetermined value is taken as the probability of selecting a chromosome for crossover operation. This value is called the *crossover rate* and is denoted by p_c. For each chromosome in the population, a uniform random number p is generated from the interval $[0, 1]$. A chromosome is selected as a parent only if $p \leq p_c$. If the number of such selected chromosomes is odd, one more chromosome is selected, or one of the selected ones, is ignored. Even this choice may be made randomly. The selected chromosomes are paired for crossover operation. For instance, consider the population

$\{(1\,0\,1\,0\,0\,1), (0\,1\,0\,0\,1\,1), (1\,1\,1\,0\,1\,1), (1\,0\,1\,0\,1\,0), (0\,0\,1\,1\,1\,0), (1\,1\,0\,0\,1\,1),$

$\quad (0\,1\,0\,1\,0\,1), (1\,0\,1\,0\,1\,1)\}$

of eight chromosomes of Example 7-1. Let the crossover rate be $p_c = 0.5$. Suppose that eight random numbers

$0.21, 0.75, 0.32, 0.89, 0.56, 0.72, 0.47,$ and 0.09

are taken from the interval $[0, 1]$. Since the first, third, seventh, and eighth random numbers are less than p_c, the corresponding chromosomes $(1\,0\,1\,0\,0\,1)$, $(1\,1\,1\,0\,1\,1)$, $(0\,1\,0\,1\,0\,1)$, and $(1\,0\,1\,0\,1\,1)$ are selected and the first and seventh chromosomes are paired with the third and eighth chromosomes, respectively, for the crossover operation.

Each pair of selected chromosomes undergoes crossover. There are several methods for crossover. One such method is the single-point operation, in which both chromosomes are cut into two segments at a randomly chosen position and the segments on the right-hand side are swapped between the chromosomes. This operation is illustrated earlier. Two-point and multi-point crossover operations are described in the literature on genetic algorithms.

Syswerda [299] has proposed a uniform crossover operation which is found to be quite effective in exploring the search space. In this operation, at every locus where genes of the two chromosomes are distinct, one of the two genes is randomly selected (with equal probability) for the first offspring and the other one is selected for the second offspring. Both of the offspring have the same gene structure as their parents at all other loci. It seems that any particular crossover method is effective for some classes of problems, but poor for others.

Mutation

Each chromosome in the population is subject to mutation. The probability that a gene of a chromosome is randomly changed to one of the possible alternatives is a predetermined value p_m. This is the mutation rate, and is the same for every gene of every chromosome of the population. To perform the mutation operation on the entire population, for each locus of each chromosome, a uniform random number p is taken from the interval [0, 1]. If $p \leq p_m$ at a locus, the existing gene is replaced by a gene randomly selected from the corresponding set of alternatives.

Generation of a new population

One of the most important features of genetic algorithms is the evolutionary operation of selecting individuals from one generation to form a population of the next generation. In the process of natural evolution, nature selects individuals of a population by their fitness for survival. The greater the fitness, the higher the chance of selection for survival. Similar survival strategies are adopted in genetic algorithms to maintain a population of fixed size in every generation in spite of the creation of new individuals.

In the literature on genetic algorithms, several methods are described for forming a new population. In one class of methods, each offspring replaces an individual of the current population. In Holland's original work on genetic algorithms, the offspring replace their parents to form a new population. Michalewicz [217] has illustrated a selection method in which the required number (population size) of individuals is randomly selected for replacement from the population, with parents replaced by their respective offspring. The probability of selection is proportional to the fitness value. In another class of selection methods, all individuals of a population and their offspring form a sample space for random selection, that is, parents and offspring compete to be selected for the next generation.

For most methods, chromosome fitness plays a significant role in selection. Some methods are deterministic in which a fixed number of individuals are selected in decreasing order of fitness value. In some methods, individuals are selected with probability proportional to fitness value. However, these probabilistic methods have a disadvantage in that the individuals with higher fitness have a higher chance of entering the next generation, adversely affecting the exploration of search space and consequently forcing the genetic algorithm to converge rapidly to a local optimum. To reduce the pressure caused by fitness-based selection, the probabilities are based on fitness values modified by a transformation. Several forms of transformation, such as linear, logarithmic, exponential, etc., are considered in the literature. Dynamic transformations, which also involve a generation number, are used to gradually decrease selection pressure over generations. Sometimes, the selection probabilities are determined on the basis of ranks of individuals in decreasing order of fitness value.

7.2.1   Genetic algorithms for system-reliability optimization

Optimization of system reliability is in general a highly complex problem in which the objective function as well as the constraints are nonlinear and the decision variables are integers. Such problems are not easily amenable to analytical treatment. The majority of reliability optimization researchers have focused their attention on the development of simple heuristics which give near-optimal solutions with less computational effort. However, heuristic methods usually require a mathematical formulation of the problem and do not provide much tradeoff between quality of solution and computational effort. A genetic algorithm can be designed for a problem without an explicit mathematical formulation, and the values of its parameters, such as population size, crossover rate, mutation rate, etc., can be appropriately chosen to balance both the quality of the solution and the computational work.

Recently, genetic algorithms have been designed to solve a variety of reliability optimization problems. Painton and Campbell [257], [258] have designed a genetic algorithm to maximize system reliability subject to a linear cost constraint. Yokota et al. [328] and Ida et al. [144] have presented a genetic algorithm to solve a redundancy allocation problem involving several failure modes. Coit and Smith [58], [59] have used a genetic algorithm for optimal redundancy allocation in a parallel-series system subject to constraints on cost and weight. For the same system, Coit and Smith [62] have applied a genetic algorithm to maximize a percentile of the probability distribution of system reliability, assuming that the component reliabilities are not deterministic. Defining penalties for infeasible and undesirable feasible solutions, Majety and Rajagopal [205] have applied a genetic algorithm to cost optimization in series-parallel and parallel-series systems subject to a minimum reliability requirement.

Dengiz et al. [74], [75] have applied a genetic algorithm to finding a cost-optimal network topology subject as a minimum requirement of network reliability. Deeter and

Smith [70] have used a genetic algorithm for a similar problem assuming that there are multiple choices for each link. Hsieh et al. [131] have used a genetic algorithm to maximize system reliability through optimal redundancy allocation, subject to resource constraints. They have also designed a genetic algorithm for reliability–redundancy allocation problems. A good description of genetic algorithms used for solving reliability optimization problems is provided by Gen and Cheng [99].

In this section, we describe the genetic algorithms of Yokota et al. [328] and Coit and Smith [58].

Reliability optimization with several failure modes

Tillman [301] considered the problem of finding the optimal redundancy allocation in a series system in which the components of each subsystem were subject to two classes of failure modes. He adopted an implicit enumeration method for solving the problem, described as follows.

A series system consists of n subsystems. Subsystem j has $x_j + 1$ components in parallel, which are subjected to two classes of failure modes: O and A. Subsystem j fails if a class-O failure mode occurs in at least one of the $x_j + 1$ components. In contrast, subsystem j fails when a class-A failure mode occurs in all $x_j + 1$ components. In general, subsystem j is subjected to s_j failure modes $1, 2, \ldots, s_j$, among which the first h_j modes $1, 2, \ldots, h_j$ belong to class O and the others belong to class A. Let q_{ju} be the probability that a component in subsystem j fails resulting in failure mode u for $1 \le u \le s_j$ and $1 \le j \le n$.

Let $Q_j^O(x_j)$ be the probability that subsystem j, consisting of $x_j + 1$ redundant components, fails due to the occurrence of one of the h_j class-O failure modes. Similarly, let $Q_j^A(x_j)$ be the probability that subsystem j fails due to the occurrence of one of the $s_j - h_j$ class-A failure modes. Then $Q_j(x_j) = Q_j^O(x_j) + Q_j^A(x_j)$ is the failure probability of subsystem j and the system reliability is

$$R_s = \prod_{j=1}^{n} [1 - Q_j(x_j)].$$

We have

$$Q_j^O(x_j) \approx \sum_{u=1}^{h_j} [1 - (1 - q_{ju})^{x_j+1}],$$

and

$$Q_j^A(x_j) \approx \sum_{u=h_j+1}^{s_j} (q_{ju})^{x_j+1}.$$

Problem 7.1

Maximize

$$R_s = \prod_{j=1}^{n}[1 - Q_j(x_j)],$$

subject to

$$\sum_{j=1}^{n} g_{ij}(x_j) \le b_i, \qquad \text{for } i = 1, 2, \dots, m,$$

$$\ell_j \le x_j \le u_j, \qquad \text{for } j = 1, 2, \dots, n,$$

x_j being a nonnegative integer.

For the purposes of illustration, Tillman [301] has specifically considered the following numerical example.

Example 7-2

Maximize

$$R_s = \prod_{j=1}^{3} \left\{ 1 - \left[1 - (1 - q_{j1})^{x_j+1}\right] - \sum_{u=2}^{4} (q_{ju})^{x_j+1} \right\},$$

subject to

$$G_1(\mathbf{x}) = (x_1 + 3)^2 + x_2^2 + x_3^2 \le 51,$$

$$G_2(\mathbf{x}) = \sum_{j=1}^{3} [x_j + \exp(-x_j)] \ge 6,$$

$$G_3(\mathbf{x}) = \sum_{j=1}^{3} [x_j \exp(-x_j/4)] \ge 3.25,$$

$$1 \le x_1 \le 4,$$

$$1 \le x_2,$$

$$0 \le x_3 \le 7,$$

x_j being an integer.

Note that for each of the three subsystems, there is one class-O failure mode and three class-A failure modes, that is, $h_j = 1$ and $s_j = 4$ for $j = 1, 2$, and 3. Failure probabilities q_{ju} are shown in Table 7.1.

The genetic algorithm

Gen et al. [102], Ida et al. [144] and Yokota et al. [328] have designed genetic algorithms for Problem 7.1. The algorithm is illustrated below using Example 7-2.

Table 7.1. Failure probabilities assigned in Example 7-2

j	u			
	1	2	3	4
1	0.01	0.05	0.10	0.18
2	0.08	0.02	0.15	0.12
3	0.04	0.05	0.20	0.10

Chromosome representation

A solution (x_1, x_2, x_3) of Example 7-2 is a vector of nonnegative integers. When the variables are bounded above, the values of the integer variables in each solution can be written as fixed-length strings of 0s and 1s using the binary system. Such a string can be taken as the chromosome representation of a solution.

The upper bounds on x_1 and x_3 are 4 and 7, respectively. Note that the first constraint and the lower limits 1 and 0 on x_1 and x_3 impose the restriction $x_2 \leq 5$. Now, the value of each variable can be expressed as a 0–1 string of length 3. Thus, solution of Example 7.2 can be represented by a 0–1 string of length 9, e.g. the solution $(x_1, x_2, x_3) = (3, 2, 1)$ is represented by the string $(0\,1\,1\ 0\,1\,0\ 0\,0\,1)$. Therefore, 0–1 strings of length 9 can be taken as chromosomes in this case. Note that some of the chromosomes correspond to infeasible solutions of the problem, e.g. the chromosomes $(0\,0\,1\,0\,0\,0\,0\,0\,1)$ and $(0\,1\,1\,0\,1\,0\,1\,0\,0)$ correspond, respectively, to the solutions $(1, 0, 1)$ and $(3, 2, 4)$ which are infeasible.

Generation of the initial population

The initial population of chromosomes is randomly generated. For the purposes of illustration, let us take the population size $s = 5$. To generate a chromosome randomly, a uniform random number p is taken from interval $[0, 1]$ for each of nine genes of the chromosome. A gene is taken as 0 if the corresponding random number does not exceed a specified value, say 0.5. Otherwise, the gene is taken as 1. The specified value may be determined on the basis of problem instance. Suppose the randomly generated initial population is

$$\mathbf{v}_1 = (0\,1\,0\,1\,0\,0\,0\,0\,1),$$
$$\mathbf{v}_2 = (0\,1\,1\,0\,1\,0\,0\,0\,1),$$
$$\mathbf{v}_3 = (0\,0\,1\,1\,0\,1\,0\,0\,1),$$
$$\mathbf{v}_4 = (0\,1\,1\,0\,0\,1\,0\,1\,1),$$
$$\mathbf{v}_5 = (1\,0\,0\,0\,1\,1\,0\,1\,0).$$

Table 7.2. Random numbers generated for mutation operation

Chromosome	Locus of gene								
	1	2	3	4	5	6	7	8	9
v_1	0.58	0.13	0.86	0.91	0.98	0.23	0.46	0.04[a]	0.63
v_2	0.69	0.53	0.28	0.05[a]	0.53	0.92	0.12	0.69	0.89
v_3	0.66	0.25	0.83	0.14	0.67	0.24	0.36	0.52	0.91
v_4	0.69	0.09[a]	0.97	0.85	0.10[a]	0.39	0.95	0.39	0.82
v_5	0.30	0.45	0.16	0.92	0.50	0.77	0.67	0.27	0.76

[a] Gene to be mutated.

The solutions corresponding to $v_1, \ldots, v_5$ are

$$x_1 = (2, 4, 1),$$

$$x_2 = (3, 2, 1),$$

$$x_3 = (1, 5, 1),$$

$$x_4 = (3, 1, 3),$$

$$x_5 = (4, 3, 2),$$

respectively.

Crossover operation

A one-point crossover operation is used in this algorithm. Suppose the crossover rate is taken as $p_c = 0.5$; this means that 50 percent of the individuals in the population are expected to be selected for crossover. Suppose, the values $0.513, 0.158, 0.817$, 0.436, and 0.719 are randomly drawn from a uniform distribution over $[0, 1]$. Since the second and fourth random numbers do not exceed the crossover rate of 0.5, the corresponding chromosomes v_2 and v_4 are selected for crossover. In the single-point crossover operation, the position for cutting the chromosomes into two segments is taken randomly. Suppose this position is 7 for the pair (v_2, v_4). Then the crossover of the pair (v_2, v_4) yields two offspring o_1 and o_2 as shown below.

$v_2 = 0\,1\,1\,0\,1\,0\,0\,0\,1$ $\longrightarrow$ $0\,1\,1\,0\,1\,0\,|\,0\,0\,1$ $\quad o_1 = 0\,1\,1\,0\,1\,0\,0\,1\,1$

$\uparrow\downarrow \quad \longrightarrow$

$v_4 = 0\,1\,1\,0\,0\,1\,0\,1\,1$ $\qquad 0\,1\,1\,0\,0\,1\,|\,0\,1\,1$ $\quad o_2 = 0\,1\,1\,0\,0\,1\,0\,0\,1$

Mutation

The mutation operation is performed on every chromosome in the population. Suppose the mutation rate is $p_m = 0.1$, that is, the probability of a gene being altered is fixed at 0.1. Suppose the random numbers taken from the interval $[0, 1]$ for all genes in the population are as shown in Table 7.2.

Table 7.3. The offspring and corresponding solutions for Example 7-2

Offspring	Solution	Fitness value
$o_1 = (0\,1\,1\,0\,1\,0\,0\,1\,1)$	$(3, 2, 3)$	$eval(o_1) = 0.629\,12$
$o_2 = (0\,1\,1\,0\,0\,1\,0\,0\,1)$	$(3, 1, 1)$	$eval(o_2) = -M$
$o_3 = (0\,1\,0\,1\,0\,0\,0\,1\,1)$	$(2, 4, 3)$	$eval(o_3) = 0.538\,10$
$o_4 = (0\,1\,1\,1\,1\,0\,0\,0\,1)$	$(3, 6, 1)$	$eval(o_4) = -M$
$o_5 = (0\,0\,1\,0\,1\,1\,0\,1\,1)$	$(1, 3, 3)$	$eval(o_5) = 0.567\,33$

The loci of genes to be mutated are:

Chromosome No.	1	2	4	4
Locus	8	4	2	5

The mutation operation on chromosome $v_1 = (0\,1\,0\,1\,0\,0\,0\,0\,1)$ alters the gene 0 in locus 8 to 1, giving the offspring $o_3 = (0\,1\,0\,1\,0\,0\,0\,1\,1)$. Similarly, mutations on the other two chromosomes, v_2 and v_4, yield

$$o_4 = (0\,1\,1\,1\,1\,0\,0\,0\,1) \quad \text{and} \quad o_5 = (0\,0\,1\,0\,1\,1\,0\,1\,1),$$

respectively. The offspring and the corresponding solutions and fitness values are given in Table 7.3.

Chromosome fitness
Let v be a chromosome and x the corresponding solution to the problem. The fitness value v is defined as

$$eval(v) = \begin{cases} R_s(x), & \text{if } x \text{ is feasible,} \\ -M, & \text{otherwise,} \end{cases}$$

where M is a large positive number.

Here, the penalty for infeasibility is $M + R_s(x)$. The fitness values of chromosomes in the initial population are

$$eval(v_1) = 0.551\,73,$$

$$eval(v_2) = 0.645\,05,$$

$$eval(v_3) = -M,$$

$$eval(v_4) = 0.658\,01,$$

$$eval(v_5) = -M.$$

Note that chromosomes v_3 and v_5 correspond to infeasible solutions $(1, 5, 1)$ and $(4, 3, 2)$, respectively.

Table 7.4. Second population for multiple failure-mode problem

	Chromosome	Solution	Fitness value
1	$v_4 = (0\,1\,1\,0\,0\,1\,0\,1\,1)$	$(3, 1, 3)$	$eval(v_4) = 0.658\,01$
2	$v_2 = (0\,1\,1\,0\,1\,0\,0\,0\,1)$	$(3, 2, 1)$	$eval(v_2) = 0.645\,05$
3	$o_1 = (0\,1\,1\,0\,1\,0\,0\,1\,1)$	$(3, 2, 3)$	$eval(o_1) = 0.629\,12$
4	$o_5 = (0\,0\,1\,0\,1\,1\,0\,1\,1)$	$(1, 3, 3)$	$eval(o_5) = 0.567\,33$
5	$v_1 = (0\,1\,0\,1\,0\,0\,0\,0\,1)$	$(2, 4, 1)$	$eval(v_1) = 0.551\,73$

Table 7.5. Final population for multiple failure-mode problem

	Chromosome	Solution	Fitness value
1	$(0\,1\,1\,0\,0\,1\,0\,1\,0)$	$(3, 1, 2)$	$0.679\,72$
2	$(0\,1\,1\,0\,0\,1\,0\,1\,1)$	$(3, 1, 3)$	$0.658\,01$
3	$(0\,1\,0\,0\,1\,0\,0\,1\,0)$	$(2, 2, 2)$	$0.652\,52$
4	$(0\,1\,1\,0\,1\,0\,0\,1\,0)$	$(3, 2, 2)$	$0.649\,88$
5	$(0\,1\,1\,0\,1\,0\,0\,0\,1)$	$(3, 2, 1)$	$0.645\,05$

Chromosome selection for a new population

The process of selecting chromosomes to form the population of the next generation is completely deterministic. It selects s ($=5$) chromosomes from the current population and the set of offspring together in decreasing order of fitness value. In this example, the five chromosomes selected for the second generation are shown in Table 7.4.

Following the above procedure, the evolution is carried out for ten generations. The population in the tenth generation is given in Table 7.5.

The best chromosome given by the algorithm is $v^1 = (0\,1\,1\,0\,0\,1\,0\,1\,0)$, with a fitness value of $0.679\,72$; that is, the best solution of the problem given by the genetic algorithm is $(3, 1, 2)$, with a corresponding system reliability of $0.679\,72$. The performance of a genetic algorithm improves with population size.

Reliability optimization of a parallel-series system

Coit and Smith [58] have designed a genetic algorithm for maximizing the reliability of a parallel-series system by means of optimal redundancy in each parallel subsystem. The problem considered by Coit and Smith is described below.

Problem 7.2

A reliability system consists of n subsystems $P_1, \ldots, P_n$ in series and subsystem P_i requires at least ℓ_i components in order to function. There are m_i types of components available to be used in P_i, and the reliability of the jth type of component in subsystem P_i is r_{ij}. Let x_{ij} denote the number of components of jth type to be used in subsystem P_i. Then vector $x_i = (x_{i1}, x_{i2}, \ldots, x_{im_i})$ represents an allocation of m_i-type components to P_i. Let $c_i(\mathbf{x}_i)$ and $w_i(\mathbf{x}_i)$ denote the total cost and total weight of P_i, respectively, for allocation $\mathbf{x}_i$. Maximize

$$R_s = \prod_{i=1}^{n} R_i(\mathbf{x}_i),$$

subject to

$$\sum_{i=1}^{n} c_i(\mathbf{x}_i) \le C,$$

$$\sum_{i=1}^{n} w_i(\mathbf{x}_i) \le W,$$

$$\ell_i \le \sum_{j=1}^{m_i} x_{ij} \le u_i \qquad \text{for } i = 1, \dots, n,$$

x_{ij} being a nonnegative integer, where $R_i(\mathbf{x}_i)$ is the reliability of subsystem P_i; u_i is the upper limit on the total number of components in P_i; and the constants C and W are the limits on the total cost and total weight of the system, respectively. The subsystem reliability $R_i(\mathbf{x}_i)$ can be written as

$$R_i(\mathbf{x}_i) = 1 - \sum_{h=0}^{\ell_i-1} \sum_{\tau \in T_i(h)} \prod_{j=1}^{m_i} \binom{x_{ij}}{t_j} r_{ij}^{t_j} (1 - r_{ij})^{x_{ij}-t_j},$$

where

$$\tau = (t_1, t_2, \dots, t_{m_i})$$

and

$$T_i(h) = \left\{ (t_1, t_2, \dots, t_{m_i}) : \sum_{j=1}^{m_i} t_j = h, t_j \le x_{ij}, \text{for } j = 1, \dots, m_i \right\}.$$

The genetic algorithm
Chromosome representation

A general solution of Problem 7.2 is represented by a vector

$$(x_{11}, x_{12}, \dots, x_{1m_1}, x_{21}, x_{22}, \dots, x_{2m_2}, \dots, x_{n1}, x_{n2}, \dots, x_{nm_n}),$$

where the segment $(x_{i1}, x_{i2}, \dots, x_{im_i})$ represents the allocation to subsystem P_i. Assume that component types for each P_i are indexed such that

$$r_{i1} \ge r_{i2} \ge \cdots \ge r_{im_i}.$$

For any feasible allocation, we have $\sum_{j=1}^{m_i} x_{ij} \le u_i$. Thus, allocation $\mathbf{x}_i = (x_{i1}, x_{i2}, \dots, x_{im_i})$ can be uniquely represented by the u_i-dimensional vector

$$\mathbf{y}_i = (1, 1, \dots, 1, 2, 2, \dots, 2, \dots, m_i, m_i, \dots, m_i, m_i + 1, m_i + 1, \dots, m_i + 1),$$

in which j appears x_{ij} times for $j = 1, 2, \ldots, m_i$, and $m_i + 1$ appears $(u_i - \sum_{j=1}^{m_i} x_{ij})$ times. If $x_{ih} = 0$ for some h, then h does not appear in vector $\mathbf{y}_i$. Similarly, if $\sum_{j=1}^{m_i} x_{ij} = u_i$, then $m_i + 1$ does not appear in $\mathbf{y}_i$. Let $k = \sum_{i=1}^{n} u_i$. Now vector $\mathbf{y} = (\mathbf{y}_1, \mathbf{y}_2, \ldots, \mathbf{y}_k)$ of dimension k represents a solution which does not violate the upper limit constraints. It is possible that a solution $\mathbf{x}$ corresponding to the k-dimensional $\mathbf{y}$-type vector may violate the lower limit constraints. Vector $\mathbf{y}$ is taken as the chromosome representation of solution $\mathbf{x}$. For the purpose of illustration, let us consider the numerical example:

$n = 3$,

$(m_1, m_2, m_3) = (6, 8, 6)$,

$(\ell_1, \ell_2, \ell_3) = (2, 2, 2)$,

$(u_1, u_2, u_3) = (4, 4, 5)$.

For example, vector $\mathbf{x} = (\mathbf{x}_1, \mathbf{x}_2, \mathbf{x}_3)$, where

$\mathbf{x}_1 = (1\,0\,0\,1\,0\,1)$,

$\mathbf{x}_2 = (1\,0\,2\,0\,1\,0\,0\,0)$,

$\mathbf{x}_3 = (0\,1\,0\,0\,1\,0)$,

is a solution to the problem. This may or may not be a feasible solution although it satisfies the bounds. The vector $\mathbf{y}$ corresponding to $\mathbf{x}$ is

$\mathbf{y} = (1467\ 1335\ 25\,777)$.

Generation of initial population
The initial population is randomly generated after fixing the population size s. A chromosome is randomly generated by the following steps for each subsystem P_i.

- Step 1: Select a number n_i randomly between ℓ_i and u_i (inclusive of both).

- Step 2: With replacement, select n_i numbers randomly from set $M_i = \{1, 2, \ldots, m_i\}$ and arrange them in nondecreasing order. (It is assumed that $r_{i1} \geq r_{i2} \geq \cdots \geq r_{im_i}$.)

- Step 3: Append the number $m_i + 1$ to this order $(u_i - n_i)$ times to get a u_i-dimensional vector $\mathbf{y}_i$ of elements $1, 2, \ldots, m_i + 1$ in nondecreasing order.

The random selections in steps 1 and 2 are to be done with equal probabilities for all possibilities. The resulting vector $\mathbf{y} = (\mathbf{y}_1, \ldots, \mathbf{y}_n)$ gives a chromosome. For example, let $(n_1, n_2, n_3) = (3, 2, 3)$ and suppose n_i numbers randomly selected from $M_i = \{1, 2, \ldots, m_i\}$ are as given in Table 7.6. Then the corresponding chromosome is $(1227\ 2399\ 11\,377)$.

Table 7.6. Numbers randomly selected from $M_i = \{1, 2, \ldots, m_i\}$

i	n_i	M_i	Elements selected
1	3	$\{1, 2, 3, 4, 5, 6\}$	1, 2, 2
2	2	$\{1, 2, 3, 4, 5, 6, 7, 8\}$	2, 3
3	3	$\{1, 2, 3, 4, 5, 6\}$	1, 1, 3

Chromosome fitness

The fitness of a chromosome is the corresponding system reliability plus a dynamic penalty that depends on the generation number and the relative degree of infeasibility. To enhance the performance of a genetic algorithm, Coit and Smith [58] have introduced an adaptive penalty for infeasibility which is based on the notion of a *near-feasibility threshold* (NFT) for each constraint. The penalty encourages the genetic algorithm to explore the feasible region as well as its NFT neighborhood, but discourages the search beyond the NFT. Let,

Δc_i excess cost for solution i
Δw_i excess weight for solution i
NFT_c near-feasibility threshold for the cost constraint
NFT_w near-feasibility threshold for the weight constraint
V_{all} the objective value without penalty for the best solution found so far
V_{feas} the objective value for the best feasible solution found so far
V_i objective value for solution i

Then, the penalized objective function for solution i is defined as

$$V_{\text{ip}} = V_i - \left[\left(\frac{\Delta w_i}{\text{NFT}_w} \right)^k + \left(\frac{\Delta c_i}{\text{NFT}_c} \right)^k \right] (V_{\text{all}} - V_{\text{feas}}),$$

where k is a predetermined severity parameter. As pointed out by Gen and Cheng [99], when the best solution found so far happens to be feasible, that is, $V_{\text{all}} = V_{\text{feas}}$, the penalty for any infeasible solution is zero. On the other hand, if the best solution found so far is infeasible, having a large objective value (without penalty), the infeasible solutions will have higher penalties. The performance of a genetic algorithm depends on the NFT values. However, it is not easy to select appropriate values for the NFT. To overcome this difficulty, Coit and Smith [58] have defined a dynamic NFT as

$$\text{NFT} = \frac{\text{NFT}_0}{1 + \lambda g},$$

where NFT_0 is the starting value (or upper bound) for the NFT, and g and λ are the generation number and a positive constant, respectively.

Crossover

All of the chromosomes in the population are ranked in nonincreasing order of fitness value. A uniform random number U, between 1.0 and the square root of the population size s, is drawn and the chromosome with the ranking closest to U^2 is selected as a parent for crossover. The number of parents to be selected is approximately $s \times p_c$. For each pair of parents, a uniform crossover operation is adopted (although other kinds of arrangements may be made too) and the genes are rearranged in increasing order in each of the k segments of the offspring. For example, consider two parents

$\mathbf{v}_1 = (1233 \ 1399 \ 23\,447)$,

$\mathbf{v}_2 = (1447 \ 1399 \ 23\,447)$.

Note that the genes in $\mathbf{v}_1$ and $\mathbf{v}_2$ differ at loci 2, 3, and 4 only. Suppose the genes 4, 3, and 7 are selected at loci 2, 3, and 4, respectively, for one of the two offspring. Then, the second offspring obtains the genes 2, 4, and 3 at loci 2, 3, and 4, respectively. The resulting offspring are

$\mathbf{o}_1 = (1437 \ 1399 \ 23\,447)$,

$\mathbf{o}_2 = (1243 \ 1399 \ 23\,447)$.

Rearrangement of genes in increasing order in each segment of $\mathbf{o}_1$ and $\mathbf{o}_2$ gives

$\mathbf{o}_1 = (1347 \ 1399 \ 23\,447)$,

$\mathbf{o}_2 = (1234 \ 1399 \ 23\,447)$.

Formulation of a new population
For the purpose of forming a population of the next generation, the best s chromosomes (with the highest fitness values) are selected from the current population of s chromosomes and the offspring are generated by crossover. Mutation is performed on some of the selected chromosomes and each parent is replaced by its offspring. The final set of s chromosomes is taken as the population of the next generation.

Mutation
Mutation is performed on some of the best s chromosomes (selected from the current population and the offspring from crossover). The number of chromosomes selected for mutation is fixed and pre-determined. Selection of chromosomes for mutation is done by simple random sampling without replacement. Each gene of a selected chromosome is subjected to mutation with a probability equal to the mutation rate p_m. When a gene in the segment corresponding to subsystem P_i is mutated, it becomes $m_i + 1$, with a probability of 0.5, and $j, 1 \leq j \leq m_i$, with a probability of $0.5/m_i$. The best chromosome is never mutated to ensure that the best solution is not adversely affected.

Cost–optimal network design
Consider a communication network with a set of nodes $N = \{1, \ldots, n\}$. A set of h_{ij} links is available to directly connect a pair of nodes i and j, for $i = 1, \ldots, n$, $j = 1, \ldots, n$, and $i \neq j$; only one of the links will be used if the nodes i and j are to be

directly connected. The links have different costs and reliabilities. Let $c_{ij}(k)$ denote the cost of link k, $1 \le k \le h_{ij}$ between nodes i and j. It is assumed that all nodes are perfectly reliable and failures of links are s-independent.

Two nodes are said to be connected if they are directly connected by a single link or there exists a sequence of nodes $v_1, v_2, \ldots, v_h$, such that the pairs (i, v_1), (v_1, v_2), $\ldots, (v_{h-1}, v_h)$, (v_h, j) are directly connected by links. Communication between two nodes is possible only when they are connected. The network is operational as long as every pair of nodes is connected. The reliability of such a network is the probability that every pair of nodes remains connected during the intended period of operation. This reliability is called *all-terminal reliability*.

Deeter and Smith [70] considered a problem given below, to design a network using the available links to minimize the total link age cost, subject to the constraint that the network reliability was not less than a specified value R_0.

Problem 7.3

Let $x_{ij} = k$ if link k, $1 \le k \le h_{ij}$ is chosen to connect nodes i and j. Let $x_{ij} = 0$ if nodes i and j are not directly connected by a link and $c_{ij}(0) = 0$. Then vector

$$\mathbf{x} = [x_{12}, x_{13}, \ldots, x_{1n}, x_{23}, \ldots, x_{2n}, \ldots, x_{(n-2)(n-1)}, x_{(n-2)n}, x_{(n-1)n}]$$

represents the network design. Let $R(\mathbf{x})$ denote the network reliability for design $\mathbf{x}$. If a design ensures that each pair of nodes is connected at time zero, then the corresponding vector $\mathbf{x}$ is said to be feasible. Let S denote the set of all such feasible vectors. The mathematical formulation of the problem is to minimize

$$z = \sum_{i=1}^{n-1} \sum_{j=i+1}^{n} c_{ij}(x_{ij}),$$

subject to

$$R(\mathbf{x}) \ge R_0 \quad \text{and} \quad \mathbf{x} \in S.$$

The genetic algorithm

Deeter and Smith [70] and Dengiz et al. [74] have developed genetic algorithms to solve Problem 7.3. We shall now implement a similar genetic algorithm to solve the following numerical example.

Let $n = 5$, $R_0 = 0.8$, and $h_{ij} = 2$ for each pair (i, j). The values of $c_{ij}(k)$ and $r_{ij}(k)$ are given in Table 7.7.

Chromosome representation

As suggested by Deeter and Smith [70] and Dengiz et al. [74], for each design, the corresponding vector $\mathbf{x}$ is taken as the chromosome representation. For example,

Table 7.7. Constants used in Problem 7.3

i	j	$(c_{ij}(1), r_{ij}(1))$	$(c_{ij}(2), r_{ij}(2))$
1	2	(15, 0.90)	(20, 0.95)
1	3	(12, 0.96)	(17, 0.97)
1	4	(8, 0.85)	(14, 0.91)
1	5	(10, 0.93)	(12, 0.97)
2	3	(18, 0.85)	(21, 0.92)
2	4	(6, 0.96)	(9, 0.98)
2	5	(16, 0.88)	(22, 0.95)
3	4	(10, 0.97)	(12, 0.99)
3	5	(8, 0.92)	(14, 0.99)
4	5	(22, 0.97)	(24, 0.99)

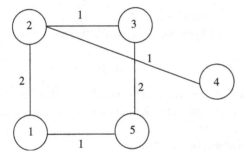

Figure 7.1. Design corresponding to chromosome **x**

consider the chromosome

$$
\begin{array}{ccccccccccc}
 & x_{12} & x_{13} & x_{14} & x_{15} & x_{23} & x_{24} & x_{25} & x_{34} & x_{35} & x_{45} \\
\mathbf{x} = (& 2 & 0 & 0 & 1 & 1 & 1 & 0 & 0 & 2 & 0 \ ).
\end{array}
$$

The design corresponding to chromosome **x** is given in Figure 7.1. The design is operational at time zero since every pair of nodes is connected. In this design, each of the pairs (1, 2), (1, 5), (2, 3), (2, 4), and (3, 5) is directly connected by a single link. Link 2 is used to connect pairs (1, 2) and (3, 5), whereas link 1 is used for the pairs (1, 5), (2, 3), and (2, 4).

Initial population

The population size is taken as $s = 10$. To generate each chromosome of the initial population, a uniform random number p is chosen from the set $\{0, 1, 2, \ldots, h_{ij}\}$ and is taken as x_{ij} for each pair (i, j), $i < j$. Suppose the initial population is shown in Table 7.8.

Table 7.8. The initial population of Problem 7.3

j	x_j
1	(1 0 0 1 1 0 0 1 0 1)
2	(0 0 1 0 1 0 1 1 0 1)
3	(0 0 2 0 1 0 0 1 0 2)
4	(1 0 0 0 1 0 0 1 0 1)
5	(0 2 0 0 0 1 0 0 1 1)
6	(0 0 1 0 1 0 1 0 1 2)
7	(0 1 0 0 1 1 0 1 0 1)
8	(0 1 0 1 0 1 0 1 0 1)
9	(0 2 0 1 0 2 0 1 0 2)
10	(1 2 0 1 0 2 0 1 0 0)

Crossover operation

Let us adopt a single-point crossover operation, with crossover rate $p_c = 0.5$. Dengiz et al. [74] have implemented this operation in their algorithm. The pairs randomly selected for this operation are

$$\{x_4, x_5\} = \{(1\,0\,0\,0\,1\,0\,0\,1\,0\,1), (0\,2\,0\,0\,0\,1\,0\,0\,1\,1)\}$$

and

$$\{x_7, x_{10}\} = \{(0\,1\,0\,0\,1\,1\,0\,1\,0\,1), (1\,2\,0\,1\,0\,2\,0\,1\,0\,0)\}.$$

The crossover operation on the first pair with a single cut at randomly selected position 3 gives the offspring

$x_{11} = (1\,0\,0\,0\,0\,1\,0\,0\,1\,1)$ with $eval(x_{11}) = 376.50$ and $R(x_{11}) = 0.7710$,

$x_{12} = (0\,2\,0\,0\,1\,0\,0\,1\,0\,1)$ with $eval(x_{12}) = 427.00$ and $R(x_{12}) = 0.7758$.

Similarly, the second pair with cutting position 8 produces the offspring

$x_{13} = (0\,1\,0\,0\,1\,1\,0\,1\,0\,0)$ with $eval(x_{13}) = 50.00$ and $R(x_{13}) = 0.0000$,

$x_{14} = (1\,2\,0\,1\,0\,2\,0\,1\,0\,1)$ with $eval(x_{14}) = 397.00$ and $R(x_{14}) = 0.9944$.

Mutation

As explained in Section 7.2.1, the mutation operation is performed on all s (=10) chromosomes of the population. The mutation operation on chromosomes $x_1, x_2, x_3, x_5, x_6, x_7$, and x_8 has yielded the new chromosomes in Table 7.9.

The genes in bold font have resulted from mutation. All genes in the chromosomes x_4, x_9, and x_{10} remain unchanged.

Table 7.9. Some new chromosomes due to mutation

j	$\mathbf{x}_j$	$eval(\mathbf{x}_j)$	$R(\mathbf{x}_j)$
15	(1 0 1 1 1 0 0 1 0 1)	397.00	0.9729
16	(0 0 1 1 1 0 1 1 0 1)	396.00	0.9633
17	(0 0 2 0 1 0 0 1 1 0)	377.33	0.6903
18	(0 2 0 0 0 1 2 0 1 2)	403.00	0.8898
19	(0 0 1 0 0 0 1 1 1 1)	389.05	0.7438
20	(0 1 0 2 1 1 0 1 0 1)	400.00	0.9903
21	(0 1 1 1 0 1 0 0 0 1)	422.00	0.9064

Chromosome fitness

Find the largest link cost $c_{max} = \max_{i,j,k} c_{ij}(k)$. The fitness of chromosome $\mathbf{x}$ is defined as

$$eval(\mathbf{x}) = n(n-1)c_{max} - \left\{ \sum_{i=1}^{n-1} \sum_{j=i+1}^{n} c_{ij}(x_{ij}) + n(n-1)c_{max} \max[R_0 - R(\mathbf{x}), 0] \right\}.$$

(7.1)

The penalty for violating the constraint $R(\mathbf{x}) \geq R_0$ is $n(n-1)c_{max} \max[R_0 - R(\mathbf{x}), 0]$. It is a maximum for any $\mathbf{x} \notin S$, since $R(\mathbf{x}) = 0$ for such $\mathbf{x}$. Maximization of the fitness defined in eq. (7.1) is equivalent to minimization of the network cost (including the penalty). For instance, the reliability for the design corresponding to chromosome $\mathbf{x} = (2\,0\,0\,1\,1\,1\,0\,0\,2\,0)$ is 0.9382, and the fitness of $\mathbf{x}$ is

$$eval(\mathbf{x}) = n(n-1)c_{max} - [c_{12}(2) + c_{15}(1) + c_{23}(1) + c_{24}(1) + c_{35}(2)$$
$$+ n(n-1)c_{max} \max(0.8 - 0.9382, 0)]$$
$$= [(5)(4)(24)] - [20 + 10 + 18 + 6 + 14$$
$$+ (5)(4)(24) \max(0.8 - 0.9382, 0)]$$
$$= 412.00.$$

The fitness values of the chromosomes in the initial population and the corresponding reliabilities are given in Table 7.10.

Chromosome selection for the next generation

The chromosome selection for the formation of the next generation is explained on pp. 183–185. The chromosomes available for selection are given in Table 7.11.

The ten best chromosomes $\mathbf{x}_5$, $\mathbf{x}_{12}$, $\mathbf{x}_{21}$, $\mathbf{x}_8$, $\mathbf{x}_{10}$, $\mathbf{x}_7$, $\mathbf{x}_9$, $\mathbf{x}_2$, $\mathbf{x}_6$, and $\mathbf{x}_1$ form the population for the second generation. This procedure is repeated 19 times to obtain the twentieth generation population, which is shown in Table 7.12.

Table 7.10. The fitness values of chromosomes in Problem 7.3

j	$eval(\mathbf{x}_j)$	$R(\mathbf{x}_j)$
1	405.00	0.9537
2	406.00	0.8224
3	386.54	0.7428
4	376.50	0.7198
5	427.00	0.8310
6	406.00	0.8106
7	412.00	0.9206
8	420.00	0.9506
9	410.00	0.9739
10	419.00	0.9210

Table 7.11. Chromosomes available for formation of the next generation

j	$\mathbf{x}_j$	$eval(\mathbf{x}_j)$	$R(\mathbf{x}_j)$
1	(1 0 0 1 1 0 0 1 0 1)	405.00	0.9537
2	(0 0 1 0 1 0 1 1 0 1)	406.00	0.8224
3	(0 0 2 0 1 0 0 1 0 2)	386.54	0.7428
4	(1 0 0 0 1 0 0 1 0 1)	376.50	0.7198
5	(0 2 0 0 0 1 0 0 1 1)	427.00	0.8310
6	(0 0 1 0 1 0 1 0 1 2)	406.00	0.8106
7	(0 1 0 0 1 1 0 1 0 1)	412.00	0.9206
8	(0 1 0 1 0 1 0 1 0 1)	420.00	0.9506
9	(0 2 0 1 0 2 0 1 0 2)	410.00	0.9739
10	(1 2 0 1 0 2 0 1 0 0)	419.00	0.9210
11	(1 0 0 0 0 1 0 0 1 1)	415.10	0.7710
12	(0 2 0 0 1 0 0 1 0 1)	401.37	0.7758
13	(0 1 0 0 1 1 0 1 0 0)	50.00	0.0000
14	(1 2 0 1 0 2 0 1 0 1)	397.00	0.9944
15	(1 0 1 1 1 0 0 1 0 1)	397.00	0.9729
16	(0 0 1 1 1 0 1 1 0 1)	396.00	0.9633
17	(0 0 2 0 1 0 0 1 1 0)	377.33	0.6903
18	(0 2 0 0 0 1 2 0 1 2)	403.00	0.8898
19	(0 0 1 0 0 0 1 1 1 1)	389.05	0.7438
20	(0 1 0 2 1 1 0 1 0 1)	400.00	0.9903
21	(0 1 1 1 0 1 0 0 0 1)	422.00	0.9064

The best chromosome in 20 generations is $\mathbf{x}_1 = (0\,1\,0\,0\,0\,1\,0\,1\,1\,0)$, Table 7.12. The design corresponding to $\mathbf{x}_1$ shown in Figure 7.2 is taken as the optimal network design. The total cost and reliability of this design are 36.00 and 0.8224, respectively.

Table 7.12. Final population for all-terminal network problem

j	x_j	$eval(x_j)$	$R(x_j)$
1	(0 1 0 0 0 1 0 1 1 0)	444.00	0.8224
2	(0 1 0 0 0 2 0 1 1 0)	441.00	0.8396
3	(0 1 0 2 0 1 0 1 0 0)	440.00	0.8671
4	(0 1 0 1 0 2 0 1 0 0)	439.00	0.8487
5	(0 2 0 1 0 1 0 1 0 0)	437.00	0.8400
6	(0 0 2 1 0 2 0 1 0 0)	437.00	0.8045
7	(0 1 0 2 0 2 0 1 0 0)	437.00	0.8852
8	(0 2 0 1 0 2 0 1 0 0)	434.00	0.8575
9	(0 1 0 1 0 1 0 0 0 1)	430.00	0.8314
10	(0 0 2 0 0 1 0 1 0 1)	428.00	0.8220

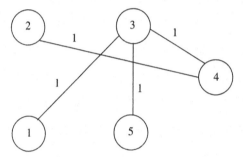

Figure 7.2. Design corresponding to chromosome **x**

7.3   The simulated annealing method

A simulated annealing algorithm is a general probabilistic method for solving combinatorial optimization problems. It involves random transitions among the solutions of the problem. Unlike iterative improvement algorithms, which improve the objective value continuously, the simulated annealing method may encounter some adverse changes in objective value in the course of its progress. Such changes are intended to lead to a global optimal solution instead of a local one.

Annealing is a physical process in which a solid is heated up to a high temperature and then allowed to cool gradually. In this process, all of the particles arrange themselves gradually into a low energy ground state level. The ultimate energy level depends on the level of the high temperature and the rate of cooling. The annealing process can be described as a stochastic model, such that at each temperature T, the solid undergoes a large number of random transitions among states of different energy levels until it attains a thermal equilibrium in which the probability of the solid appearing in a state with energy level E is given by

$$\Pr(X = E) = \frac{1}{Z(T)} \exp\left(-\frac{E}{K_B T}\right),$$

where X denotes the random energy level of the solid, $Z(T)$ is a normalization factor, and K_B is the Boltzmann constant.

The above probability distribution is called the *Boltzmann distribution*. As the temperature T decreases, equilibrium probabilities associated with higher energy level states decrease. When the temperature approaches zero, only the states with the lowest energy levels will have nonzero probability. If the cooling is not sufficiently slow, thermal equilibrium will not be attained at any temperature and consequently the solid will finally have a metastable condition. Details of the process are given in Arts and Korst [14].

To simulate the random transitions among the states and the attainment of thermal equilibrium at a fixed temperature T, Metropolis et al. [216] developed a method in which a transition from one state to another occurs due to random perturbation in the state. If perturbation results in a reduction of the energy level, the transition to the new state is accepted. If the perturbation increases the energy level by ΔE (>0), then the transition to the new state is accepted with probability $\exp[-\Delta E/(K_B T)]$. This method is called the *Metropolis algorithm*. The criterion for the acceptance of the transition is called the *Metropolis criterion*.

Kirkpatrick [157], Kirkpatrick et al. [158], [159] developed a simulated annealing algorithm for solving combinatorial optimization problems, and this can be described as follows.

Problem 7.4

Minimize $z = C(\mathbf{x})$, subject to $\mathbf{x} \in S$, where S is a finite or countably infinite set of feasible solutions of the problem, and $C(\mathbf{x})$ is the objective value for solution $\mathbf{x}$. A maximization problem can also be formulated in the above form by changing the sign of the objective function.

In a simulated annealing for Problem 7.4, the solutions in S are analogous to the states of the solid, and the objective value $C(\mathbf{x})$ is analogous to the energy level of a state. A control parameter T plays the role of temperature. Simulated annealing applies the Metropolis algorithm for a decreasing sequence of T values. The application of the Metropolis algorithm for a fixed value of T causes a sequence of transitions among the solutions, which constitute a Markov chain and the sequence is long enough to ensure equilibrium (a stationary probability distribution of states). By using the Metropolis criterion, we move from solution $\mathbf{x}$ to another solution $\mathbf{y}$ with probability

$$\Pr(\mathbf{x}, \mathbf{y}) = \begin{cases} 1, & \text{if } \Delta C < 0, \\ \exp(-\Delta C/T), & \text{otherwise,} \end{cases}$$

where $\Delta C = C(\mathbf{y}) - C(\mathbf{x})$. The simulated annealing is stopped by a criterion which usually involves the T control value.

Cooling schedules

The design of the simulated annealing requires determination of

1. an initial value of control parameter T,

2. a final value of T (or a stopping criterion),

3. the length of the Markov chain for each value of T, and

4. a rule for decreasing T.

The determination of these parameters and criteria provides a *cooling schedule*. Several approaches are available in the literature for determining a cooling schedule. For a good description of such approaches, refer to van Laarhoven and Arts [315]. Some simple cooling schedules are discussed here.

The initial value of T

The initial value T_0 of the control parameter T is determined such that the probability of accepting a transition, $\exp(-\Delta C / T_0)$, is close to 1 for any transition in which the change ΔC in the objective value is positive. For such T_0, any adverse transition is almost surely accepted. For determining T_0, Kirkpatrick et al. [158] suggest the following simple procedure. Select a large value for T_0 and perform several transitions. If the fraction of accepted transitions is less than a fixed value p, multiply T_0 by two. Repeat this procedure until the fraction exceeds p.

The final value of T

The value of T is gradually decreased as the algorithm progresses. Several researchers have used a lower limit T_f on T in the criterion for stopping the algorithm. When the control parameter gradually decreases to, or is below, T_f, the algorithm is stopped. There are several other criteria for stopping simulated annealing algorithms. The algorithm may be terminated if the final solutions obtained in a specified number of successive Markov chains coincide. In some cooling schedules, the algorithm is stopped when the fraction of accepted transitions falls below a specified value.

Length of the Markov chain

Ideally, the length L_k of the Markov chain at $T = T_k$ must be large enough to ensure an equilibrium condition at that parameter value. However, such an approach leads to enormous lengths for small values of T. For this reason, in some cooling schedules L_k is not allowed to exceed a specified number, which is usually a polynomial function of the problem size. Some researchers have adopted a fixed length for all T based on the number of variables. In some schedules, the length at any value of T is limited by the number of rejections at that T. However, this leads to smaller lengths as the temperature decreases.

A rule for decreasing T

A simple rule for decreasing the control parameter is

$$T_{k+1} = \alpha T_k, \qquad \text{for } k = 0, 1, 2, \ldots,$$

where α is a constant between 0 and 1. This rule is proposed by Kirkpatrick et al. [158], with $\alpha = 0.95$. Any two successive values of T must be close enough to avoid re-establishment of equilibrium through a large number of transitions.

The simulated annealing algorithm

There are several variations of simulated annealing, which arise due to different cooling schedules and stopping criteria. The following is a general description of simulated annealing.

- Step 0: Select T_0 (and select T_f if the stopping criterion involves T_f). Randomly obtain a solution $\mathbf{x}_0$ from S, and evaluate $f_0 = C(\mathbf{x}_0)$. Let $k = 0$.

- Step 1: Repeat the following procedure a sufficient number of times to ensure a near-equilibrium condition.

  Randomly select a transition $\mathbf{x}_k \to \mathbf{y}$ ($\in S$) and compute $\Delta C = C(\mathbf{y}) - C(\mathbf{x}_k)$. If $\Delta C \leq 0$, accept the transition. If $\Delta C > 0$, accept the transition with probability $\text{Pr}_k(\Delta C) = \exp(-\Delta C / T_k)$, and reject it with probability $1 - \text{Pr}_k(\Delta C)$. If the transition is accepted, update $\mathbf{x}_k = \mathbf{y}$ and $f_k = C(\mathbf{y})$. (To accept or reject the transition with $\Delta C > 0$, first generate a random number p from $(0, 1)$. If $p \leq \text{Pr}_k(\Delta C)$, accept the transition; otherwise, reject it.)

- Step 2: Let $k = k+1$. Find T_k from T_{k-1}, based on the rule for decreasing the control parameter T. If the stopping criterion is satisfied, stop. Otherwise, let $\mathbf{x}_k = \mathbf{x}_{k-1}$, and $f_k = f_{k-1}$, and go to step 1.

  A transition $\mathbf{x}_k \to \mathbf{y}$ is usually selected in such a way that $\mathbf{y}$ is in the neighborhood of $\mathbf{x}_k$.

7.3.1 Reliability optimization by simulated annealing

Two examples are provided here to illustrate simulated annealing. The first example deals with assignment in a ten-component complex system, whereas the second example is the redundancy allocation problem described in Example 3-1.

Example 7-3

Consider a ten-component assignment problem in a complex system as shown in Figure 7.3. Let $r_1, \ldots, r_{10}$ be the reliabilities of components $1, \ldots, 10$, respectively. As mentioned in Chapter 9, the system reliability depends on the assignment of components to positions. Consider a component assignment $\pi = (\pi_1, \ldots, \pi_{10})$ in

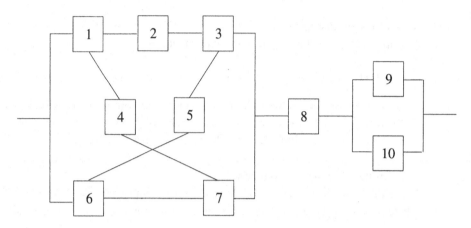

Figure 7.3. A ten-component complex structure

which component π_j is assigned to position j. Let $a_j = r_{\pi_j}$ for $j = 1, \ldots, 10$, that is, a_j is the reliability of the component assigned to position j and π can be written as

$$R_s(\pi) = [(a_1a_2a_3 + a_1a_7 + a_3a_6 + a_6a_7 - a_1a_2a_3a_7 - a_1a_2a_3a_6$$

$$- a_1a_6a_7 - a_3a_6a_7 + a_1a_2a_3a_6a_7)a_4a_5$$

$$+ (a_1a_2a_3 + a_1a_7 + a_6a_7 - a_1a_2a_3a_7 - a_1a_6a_7)a_4(1 - a_5)$$

$$+ (a_1a_2a_3 + a_3a_6 + a_6a_7 - a_1a_2a_3a_6 - a_3a_6a_7)(1 - a_4)a_5$$

$$+ (a_1a_2a_3 + a_6a_7 - a_1a_2a_3a_6a_7)(1 - a_4)(1 - a_5)]a_8[1 - (1 - a_9)(1 - a_{10})].$$

The problem is to maximize $R_s(\pi)$ over all assignments. Transform this problem into the combinatorial optimization Problem 7.4 by defining $C(\pi) = -R_s(\pi)$ and denoting the set of all ten assignments by S. Following is a simulated annealing application for this problem.

Suppose the reliabilities of components $1, \ldots, 10$ are

$$\mathbf{r} = (0.98, 0.95, 0.94, 0.93, 0.88, 0.80, 0.77, 0.69, 0.66, 0.64).$$

The design parameters are set as

$$T_0 = 0.002,$$

$$T_f = 0.000\,05,$$

$$L_k = 2n, \qquad (n = 5),$$

$$T_{k+1} = (0.9)T_k, \qquad \text{for } k = 0, 1, 2, \ldots.$$

The sequence $(1, \ldots, 10)$ is taken as the initial solution. The initial solution can be any random sequence of $1, 2, \ldots, 10$. We consider only the transitions which are caused by an interchange of two components. In the initial iteration (with $T = T_0$), any transition that increases the objective value $C(\pi)$ by $0.000\,102\,6$ is accepted by the

Table 7.13. Results of Example 7-3 by the simulated annealing method

k	Parameter T_k	Final solution in the iteration	Reliability
1	0.002 000	(4, 7, 2, 6, 9, 8, 5, 1, 10, 3)	0.904 805 97
2	0.001 800	(3, 10, 5, 8, 9, 6, 2, 1, 7, 4)	0.916 147 97
3	0.001 620	(6, 9, 5, 8, 7, 3, 2, 1, 10, 4)	0.927 541 02
4	0.001 458	(6, 7, 9, 10, 8, 3, 2, 1, 5, 4)	0.931 997 10
5	0.001 312	(6, 9, 7, 8, 10, 2, 3, 1, 5, 4)	0.935 345 62
6	0.001 181	(5, 9, 7, 8, 10, 2, 3, 1, 6, 4)	0.934 066 60
7	0.001 063	(5, 9, 6, 8, 10, 2, 3, 1, 7, 4)	0.933 772 39
8	0.000 957	(5, 8, 6, 9, 10, 3, 2, 1, 7, 4)	0.934 581 78
9	0.000 861	(7, 10, 6, 8, 9, 3, 2, 1, 5, 4)	0.935 374 31
10	0.000 775	(7, 8, 6, 10, 9, 2, 3, 1, 5, 4)	0.935 775 34
11	0.000 697	(6, 10, 7, 8, 9, 2, 3, 1, 5, 4)	0.935 305 87
12	0.000 628	(6, 8, 7, 10, 9, 2, 3, 1, 5, 4)	0.935 555 52
13	0.000 565	(7, 8, 6, 10, 9, 2, 3, 1, 5, 4)	0.935 775 34
14	0.000 508	(7, 8, 6, 9, 10, 2, 3, 1, 5, 4)	0.935 675 06
15	0.000 458	(6, 8, 7, 10, 9, 2, 3, 1, 5, 4)	0.935 555 52
16	0.000 412	(6, 9, 7, 10, 8, 2, 3, 1, 5, 4)	0.935 453 70
17	0.000 371	(6, 9, 7, 8, 10, 3, 2, 1, 5, 4)	0.935 723 37
18	0.000 334	(6, 10, 7, 8, 9, 3, 2, 1, 5, 4)	0.935 616 57
19	0.000 300	(6, 9, 7, 8, 10, 3, 2, 1, 5, 4)	0.935 723 37
20	0.000 270	(7, 8, 6, 10, 9, 3, 2, 1, 5, 4)	0.935 515 88
21	0.000 243	(7, 8, 6, 9, 10, 3, 2, 1, 5, 4)	0.935 555 52
22	0.000 219	(7, 8, 6, 9, 10, 3, 2, 1, 5, 4)	0.935 555 52
23	0.000 197	(7, 8, 6, 9, 10, 3, 2, 1, 5, 4)	0.935 555 52
24	0.000 177	(6, 8, 7, 9, 10, 3, 2, 1, 5, 4)	0.935 775 34
25	0.000 160	(6, 8, 7, 10, 9, 3, 2, 1, 5, 4)	0.935 675 06
26	0.000 144	(6, 8, 7, 9, 10, 3, 2, 1, 5, 4)	0.935 775 34
27	0.000 129	(6, 9, 7, 8, 10, 3, 2, 1, 5, 4)	0.935 723 37
28	0.000 116	(6, 10, 7, 9, 8, 3, 2, 1, 5, 4)	0.935 457 18
29	0.000 105	(6, 8, 7, 10, 9, 3, 2, 1, 5, 4)	0.935 675 06
30	0.000 094	(6, 8, 7, 10, 9, 3, 2, 1, 5, 4)	0.935 675 06
31	0.000 085	(6, 8, 7, 9, 10, 3, 2, 1, 5, 4)	0.935 775 34
32	0.000 076	(6, 8, 7, 10, 9, 3, 2, 1, 5, 4)	0.935 675 06
33	0.000 069	(6, 9, 7, 8, 10, 3, 2, 1, 5, 4)	0.935 723 37
34	0.000 062	(6, 8, 7, 9, 10, 3, 2, 1, 5, 4)	0.935 775 34
35	0.000 056	(6, 8, 7, 9, 10, 3, 2, 1, 5, 4)	0.935 775 34
36	0.000 050	(6, 8, 7, 9, 10, 3, 2, 1, 5, 4)	0.935 775 34

Metropolis criterion with a probability of 0.95. Such transitions are accepted with a probability of only 0.128 in the final iteration (with $T \approx T_f$). The total number of iterations carried out is 36, with each iteration involving transitions at a fixed value of

T. The final solutions of these iterations are given in Table 7.13. The best solution is $\pi^* = (6, 8, 7, 9, 10, 3, 2, 1, 5, 4)$, with a corresponding system reliability $R_s(\pi^*) = 0.935\,775\,34$. This solution is taken as the global optimal solution. It is found from complete enumeration of all ten solutions that π^* is an optimal solution. Since simulated annealing is a probabilistic algorithm, it is applied 20 times to see the consistency of its performance, and yield π^* every time. However, of the 20 applications π^* was not obtained twice for Markov chain length $L_k = n$. The performance of simulated annealing for $L_k = n$ improves when α is increased.

Example 7-4
Consider Example 3-1. Suppose the upper limits on the variables $x_1, \ldots, x_5$ are

$(u_1, u_2, u_3, u_4, u_5) = (5, 5, 4, 4, 3)$.

Without loss of generality, we can take 1 as the lower limit on each variable. Now, the problem is to maximize the reliability of a five-component series system subject to three separable constraints, and lower and upper limits.

The problem can be viewed as the minimization of $C(\mathbf{x}) = -R_s(\mathbf{x})$, subject to the three constraints and limits on the variables. Here, the simulated annealing design parameters are taken as

$T_0 = 0.02,$

$T_f = 0.008,$

$L_k = 2n, \qquad (n = 5),$

$T_{k+1} = (0.9)T_k, \qquad \text{for } k = 0, 1, 2, \ldots .$

The initial solution is the randomly generated feasible solution $\mathbf{x} = (4, 1, 1, 1, 3)$, with system reliability $R_s(\mathbf{x}) = 0.488\,697\,30$. In this implementation, the transitions occur due to:

1. a unit increase in one of the variables, and

2. a simultaneous unit increase in one variable and a unit decrease in another variable.

Either option is equally likely to be selected. If option 1 is chosen, one of the variables is randomly selected for an increment with equal probability. When option 2 is chosen, one of the 20 ordered pairs of variables is randomly selected. If pair (i, j) is selected, x_i is to be reduced by 1, and x_j is to be increased by the same value.

The total number of iterations performed is nine. The final solutions of these iterations are given in Table 7.14. The best solution is $\mathbf{x}^* = (3, 2, 2, 3, 3)$, with a corresponding system reliability $R_s(\mathbf{x}^*) = 0.904\,467\,30$. This solution is taken as the global optimal solution. It is known that $\mathbf{x}^*$ is an optimal solution.

Table 7.14. Results of Example 7-4 by the simulated annealing method

k	Parameter T_k	Final solution in the iteration	Reliability
1	0.020 00	(3, 2, 2, 3, 3)	0.904 467 30
2	0.018 00	(3, 2, 3, 3, 2)	0.869 228 31
3	0.016 20	(2, 2, 2, 4, 3)	0.900 776 91
4	0.014 58	(3, 2, 2, 4, 2)	0.886 478 86
5	0.013 12	(2, 2, 2, 4, 3)	0.900 776 91
6	0.011 81	(3, 2, 2, 3, 3)	0.904 467 30
7	0.010 63	(2, 3, 2, 3, 3)	0.892 416 19
8	0.009 57	(3, 2, 2, 3, 3)	0.904 467 30
9	0.008 61	(3, 2, 2, 3, 3)	0.904 467 30

7.3.2 The nonequilibrium simulated annealing algorithm

Although simulated annealing gives satisfactory solutions for combinatorial optimization problems, its major disadvantage is the amount of computational effort involved. In order to improve the rate of convergence and reduce the computational time, Cardoso et al. [44] introduced a nonequilibrium simulated annealing algorithm (NESA) by modifying the algorithms of Metropolis et al. [216] and Glauber [109]. In NESA, there is no need to reach an equilibrium condition through a large number of transitions at any fixed temperature. The temperature is reduced as soon as an improved solution is obtained. Ravi et al. [275] have recently improved NESA by incorporating a simplex-like heuristic in the method. They have applied this variant of NESA, denoted as I-NESA, to reliability optimization problems. It consists of two phases: phase I uses a NESA and collects solutions obtained at regular NESA progress intervals, and phase II starts with the set of solutions obtained in phase I and uses a heuristic procedure to improve the best solution further. Both phases are described below.

Phase I
- Step 0: Let $k = 0$ and $k_{max} = 1000$ (k_{max} must be selected such that $n \ll k_{max}/10$). Choose an initial value T_0 for the control parameter T. Let $\beta = 0.06$.

- Step 1: Randomly generate a feasible solution $\mathbf{x}_k$, and evaluate $f_k = C(\mathbf{x}_k)$. Let $l = 0$ and $l_{max} = 100n$.

- Step 2: Generate a new solution $\mathbf{y}$ in the neighborhood of $\mathbf{x}_k$ as

$$y_j = x_{kj} + (u_j - \ell_j)(2U - 1)^3, \qquad \text{for } j = 1, \ldots, n,$$

  where x_{kj} is the value of the jth variable in $\mathbf{x}_k$ and U is a random number from (0, 1).

- Step 3: If $\mathbf{y}$ is infeasible, go to step 2. Otherwise, evaluate $h = C(\mathbf{y})$ and $\Delta = h - f_k$.

- Step 4: If $\Delta < 0$, let $\mathbf{x}_k = \mathbf{y}$ and $f_k = h$, and go to step 7. Otherwise (when $\mathbf{y}$ is feasible but not better than $\mathbf{x}_k$), compute

$$Pr = \frac{1}{1 + \exp(\Delta / T_k)}.$$

- Step 5: Take a random number U from $(0, 1)$. If $U < Pr$, update $\mathbf{x}_k = \mathbf{y}$, $f_k = h$, and $l = l + 1$.

- Step 6: If $l \le l_{max}$, go to step 2.

- Step 7: Store $\mathbf{x}_k$ and f_k. Compute $T_{k+1} = (1 - \beta)T_k$ and update $k = k + 1$. If $k \le k_{max}$ or $(T_k > 10^{-4})$, go to step 1.

- Step 8: Select $\mathbf{x}_k$ (along with the corresponding f_k) for every k that is a multiple of ten, from k_{max} solutions. Denote the number of selected solutions by q and go to phase II.

Phase II

Phase II uses the following terms for counting and termination purposes:

ITR	iteration number
MAXITR	maximum number of iterations allowed
IFAIL	number of infeasible solutions obtained successively in a single iteration
MAXFAIL	upper limit on *IFAIL*

- Step 0: Let $ITR = 0$ and $IFAIL = 0$. Choose positive integer values for *MAXFAIL* and *MAXITR*.

- Step 1: Sort the set of q solutions obtained from phase I in ascending order of the objective values. Denote the first solution $\mathbf{x}_{[1]}$ in the order as $\mathbf{z}_1$.

- Step 2: Randomly choose n distinct points from $\{\mathbf{x}_{[2]}, \mathbf{x}_{[3]}, \ldots, \mathbf{x}_{[q]}\}$ and denote them by $\mathbf{z}_2, \mathbf{z}_3, \ldots, \mathbf{z}_{n+1}$.

- Step 3: Obtain

$$\mathbf{z}' = \frac{1}{n-1} \sum_{i=1}^{n} \mathbf{z}_i \text{ and } \hat{\mathbf{z}} = 2\mathbf{z}' - \mathbf{z}_{n+1}.$$

- Step 4: If $\hat{\mathbf{z}}$ is infeasible, let $IFAIL = IFAIL + 1$ and go to step 5. If $\hat{\mathbf{z}}$ is feasible, compute $C(\hat{\mathbf{z}})$ and go to step 6.

- Step 5: If $IFAIL \le MAXFAIL$, go to step 2. Otherwise, go to step 8.

- Step 6: If $C(\hat{\mathbf{z}}) > C(\mathbf{x}_q)$, go to step 2. Otherwise, replace $\mathbf{x}_q$ with $\hat{\mathbf{z}}$ and $C(\mathbf{x}_q)$ with $C(\hat{\mathbf{z}})$.

- Step 7: Let $ITR = ITR + 1$ and $IFAIL = 0$. If $ITR \le MAXITR$, go to step 1.

- Step 8: Take z_1 as the required solution and stop.

Ravi et al. [275] have demonstrated this method for three different reliability optimization problems.

7.4   The tabu search method

Tabu search is a metaheuristic that guides a heuristic method to expand its search beyond the local optimality. It is an artificial intelligence technique which utilizes memory at every stage (information about the solutions visited up to that stage) in order to provide an efficient search for optimality. It is based on ideas proposed by Fred Glover. An excellent description of tabu search methodology can be found in Glover and Laguna [110].

A tabu search for any complex optimization problem combines the merits of artificial intelligence with those of optimization procedures. Tabu search allows the heuristic to cross boundaries of feasibility or local optimality, which are major impediments in any local search procedure. The most prominent feature of tabu search is the design and use of memory-based strategies at every stage, for exploration in the neighborhood of a solution. Tabu search is very useful for solving large complex optimization problems that are very difficult to solve by exact methods.

Memory
Memory is a vital component of the tabu search procedure. The memory used in tabu search can be divided into two types: (1) short-term memory, and (2) longer-term memory. It can be used to compare the merits of solutions visited during the search and to measure the effect of choices made. The two types of memory-based strategies adopted in tabu search are very different, and appear in two forms: *explicit* and *attributive*. The explicit memory contains complete solutions, particularly the elite solutions, visited during the search. The attributive memory records information about attributes associated with moves in the search. For example, if the value of a variable x_j changes, the attribute to be recorded may be index j. It may also be $j+$ ($j-$) if x_j is increased (decreased) by the move.

Short-term memory
The most widely used short-term memory is of the attributive type. It records attributes associated with recent moves in the search. A tabu search that involves strategies associated only with short-term memory is a simple search procedure which proceeds like a local search method.

In any iteration of the local search procedure, a neighborhood $N(\mathbf{x})$ of a solution $\mathbf{x}$ is explored, and the move that yields the best point in $N(\mathbf{x})$ is selected. It stops when no point in $N(\mathbf{x})$ is better than the current best solution. In contrast, in every iteration of a simple tabu search procedure, a truncated neighborhood $N'(\mathbf{x})$ of $\mathbf{x}$ is explored, and the move that gives the best point in $N'(\mathbf{x})$ is selected. At any stage, the short-term memory records some attributes as *tabu active*. An attribute retains its acquired tabu-active status for a certain number of iterations, which is referred to as the *tabu tenure*. As long as an attribute is tabu active, it forbids consideration of moves that contain it. Thus, the recency-based memory (short-term memory) forbids exploration of some parts of neighborhood $N(\mathbf{x})$, that is, the memory restricts exploration in the neighborhood of the current solution. A *tabu list* refers to the tabu status of all attributes. This list changes from iteration to iteration. In other words, the tabu status of any attribute may oscillate between tabu-active and tabu-inactive alternatives during the progress of the search.

Tabu methods may sometimes be too restrictive to provide effective guidance in the search. In such cases, it is necessary to find a way to override the tabu restrictions. For this purpose, *aspiration criteria* are introduced in tabu search: these refer to aspiration by attribute or aspiration by move. A simple and widely used aspiration criterion is to override the tabu-active status of an attribute (or some attributes) if such an operation yields a solution that is better than the best solution obtained so far. Another criterion is the *default aspiration criterion*, which allows selection of the move with the least tabu restrictions when all current moves have tabu-active attributes.

Longer-term memory

Advanced versions of tabu search include longer-term memory, which is complementary to short-term memory. Longer-term memory enhances the strength of the search by guiding and expanding the search. It collects useful information from the solution trajectory. *Frequency-based* memory is a longer-term memory that consists of information about how often attributes change and how often any specific attribute is a member of the solutions visited. This information is used to increase or decrease the scope of inclusion of any attribute in a move through penalties and incentives. For example, if an attribute has a high frequency of occurrence in high quality solutions, then the strategy must increase the scope of inclusion of the attribute in a move. Most applications of tabu search that involve longer-term memory use a simple linear function of the frequency measure to define penalties and incentives. The explicit memory that records elite solutions and their unexplored neighborhoods is also a part of longer-term memory. In a tabu search involving longer-term memory, the modified neighborhood (to be explored) of the current solution consists of some original neighborhood points and the elite solutions recorded during the progress of the search. Sometimes, components of elite solutions are integrated to direct the search more effectively.

Two important components of rigorous tabu search are intensification and diversification strategies. The intensification strategies involve modification of choice rules to encourage solution features that are previously found good. In attractive regions, they also initiate a rigorous search using strategies based on short-term memory. The diversification strategies are useful to extend the search into new regions. Although short-term memory provides some diversification in tabu search, certain strategies associated with longer-term memory increase the diversity. Some strategies use combinations of attributes to generate more diverse solutions. The diversification strategies are very useful when the solution space consists of several ridges and humps (with respect to the objective function value).

Guidelines

There is no fixed and specific sequence of operations to be performed for all tabu search procedures. The memory structures and the associated search strategies in tabu search vary with problem type. The implementation of a tabu search requires ingenuity, good knowledge of the nature of the problem, and numerical experimentation. Such an exercise is quite difficult in general, but highly rewarding in terms of solution quality for complex problems. The following are a few general guidelines in tabu search implementation:

1. For tabu restrictions based on attributive memory, select attributes whose tabu status does not impose too much restriction on the choice of move.

2. Maintain separate tabu lists for different attribute types, with tabu tenure for each list depending on the nature of the corresponding attribute type.

3. Define an appropriate aspiration criterion.

4. Define intensification and diversification strategies.

5. Adopt appropriate penalties and incentives (as functions of frequency-based memory) to guide the search more effectively.

7.4.1   Reliability optimization by tabu search

A simple tabu search based on short-term memory is implemented for redundancy allocation and reliability–redundancy allocation problems in a ten-component complex reliability system. Even the simplest tabu search holds some promise for improvement in solution quality in large reliability optimization problems.

Example 7-5

Consider a redundancy allocation problem to maximize

$$R_s = f(x_1, \ldots, x_{10}),$$ (7.2)

Table 7.15. Constants of stage j for Example 7-5

	1	2	3	4	5	6	7	8	9	10
r_j	0.9	0.7	0.9	0.7	0.7	0.9	0.7	0.8	0.8	0.8
p_j	4.4	5.0	3.2	3.1	4.4	2.9	2.9	3.1	2.5	2.8
c_j	7.8	6.3	8.1	9.2	7.3	6.2	6.5	8.1	9.2	7.9
w_j	8.6	6.1	8.4	7.5	6.7	5.0	5.1	6.8	5.4	8.9

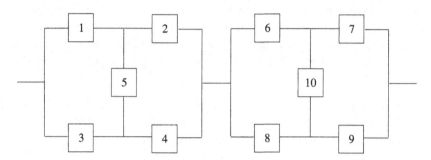

Figure 7.4. A system with two bridge structures

subject to

$$\sum_{j=1}^{10} p_j x_j^2 \le P,$$

$$\sum_{j=1}^{10} c_j \left[x_j + \exp(x_j/4) \right] \le C,$$

$$\sum_{j=1}^{10} w_j x_j \exp(x_j/4) \le W,$$

$$x_j = 1, 2, 3, \ldots, \qquad \text{for all } j,$$

where $f(x_1, \ldots, x_{10})$ is the reliability of a system which consists of two bridge structures connected in series as shown in Figure 7.4; and x_j is the number of redundant components at stage j, $j = 1, \ldots, 10$. The component reliabilities and the constants in the constraints are given in Table 7.15. Let $(P, C, W) = (300, 400, 400)$.

As explained in previous chapters, several methods are available to solve this nonlinear integer programming problem. Among them, the GAG1 method, explained in Chapter 3, is a simple heuristic algorithm that takes into account the nature of the objective function and the constraints. In every iteration, based on a criterion, the GAG1 algorithm selects a move in which redundancy level x_j, at stage j, is increased by 1. It continues until no increment is feasible at any stage.

For any solution $\mathbf{x}$, consider the neighborhood

$$N(\mathbf{x}) = \{(y_1, \ldots, y_{10}): y_k = x_k + 1, \text{ for some } k, \text{ and } y_j = x_j, \text{ for all } j \text{ other than } k\}.$$

In each iteration, the GAG1 algorithm makes a move from a feasible solution $\mathbf{x}$ to one of the feasible solutions in the neighborhood $N(\mathbf{x})$. To select the best move at $\mathbf{x}$ (with respect to a specific criterion), all feasible solutions must be explored. It involves a considerable amount of computational effort when the number of stages in the system is large.

Implementation of tabu search

To illustrate the implementation of tabu search, a simple version based on short-term memory is incorporated in the GAG1 algorithm. We can introduce tabu restrictions on the selection of moves to reduce the computational work and to guide the search for a better final solution. Consider the indices $1, 2, \ldots, 10$ of variables $x_1, x_2, \ldots, x_{10}$ as attributes. Whenever a variable x_j is increased, the corresponding attribute j becomes tabu active. The tabu tenure for each attribute is chosen as $t = 5$. When an attribute becomes tabu active in an iteration, it will have the same status for five successive iterations. To monitor the tabu status of each attribute, a ten-dimensional vector $\mathbf{tbs} = [tbs(1), tbs(2), \ldots, tbs(10)]$ is defined, such that, in any iteration $tbs(j)$ represents the number of following iterations in which attribute j will remain tabu active. Note that $0 \le tbs(j) \le t$ for all j. Suppose an attribute j becomes tabu active in some iteration, that is, x_j is increased. Then $tbs(j)$ is changed from 0 to t and the tabu status of each attribute $i \ne j$ is updated as

$$tbs(i) = \max\{0, tbs(i) - 1\}.$$

When the number of variables in the problem is large, the tabu status of attributes can also be stored in the form of a t-dimensional vector to save computer memory and execution time. Now, the tabu restriction is to forbid any move that increases the value of a variable associated with a tabu-active attribute. In any iteration (except the first t iterations), tabu restriction forbids exploration of t moves. An aspiration criterion is adopted here as follows. If none of the tabu-free moves gives a feasible solution, the tabu restriction is overridden to explore the forbidden moves. The forbidden move that gives the best feasible solution is selected. Here, one can also select a move overriding all attributes with the least tabu status if such a move gives a feasible solution.

The sequence of solutions visited by tabu search in this example is listed in Table 7.16. At the beginning of the ninth iteration, the current solution is

$$(2, 3, 1, 2, 1, 2, 3, 2, 2, 1);$$

the residual resources are 154.50, 133.51, and 188.25; and the tabu list is

$$\mathbf{tbs} = (2, 3, 0, 0, 0, 0, 4, 5, 1, 0).$$

Table 7.16. Sequence of solutions of Example 7-5 from tabu search

Iteration	Allocation $(x_1, \ldots, x_{10})$	System reliability
0	(1, 1, 1, 1, 1, 1, 1, 1, 1, 1)	0.809 784
1	(1, 2, 1, 1, 1, 1, 1, 1, 1, 1)	0.868 626
2	(1, 2, 1, 1, 1, 1, 2, 1, 1, 1)	0.913 030
3	(1, 2, 1, 1, 1, 2, 2, 1, 1, 1)	0.932 150
4	(1, 2, 1, 2, 1, 2, 2, 1, 1, 1)	0.954 971
5	(1, 2, 1, 2, 1, 2, 2, 1, 2, 1)	0.968 690
6	(2, 2, 1, 2, 1, 2, 2, 1, 2, 1)	0.979 289
7	(2, 3, 1, 2, 1, 2, 2, 1, 2, 1)	0.986 574
8	(2, 3, 1, 2, 1, 2, 3, 1, 2, 1)	0.991 458
9	(2, 3, 1, 2, 1, 2, 3, 2, 2, 1)	0.993 854
10	(2, 3, 2, 2, 1, 2, 3, 2, 2, 1)	0.995 383
11	(2, 3, 2, 3, 1, 2, 3, 2, 2, 1)	0.997 257
12	(2, 3, 2, 3, 1, 2, 3, 2, 3, 1)	0.998 173
13	(2, 3, 2, 3, 1, 3, 3, 2, 3, 1)	0.998 544
14	(2, 4, 2, 3, 1, 3, 3, 2, 3, 1)	0.999 107
15	(2, 4, 2, 3, 1, 3, 4, 2, 3, 1)	0.999 408
16	(3, 4, 2, 3, 1, 3, 4, 2, 3, 1)	0.999 569
17	(3, 4, 2, 4, 1, 3, 4, 2, 3, 1)	0.999 727

Here, the GAG1 method would have increased x_2 by 1. Since attribute 2 is still tabu active, this move is forbidden. In this iteration, the tabu search explores the moves that increase one of the variables x_3, x_4, x_5, x_6, and x_{10}, and selects the move in which x_3 is increased. Therefore, the resulting solution in the ninth iteration is

$$(2, 3, 2, 2, 1, 2, 3, 2, 2, 1).$$

The final solution obtained by the tabu search is

$$\mathbf{x}^* = (3, 4, 2, 4, 1, 3, 4, 2, 3, 1),$$

and the corresponding system reliability is $R_s(\mathbf{x}^*) = 0.9997$.

Example 7-6
Consider the reliability–redundancy allocation problem of maximizing

$$R_s = f(x_1, \ldots, x_{10}, r_1, \ldots, r_{10}), \tag{7.3}$$

Table 7.17. Constants of stage j for Example 7-6

	1	2	3	4	5	6	7	8	9	10
p_j	2.26	4.11	1.50	4.54	1.30	3.13	1.98	4.08	3.18	1.04
w_j	8.13	8.85	5.29	7.52	5.23	8.48	7.41	7.73	8.06	6.15
$\alpha_j \times 10^{-5}$	3.975	7.766	5.692	4.735	5.777	4.585	6.144	4.112	7.939	5.771

subject to

$$\sum_{j=1}^{10} p_j x_j^2 \le P,$$

$$\sum_{j=1}^{10} c_j(r_j)\left[x_j + \exp\left(\frac{x_j}{4}\right)\right] \le C,$$

$$\sum_{j=1}^{10} w_j x_j \exp\left(\frac{x_j}{4}\right) \le W,$$

$$0 \le r_j \le 1, \qquad \text{for } j = 1, \cdots, n,$$

x_j being a positive integer; where $f(x_1, \ldots, x_{10}, r_1, \ldots, r_{10})$ is the reliability of the system described in Example 7-5; and r_j and x_j are the reliability of a component and the number of redundant components at stage j, respectively. The cost function $c_j(r_j)$ is given by

$$c_j(r_j) = \alpha_j \left(\frac{-1000}{\ln r_j}\right)^{1.5},$$

and $(P, C, W) = (200, 400, 500)$. All other constants in the constraints are given in Table 7.17.

The THK method explained in Chapter 8 can be used to solve this problem. Considering maximum system reliability over all feasible choices of $\mathbf{x} = (x_1, \ldots, x_{10})$ for each selected vector $\mathbf{r} = (r_1, \ldots, r_{10})$, the method treats this problem as a nonlinear programming problem and uses the Hooke–Jeeves (H–J) direct search method to obtain an optimal $\mathbf{r}$.

In its exploratory move, the H–J method investigates the effect of increasing each variable sequentially. If the increase in r_1 is advantageous, the increment is retained. Otherwise, a decrease in r_1 is considered. Similarly, if the decrease in r_1 is advantageous, the decrement in r_1 is retained. Otherwise, no change is made in r_1. This procedure is repeated sequentially for variables $r_2, \ldots, r_{10}$.

A simple tabu search is implemented in the H–J method as follows. The indices $1, 2, \ldots, 10$ of variables $r_1, \ldots, r_{10}$ are taken as the attributes. Since the changes in two variables r_j are positive or negative when we move from one solution to another, two tabu lists, **tb1** and **tb2**, are used to guide the search. Initially, $tb1(j) = tb2(j) = 0$ for each attribute j. The tabu tenure for each attribute is chosen as $t = 1$. If a variable r_j gets an increment in an iteration, the associated attribute j becomes tabu active with $tb1(j) = t$. If a variable r_j gets a decrement in an iteration, the associated attribute j becomes tabu active with $tb2(j) = t$. The lists **tb1** and **tb2** are updated exactly as **tbs** is done in the previous example.

Tabu restrictions are defined on the basis of the tabu status of attributes as follows. If an attribute j is tabu active with $tb1(j) > 0$, then a decrease is not considered in r_j. Similarly, if an attribute k is tabu active with $tb2(k) > 0$, then an increment is not considered in r_k. If the exploratory move does not provide a better solution under the tabu restrictions, the tabu restrictions are overridden to explore other forbidden solutions. Suppose $tb1(1) = 1$, $tb1(3) = 1$, and $tb2(4) = 1$, and all other attributes are tabu inactive in an exploratory move subject to tabu restrictions. Suppose the tabu restrictions have not provided any better solution. Then explore the solutions which can be obtained by sequentially making forbidden changes in the variables r_1, r_3, and r_4. First, decrease the value of r_1. If it gives an improvement (without violating the constraints) retain the increase in r_1. Next, repeat the same for r_3. Determine whether an increase in r_4 yields an improvement without violating the constraints. If so, retain the increment in r_4. If the resulting solution is better than the current best solution, it is taken as the next solution. The method stops when the increment h (in r_j) reaches its minimum value and the exploratory move does not give any improvement in the objective function even with the aspiration criterion.

The solution yielded by tabu search in the THK method is

$$\mathbf{r}^* = (0.7984, 0.7031, 0.6625, 0.7, 0.6875, 0.7125, 0.7, 0.7, 0.7, 0.7),$$

$$\mathbf{x}^* = (2, 2, 3, 3, 3, 3, 3, 3, 2, 3),$$

with a corresponding system reliability of 0.992 807. This is better than the solution

$$\mathbf{r}^* = (0.7516, 0.7266, 0.6875, 0.75, 0.65, 0.7, 0.7, 0.7, 0.7, 0.7),$$

$$\mathbf{x}^* = (3, 2, 3, 2, 3, 3, 3, 3, 2, 3),$$

with system reliability 0.991 514, given by the THK method without any tabu restrictions. The gain may increase if the tabu search is also implemented using the heuristic method of Aggarwal [5] for obtaining optimal redundancy allocation for each choice of reliability vector.

7.5   Discussion

Genetic algorithms are heuristic methods which imitate the natural evolution process. They are very useful for solving complex discrete optimization problems and do not require sophisticated mathematical treatment. They can be easily designed and implemented on a computer for a wide spectrum of discrete problems.

Genetic algorithms have been designed for solving redundancy allocation problems in reliability systems. As demonstrated through the examples, the chromosome definition and the selection of parameters provide a lot of flexibility in adopting a genetic algorithm for any particular type of problem. However, there is some difficulty in determining appropriate values for the parameters and penalties for infeasibility. If these values are not selected properly, a genetic algorithm may rapidly converge to a local optimum or slowly converge to the global optimum. Large population size and more generations not only enhance the solution quality but also increase the computational effort. Experiments are usually recommended to obtain appropriate genetic algorithm parameters for solving a specific type of problem. The definition of the chromosome is a critical element of genetic algorithm design and it usually requires some ingenuity.

An important advantage of a genetic algorithm is its presentation of several good solutions (mostly optimal or near optimal). The multiple solutions yielded by a genetic algorithm provide a lot of flexibility for decision making.

Simulated annealing is a global optimization technique that can be used for solving large-scale combinatorial optimization problems. It may be noted that, unlike many discrete optimization methods, simulated annealing does not exploit any special structure that exists in the objective function or in the constraints. However, simulated annealing is relatively more effective when the problem is highly complex without having any special structure. Redundancy allocation problems in reliability systems are nonlinear integer programming problems of this type. Thus, simulated annealing can be quite useful for solving complex reliability optimization problems, including network optimization, as has been demonstrated by Atiqullah and Rao [17]. Simulated annealing can also be used to solve problems involving continuous variables.

Although several approaches are available in the literature for designing a simulated annealing method, the design still requires ingenuity and sometimes much experimentation. A major disadvantage of simulated annealing is that it requires a large amount of computational effort (with a large number of function evaluations and tests for solution feasibility). However, it has great potential for providing an optimal or near-optimal solution. The numerical examples in Chapter 7 encourage the application of simulated annealing to optimization in reliability systems.

Tabu search is very useful for solving large-scale complex optimization problems. The salient feature of this method is the utilization of memory (information about previous solutions) to guide the search beyond local optimality. There is no fixed sequence of operations in tabu search and its implementation is problem-specific. Thus, tabu search can be described as a metaheuristic rather than a method. A simple tabu search method that uses only short-term memory is quite easy to implement. Most often such methods yield good solutions when attributes, tabu tenure, and aspiration criteria are appropriately defined. In this chapter, a simple tabu search is implemented (mainly for demonstration purposes) for solving redundancy and reliability–redundancy allocation problems, and has given encouraging results. Numerical experimentation has revealed that even a simple tabu search can be quite effective for large complex problems. However, a simple tabu search can be ineffective for small and large series systems. The advanced version of tabu search is likely to give much better results in reliability optimization problems. One major disadvantage of tabu search is the difficulty involved in defining effective memory structures and memory-based strategies which are problem-dependent. This task really requires good knowledge of the nature of the problem, ingenuity, and some numerical experimentation. A well-designed tabu search can offer excellent solutions in large-scale system-reliability optimization.

EXERCISES

7.1   Most classical optimization methods are uni-directional search methods. How about genetic algorithms? What's the difference between genetic algorithms and most of the classical optimization methods?

7.2   A genetic algorithm is called a probabilistic method. What are probabilistic methods? Discuss other probabilistic methods.

7.3   When a genetic algorithm is applied, chromosome representation is a difficult part of some problems. Let one example of chromosome representation in a one-machine–six-jobs scheduling problem be (1, 2, 3, 4, 5, 6). This chromosome representation means that job 1 precedes job 2, job 2 precedes job 3, etc. Is the above chromosome representation correct? If it is incorrect, explain the reason and discuss the correct representation. (Hint: perform crossover and a mutation operation.)

7.4   The usual way to select a new population following the creation of offspring is by fitness-based selection. However, these approaches have one critical disadvantage. Find that disadvantage, and explain how to overcome it.

7.5   Explain advantages and disadvantages of the genetic algorithm. Discuss the areas of application of the genetic algorithm.

7.6   In tabu search and simulated annealing, "move" can transit a solution to another (neighborhood). Find and explain the operations of the genetic algorithm which are similar to "move".

7.7   Develop a general genetic algorithm and draw the corresponding flowchart.

7.8   In Problem 7.2, let $(m_1, m_2, m_3) = (6, 7, 6)$, $(l_1, l_2, l_3) = (1, 1, 1)$, $(u_1, u_2, u_3) = (5, 6, 5)$, and the vector $\mathbf{x} = (x_1, x_2, x_3)$, where $x_1 = (1\,0\,0\,0\,0\,3)$, $x_2 = (0\,2\,0\,0\,1\,1\,2)$, and $x_3 = (0\,1\,1\,1\,0\,0)$. Find vector $\mathbf{y}$.

7.9   In Problem 7.3, let $x_1 = (1\,0\,0\,1\,1\,0\,0\,1\,0\,1)$ and $x_2 = (1\,0\,0\,1\,0\,1\,0\,0\,1\,1)$. If the crossover operation with a single cut is selected in position 5, the mutation operation is performed in position 6 of x_2, and the value of the gene changes from 1 to 0, find the offspring and the fitness value.

7.10 Heuristic methods usually converge quickly to a local optimal solution, which may be far from the global optimal solution. In order to overcome this drawback, the simulated annealing methods allow an uphill move. Describe how simulated annealing techniques make it possible to allow an uphill move.

7.11 Briefly explain the cooling schedules and stopping criteria used in simulated annealing methods. Why are cooling schedules necessary for simulated annealing methods? In a rule for decreasing T ($T_{k+1} = \alpha T_k$), what happens if $\alpha = 1$?

7.12 In Example 7-1, find each acceptance probability when $T_0 = 0.002$ and $T_f = 0.000\,05$ ($\Delta C = 0.000\,1026$). Which has a higher probability? What does this result mean?

7.13 In Example 7-2, suppose that position 3, for unit increase, and position 5, for unit decrease, are randomly selected in the first iteration, and positions 4 and 3 are selected in the second iteration. Let $x_0 = (3, 2, 2, 3, 3)$ and the randomly generated probability be $p_1 = 0.15$ and $p_2 = 0.3$. Perform two iterations of simulated annealing methods.

7.14 Define neighborhoods, moves, and tabu lists. In simulated annealing methods, neighborhoods are selected by the Metropolis criteria. How are neighborhoods obtained in tabu search methods?

7.15 It is broadly known that tabu search techniques are highly dependent on an initial solution. In other words, if a good solution is used as an initial solution, then tabu search techniques generate a good solution. Improve tabu search techniques by using the above motivation.

7.16 There are three kinds of moves in tabu search techniques: moves which are not in the tabu list (tabu free), moves which are in the tabu list and satisfy the aspiration function, and moves which are in the tabu list but do not satisfy the aspiration function. Find admissible moves among these moves. Which admissible move is selected as the move that generates the next neighbor?

7.17 In Example 7-3, let $\mathbf{x} = (2, 3, 2, 2, 1, 2, 3, 2, 2, 1)$ and $\mathbf{tbs} = (2, 3, 5, 0, 0, 0, 4, 5, 1, 0)$ in the tenth iteration. Show three classifications of moves in this problem. Find the next neighbor, and show the reason why you chose that neighbor.

8 Reliability–redundancy allocation

8.1 Introduction

A design engineer has several options to improve the reliability of a system with a given basic design. Two important criteria are: (1) to increase the component reliabilities, and (2) to provide redundancy at various stages. Any increment in component reliability causes an increase in the component cost. However, the cost will increase exponentially as the reliability is increased beyond a certain limit. When it is feasible to provide component redundancy, which also requires an additional cost, it may be advantageous to increase the component reliability to some level and provide redundancy at that level. In other words, the design engineer can make a tradeoff between component reliability increments and component or subsystem redundancies.

The mathematical formulation becomes a mixed integer nonlinear programming problem (MINLP) in which the continuous variables represent the component reliabilities and the integer variables represent the levels of redundancy. This problem is no less difficult than the nonlinear integer programming optimal redundancy allocation problems discussed in previous chapters. Also, the discrete optimization methods explained earlier are not directly applicable to MINLP. However, some methods, such as branch-and-bound, generalized Benders decomposition, outer approximation, etc., are useful in solving MINLPs; see Floudas [94]. Researchers in the field of system reliability adopt several approaches to solve mixed integer programming formulations of reliability–redundancy allocation problems. Misra and Ljubojevic [229] were the first to introduce this kind of reliability optimization problem. They considered the problem of simultaneously determining optimal component reliabilities and optimal redundancy levels at n stages of a series system subject to a cost constraint. They assumed that component cost is an exponential function of component reliability at each stage. Treating the redundancy levels as continuous variables, they solved the resulting nonlinear programming problem using the Lagrange multipliers method and using a trial-and-error approach to round off the values of redundancy levels to integers. This approach is heuristic in nature, and it can also be adopted for a general reliability system with a single resource constraint. The general problem can be described as follows.

Problem 8.1

Let r_j and x_j denote the component reliability and the number of redundancies at stage j, respectively. Let $f(x_1, \ldots, x_n, r_1, \ldots, r_n)$ represent the corresponding system reliability. Suppose there are m resources and the available amount of resource i is b_i

for $i = 1, \dots, m$. Let $g_{ij}(x_j, r_j)$ denote the consumption of resource i at stage j for component reliability r_j and redundancy level x_j. Then the reliability–redundancy allocation problem can be mathematically represented so as to maximize

$$R_s = f(\mathbf{x}, \mathbf{r}),$$

subject to

$$\sum_{j=1}^{n} g_{ij}(x_j, r_j) \le b_i, \qquad \text{for } i = 1, \dots, m, \tag{8.1}$$

$$l_j \le x_j \le u_j, \qquad \text{for } j = 1, \dots, n, \tag{8.2}$$

$$r_j^l \le r_j \le r_j^u, \qquad \text{for } j = 1, \dots, n, \tag{8.3}$$

x_j being a nonnegative integer, where $\mathbf{x} = (x_1, \dots, x_n)$ and $\mathbf{r} = (r_1, \dots, r_n)$.

Distinct approaches are adopted in the literature to solve this mixed integer optimization problem. The main approaches are: (1) a combination of a search method for component reliability allocation and a heuristic method for redundancy allocation (for fixed component reliabilities), (2) a combination of a branch-and-bound method and a Lagrange multipliers method, (3) a surrogate multipliers method, and (4) evolutionary algorithms. All of these approaches are described in later sections. Following are two examples which have been commonly used to illustrate the reliability–redundancy allocation methods.

Example 8-1

Tillman et al. [304] developed a heuristic method for solving Problem 8.1 and illustrated it using the following example. Maximize

$$f(\mathbf{x}, \mathbf{r}) = \prod_{j=1}^{n} \left[1 - (1 - r_j)^{x_j} \right],$$

subject to

$$\sum_{j=1}^{n} p_j x_j^2 \le P,$$

$$\sum_{j=1}^{n} c_j(r_j) \left[x_j + \exp\left(\frac{x_j}{4}\right) \right] \le C,$$

$$\sum_{j=1}^{n} w_j x_j \exp\left(\frac{x_j}{4}\right) \le W,$$

$$0 \le r_j \le 1, \qquad \text{for } j = 1, \dots, n,$$

x_j being a positive integer.

Table 8.1. Constants for Example 8-1

j	$\alpha_j \times 10^5$	p_j	w_j	β_j
1	2.330	1	7	1.5
2	1.450	2	8	1.5
3	0.541	3	8	1.5
4	8.050	4	6	1.5
5	1.950	2	9	1.5

$(P, C, W) = (110, 175, 200)$ and $t = 1000$.

This is a reliability–redundancy allocation problem in which the reliability of a series system is to be maximized subject to three constraints. It may be noted that the second constraint involves both integer and continuous variables, whereas the other two constraints involve only integer variables. It is assumed that for component reliability r_j at stage j, the component cost is

$$c_j(r_j) = \alpha_j \left(\frac{-t}{\ln r_j} \right)^{\beta_j}.$$

The constants α_j and β_j represent the inherent characteristics of each component at the jth stage, and t is the duration for which the system is required to operate. The ratio $-(\ln r_j)/t$ gives the component failure rate when the failure distribution is a negative exponential. The constants in the constraints are given in Table 8.1.

This example has been extensively used to demonstrate the various methods for solving reliability–redundancy allocation problems.

Example 8-2

Consider a five-stage complex system, as shown in Figure 1.10. Let x_j denote the number of components in parallel at stage j, and let r_j denote the reliability of each of them. Then, the reliability $R_j(x_j)$ of stage j is

$$R_j(x_j) = 1 - (1 - r_j)^{x_j},$$

and the corresponding system reliability is

$$\begin{aligned}
R_s = {} & R_1(x_1)R_2(x_2) + R_3(x_3)R_4(x_4) + R_1(x_1)R_4(x_4)R_5(x_5) + R_2(x_2)R_3(x_3)R_5(x_5) \\
& - R_1(x_1)R_2(x_2)R_3(x_3)R_4(x_4) - R_1(x_1)R_2(x_2)R_3(x_3)R_5(x_5) \\
& - R_1(x_1)R_2(x_2)R_4(x_4)R_5(x_5) - R_1(x_1)R_3(x_3)R_4(x_4)R_5(x_5) \\
& - R_2(x_2)R_3(x_3)R_4(x_4)R_5(x_5) + 2R_1(x_1)R_2(x_2)R_3(x_3)R_4(x_4)R_5(x_5).
\end{aligned}$$

The problem is to find optimal values of $x_1, \ldots, x_5, r_1, \ldots, r_5$ satisfying the constraints given in Example 8-1. This example was considered by Hikita et al. [126] to illustrate their surrogate constraints method.

8.2   The method of Tillman, Hwang, and Kuo

Tillman et al. [304] developed a heuristic method (THK) for solving reliability–redundancy allocation, Problem 8.1. The method goes through a sequence of component reliability vectors using a search method and derives, for each vector $\mathbf{r} = (r_1, \ldots, r_n)$ in the sequence, an optimal or near-optimal redundancy allocation using a heuristic method. For a fixed vector $\mathbf{r}$ of component reliabilities, let $\mathbf{x}^* = (x_1^*, \ldots, x_n^*)$ be an optimal redundancy allocation and let $T(\mathbf{r})$ denote the corresponding system reliability, that is,

$$T(\mathbf{r}) = f(\mathbf{x}^*, \mathbf{r})$$

$$= \max_{\mathbf{x}} \{ f(\mathbf{x}, \mathbf{r}) : (\mathbf{x}, \mathbf{r}) \text{ satisfies eqs. (8.1) and (8.3)} \}.$$

The heuristic method of Aggarwal et al. [7] was used by Tillman et al. [304] for deriving an optimal redundancy allocation $\mathbf{x}^*$ that gives $T(\mathbf{r})$ for any fixed $\mathbf{r}$. The value of $T(\mathbf{r})$ is taken as zero when no redundancy allocation $\mathbf{x} = (x_1, \ldots, x_n)$ is feasible for $\mathbf{r}$. The function $T(\mathbf{r})$ is maximized with respect to $\mathbf{r}$ using the Hooke–Jeeves method (Appendix 2). Although the THK method (Figure 8.1) was demonstrated by Tillman et al. [304] for solving a problem in a series system, it can also be applied for any general reliability system.

To solve Example 8-1 by the THK method, the initial value of step size h and the lower limit on h in the Hooke–Jeeves method are taken as 0.05 and 0.003 13, respectively. The solution yielded by the THK method is

$$(x_1, x_2, x_3, x_4, x_5) = (3, 3, 2, 2, 3),$$

$$(r_1, r_2, r_3, r_4, r_5) = (0.784 \, 38, 0.825 \, 00, 0.900 \, 00, 0.775 \, 00, 0.778 \, 13),$$

and the corresponding system reliability is 0.915 363. For this solution, the residual resources $b_i - \sum g_{ij}(x_j, r_j)$ are 37.0, 0.1474, and 1.4118 for $i = 1, 2,$ and 3, respectively. The initial solution, the improved solutions at intermediate stages of the algorithm, and the final solution are given in Table 8.2.

We can also use other direct search methods, such as the flexible polygon search of Nelder and Mead [251], the conjugate direction method of Powell [264] etc., instead of the direct search method of Hooke and Jeeves [130]. A good description of these methods can be found in Reklaitis et al. [276]. Similarly, we can use any good heuristic method for redundancy allocation as an alternative to the method of Aggarwal et al. [7].

8.3   The method of Gopal, Aggarwal, and Gupta

Gopal et al. [113] developed a heuristic method (GAG2) that starts with 0.5 as the component reliability at each stage of the system and increases the component reliability at one of the stages by a specified value h in every iteration. The selection of a stage

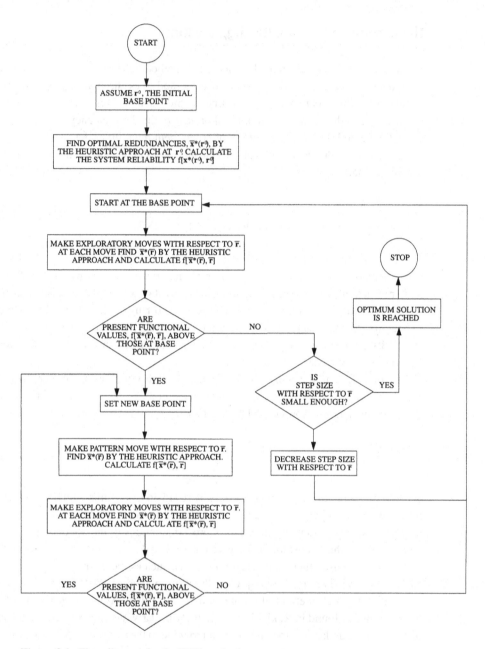

Figure 8.1. Flow diagram for the THK method

for improving component reliability is based on a *sensitivity factor*. For any particular choice of component reliabilities $r_1, \ldots, r_n$, an optimal or near-optimal redundancy allocation $x_1, \ldots, x_n$ is derived by the heuristic method of Gopal et al. [112]. In fact, any heuristic redundancy allocation method can be used for the same purpose. When

Table 8.2. Output from the THK method

x	r	System reliability
(3, 3, 3, 2, 2)	(0.600 00, 0.600 00, 0.600 00, 0.600 00, 0.600 00)	0.578 610
(3, 3, 3, 2, 2)	(0.650 00, 0.650 00, 0.650 00, 0.650 00, 0.650 00)	0.675 150
(3, 3, 3, 2, 2)	(0.750 00, 0.750 00, 0.750 00, 0.750 00, 0.750 00)	0.838 348
(3, 3, 3, 2, 2)	(0.800 00, 0.800 00, 0.800 00, 0.800 00, 0.800 00)	0.870 637
(3, 3, 3, 2, 2)	(0.825 00, 0.800 00, 0.825 00, 0.775 00, 0.800 00)	0.894 444
(3, 2, 3, 2, 3)	(0.825 00, 0.812 50, 0.837 50, 0.775 00, 0.787 50)	0.898 475
(3, 2, 3, 2, 3)	(0.825 00, 0.825 00, 0.850 00, 0.775 00, 0.775 00)	0.901 887
(3, 3, 2, 2, 3)	(0.812 50, 0.825 00, 0.875 00, 0.775 00, 0.762 50)	0.911 035
(3, 3, 2, 2, 3)	(0.800 00, 0.825 00, 0.900 00, 0.775 00, 0.750 00)	0.912 875
(3, 3, 2, 2, 3)	(0.800 00, 0.825 00, 0.900 00, 0.775 00, 0.753 13)	0.913 412
(3, 3, 2, 2, 3)	(0.796 88, 0.825 00, 0.900 00, 0.775 00, 0.759 38)	0.914 094
(3, 3, 2, 2, 3)	(0.790 63, 0.825 00, 0.900 00, 0.775 00, 0.768 75)	0.914 809
(3, 3, 2, 2, 3)	(0.784 38, 0.825 00, 0.900 00, 0.775 00, 0.778 13)	0.915 363

the increments in component reliabilities do not give any higher system reliability, the increment h is reduced and the procedure is repeated with the new increment h. This process stops when h falls below a specified limit h_0. Let $c_j(r_j)$ denote component cost at stage j when component reliability at that stage is r_j. The following is a stepwise description of the GAG2 method.

GAG2 algorithm

- Step 0: Initialize $(r_1, \ldots, r_n) = (0.5, \ldots, 0.5)$. Derive a heuristic optimal redundancy allocation $\mathbf{x}^* = (x_1^*, \ldots, x_n^*)$ by a heuristic method for component reliabilities $r_1, \ldots, r_n$. Let $\mathbf{r}^0 = (r_1, \ldots, r_n)$ and $\mathbf{x}^0 = \mathbf{x}^*$. Compute the corresponding system reliability $f(\mathbf{x}^0, \mathbf{r}^0)$. Let $R_s^0 = f(\mathbf{x}^0, \mathbf{r}^0)$.

- Step 1: Evaluate the sensitivity factor

$$s_j(\mathbf{x}^*, \mathbf{r}) = \frac{f(x_1^*, \ldots, x_n^*, r_1, \ldots, r_j + h, \ldots, r_n) - f(\mathbf{x}^*, \mathbf{r})}{c_j(r_j + h) - c_j(r_j)}$$

for each stage j and arrange these values in decreasing order. Suppose $[s_{v_1}(\mathbf{x}^*, \mathbf{r}), \ldots, s_{v_n}(\mathbf{x}^*, \mathbf{r})]$ is the decreasing order of $s_j(\mathbf{x}^*, \mathbf{r})$. Let $k = 0$.

- Step 2: Let $k = k + 1$. If $k > n$, go to step 5.

- Step 3: Let $r_{v_k} = r_{v_k} + h$. Obtain an optimal redundancy allocation $(x_1^*, \ldots, x_n^*)$ for component reliabilities $(r_1, \ldots, r_n)$. If no feasible allocation exists, go to step 4. Otherwise, compute the corresponding system reliability $f(x_1^*, \ldots, x_n^*, r_1, \ldots, r_n)$. If $R_s^0 < f(x_1^*, \ldots, x_n^*, r_1, \ldots, r_n)$, then let $x_j^0 = x_j^*$ and $r_j^0 = r_j$, for $j = 1, \ldots, n$ and $R_s^0 = f(x_1^*, \ldots, x_n^*, r_1, \ldots, r_n)$. Go to step 2.

Table 8.3. Output from the GAG2 method for Example 8-1

	h	x^*	r	R_S
1	0.05	$(3, 3, 3, 2, 2)$	$(0.50, 0.50, 0.50, 0.50, 0.50)$	0.376 831
2	0.05	$(3, 3, 2, 2, 3)$	$(0.50, 0.50, 0.55, 0.50, 0.50)$	0.400 697
3	0.05	$(3, 3, 2, 2, 3)$	$(0.50, 0.50, 0.60, 0.50, 0.50)$	0.422 051
4	0.05	$(3, 3, 2, 2, 3)$	$(0.50, 0.50, 0.65, 0.50, 0.50)$	0.440 892
5	0.05	$(3, 3, 2, 2, 3)$	$(0.50, 0.55, 0.65, 0.50, 0.50)$	0.457 961
6	0.05	$(3, 3, 2, 2, 3)$	$(0.50, 0.55, 0.70, 0.50, 0.50)$	0.474 923
7	0.05	$(3, 3, 2, 2, 3)$	$(0.50, 0.55, 0.70, 0.50, 0.55)$	0.493 309
8	0.05	$(3, 3, 2, 2, 3)$	$(0.55, 0.55, 0.70, 0.50, 0.55)$	0.512 407
9	0.05	$(3, 2, 2, 3, 3)$	$(0.55, 0.60, 0.70, 0.50, 0.55)$	0.552 506
10	0.05	$(3, 2, 2, 3, 3)$	$(0.55, 0.65, 0.70, 0.50, 0.55)$	0.577 172
11	0.05	$(3, 2, 2, 3, 3)$	$(0.55, 0.65, 0.75, 0.50, 0.55)$	0.594 613
12	0.05	$(3, 2, 2, 3, 3)$	$(0.55, 0.65, 0.75, 0.50, 0.60)$	0.612 360
13	0.05	$(3, 2, 2, 3, 3)$	$(0.60, 0.65, 0.75, 0.50, 0.60)$	0.630 635
14	0.05	$(3, 2, 2, 3, 3)$	$(0.60, 0.70, 0.75, 0.50, 0.60)$	0.653 992
15	0.05	$(3, 2, 2, 3, 3)$	$(0.60, 0.70, 0.80, 0.50, 0.60)$	0.669 688
16	0.05	$(3, 2, 2, 3, 3)$	$(0.60, 0.70, 0.80, 0.50, 0.65)$	0.684 802
17	0.05	$(3, 2, 2, 3, 3)$	$(0.65, 0.70, 0.80, 0.50, 0.65)$	0.700 258
18	0.05	$(3, 2, 2, 3, 3)$	$(0.65, 0.75, 0.80, 0.50, 0.65)$	0.721 420
19	0.05	$(3, 2, 2, 3, 3)$	$(0.65, 0.75, 0.80, 0.55, 0.65)$	0.749 349
20	0.05	$(3, 2, 2, 3, 3)$	$(0.65, 0.75, 0.80, 0.55, 0.70)$	0.761 778
21	0.05	$(3, 2, 2, 3, 3)$	$(0.65, 0.75, 0.80, 0.60, 0.70)$	0.784 513
22	0.05	$(3, 2, 2, 3, 3)$	$(0.70, 0.75, 0.80, 0.60, 0.70)$	0.797 525
23	0.05	$(3, 2, 2, 3, 3)$	$(0.70, 0.75, 0.85, 0.60, 0.70)$	0.812 063
24	0.05	$(3, 2, 2, 3, 3)$	$(0.70, 0.80, 0.85, 0.60, 0.70)$	0.831 552
25	0.05	$(3, 2, 2, 3, 3)$	$(0.70, 0.80, 0.85, 0.65, 0.70)$	0.850 320
26	0.05	$(3, 2, 2, 3, 3)$	$(0.70, 0.80, 0.85, 0.65, 0.75)$	0.860 261
27	0.05	$(3, 2, 2, 3, 3)$	$(0.75, 0.80, 0.85, 0.65, 0.75)$	0.870 318
28	0.05	$(3, 2, 2, 3, 3)$	$(0.75, 0.85, 0.85, 0.65, 0.75)$	0.886 183
29	0.05	$(3, 2, 2, 3, 3)$	$(0.75, 0.85, 0.85, 0.70, 0.75)$	0.900 881
30	0.05	$(3, 2, 2, 3, 3)$	$(0.75, 0.85, 0.90, 0.70, 0.75)$	0.912 401
31	0.05	$(3, 2, 2, 3, 3)$	$(0.75, 0.85, 0.90, 0.70, 0.80)$	0.919 469
32	0.05	$(3, 2, 2, 3, 3)$	$(0.80, 0.85, 0.90, 0.70, 0.80)$	0.926 591
33	0.0125	$(3, 2, 2, 3, 3)$	$(0.80, 0.8625, 0.90, 0.70, 0.80)$	0.929 998
34	0.001 56	$(3, 2, 2, 3, 3)$	$(0.80, 0.8625, 0.901 56, 0.70, 0.80)$	0.930 289

- Step 4: Let $r_{v_k} = r_{v_k} - h$, and go to step 2.

- Step 5: If the value of R_s^0 increases at least once in n iterations of the loop involving steps 2–4, go to step 1.

- Step 6: Let $h = h/2$. If h is less than the specified value h_0, take $(x_1^0, \ldots, x_n^0, r_1^0, \ldots, r_n^0)$ as the required solution, R_s^0 as the corresponding system reliability, and stop. Otherwise, go to step 1.

Example 8-1 is now solved by the GAG2 method using the heuristic redundancy allocation method (GAG1) of Gopal et al. [112] for any fixed component reliabilities. For an initial value $h = 0.05$ and a lower limit $h_0 = 0.001$, the final solution yielded by the GAG2 method is

$$\mathbf{x}^0 = (3, 2, 2, 3, 3),$$

$$\mathbf{r}^0 = (0.80, 0.8625, 0.901\,56, 0.70, 0.80),$$

and the corresponding system reliability is $R_s^0 = 0.930\,289$. For this solution, the residual resources $b_i - \sum g_{ij}(x_j, r_j)$ are 27.0, 0.0265, and 7.5189 for $i = 1, 2$, and 3, respectively. The initial solution, improved solutions at intermediate stages of the algorithm, and the final solution are given in Table 8.3. For each solution listed in the table, the corresponding value of h is also given.

It may be noted that for this example, the GAG2 method has yielded a better solution than the THK method. However, it is difficult to compare the methods without sufficient numerical experimentation on a variety of problems. Although the performance of the THK method may be affected by the choice of initial component reliabilities, the method usually yields a satisfactory solution since the search for component reliabilities is based on the Hooke–Jeeves search method.

8.4   The method of Kuo, Lin, Xu, and Zhang

Kuo et al. [178] developed a branch-and-bound technique (KLXZ) to solve nonlinear integer and mixed integer programming formulations of reliability optimization problems using Lagrange multipliers. In this technique, an upper bound is derived at every node by solving a nonlinear programming problem by the Lagrange multipliers method. Consider the reliability–redundancy allocation Problem 8.1. In the branch-and-bound technique, each node is associated with an optimization problem, which is the same as Problem 8.1 with some x_js having fixed integer values. The initial node 1 is associated with Problem 8.1. The integer restriction is relaxed for all x_js and an optimal solution $(\mathbf{x}^*, \mathbf{r}^*)$ of the relaxed version of Problem 8.1 is obtained by the Lagrange multipliers method. The optimal value $f(\mathbf{x}^*, \mathbf{r}^*)$ is an upper bound on the objective value of the problem associated with node 1. If x_j^* is the integer for all j, take $(\mathbf{x}^*, \mathbf{r}^*)$ as the required optimal solution of Problem 8.1 and stop. Otherwise, a variable x_j for which x_j^* is not an integer is considered and two nodes (2 and 3) are generated from node 1 with respect to x_j. One of the two nodes corresponds to the optimization Problem 8.1 with x_j restricted to be $\lfloor x_j^* \rfloor$ and the other node corresponds to Problem 8.1 with x_j restricted to be $\lfloor x_j^* \rfloor + 1$, where $\lfloor x_j^* \rfloor$ denotes the largest integer less than or equal to x_j^*. A similar procedure is followed for branching at every selected node. If the optimal solution $(\mathbf{x}^*, \mathbf{r}^*)$ of the relaxed version of the problem associated with a newly generated node contains all integer x_js, then it is feasible for Problem 8.1. In such a case, $f(\mathbf{x}^*, \mathbf{r}^*)$ is compared with the objective value of the current best solution of Problem 8.1, and the node is fathomed. If all x_j^*s are not integers, $f(\mathbf{x}^*, \mathbf{r}^*)$ is taken as the upper bound associated with the node. The node with the highest upper bound is always selected for branching. Let Q_k denote the optimization problem associated with node k, and β_k denote the optimal objective value of the relaxed version of Q_k. The branching method of Kuo et al. [178] is described in the following algorithm.

The KLXZ algorithm

- Step 0: Take the original Problem 8.1 as Q_1 for the initial node 1 and obtain an optimal solution $(\mathbf{x}^*, \mathbf{r}^*)$ of the relaxed version of Q_1 (ignoring the integer restriction on x_j). If x_j^* is the integer for all j, then take $(\mathbf{x}^*, \mathbf{r}^*)$ as an optimal solution of Problem 8.1 and stop. Otherwise, go to step 1.

- Step 1: Let $N = \phi, s = 1, k = 1$ and $f^* = -\infty$. Select the node k for branching, and go to step 2.

- Step 2: Select a variable x_j for which x_j^* is not an integer and generate two nodes $s+1$ and $s+2$ from the selected node k. Associate with node $s+1$ the problem Q_{s+1}, which is Q_k with the additional restriction $x_j = \lfloor x_j \rfloor$, and derive the corresponding upper bound β_{s+1} by solving the relaxed version of Q_{s+1}. Similarly, associate with node $s + 2$ the problem Q_{s+2}, which is Q_k with the additional restriction $x_j = \lfloor x_j \rfloor + 1$, and derive the corresponding bound β_{s+2}.

- Step 3: If all x_js are integers in the optimal solution of the relaxed version of Q_{s+1} and $\beta_{s+1} > f^*$, let $f^* = \beta_{s+1}$ and $(x^0, r^0) = (\mathbf{x}^*, \mathbf{r}^*)$. If at least one x_j^* is not an integer, let $N = N \cup \{s + 1\}$. Repeat the same with respect to node $s + 2$.

- Step 4: If $N = \phi$, stop. Otherwise, select from set N a node, say k, with maximum bound β_k. Let $(\mathbf{x}^*, \mathbf{r}^*)$ denote the optimal solution of the relaxed version of the problem Q_k associated with node k. Go to step 2.

Kuo et al. [178] solved Example 8-1 using this method. The solution derived by them is $\mathbf{x}^* = (3, 3, 2, 3, 2)$, $\mathbf{r}^* = (0.779\,60, 0.800\,65, 0.902\,27, 0.710\,44, 0.859\,47)$, and the corresponding system reliability is $0.929\,75$.

8.5   The method of Xu, Kuo, and Lin

Xu et al. [326] developed a heuristic method (XKL) for the reliability–redundancy allocation problem. This method iteratively derives a sequence of feasible solutions. Two ways in which a solution is derived from the previous solution are:

1. One of the x_js is increased by 1 and an optimal **r** is obtained corresponding to the new **x** by solving a nonlinear programming problem.

2. One of the x_js is increased by 1 and another of them is reduced by 1 and an optimal **r** is obtained for the new **x** by solving a nonlinear programming problem.

The selection of variable x_j for increase or decrease is based on a sensitivity function. Xu et al. [326] assumed that the functions $f(\mathbf{x}, \mathbf{r})$ and $g_{ij}(x_j, r_j)$ are differentiable and monotonic nondecreasing functions. Let $\mathbf{x}^0 = (x_1^0, \dots, x_n^0)$ be a fixed vector of redundancy levels.

Problem 8.2

Consider the nonlinear programming problem $P(\mathbf{x}^0)$ to maximize

$$z = f(\mathbf{x}^0, \mathbf{r}),$$

subject to

$$\sum_{j=1}^{n} g_{ij}(x_j^0, r_j) \le b_i, \qquad \text{for } i = 1, \dots, m,$$

$$0 \le r_j \le 1, \qquad \text{for } j = 1, \dots, n.$$

A nonlinear programming problem of the type shown in Problem 8.2 has to be solved in every iteration of the XKL method. The following algorithm is a stepwise description of the XKL method.

XKL algorithm

- Step 0: Start with a reasonable redundancy allocation $\mathbf{x}^0$. Find an optimal solution $\mathbf{r}^0$ of Problem 8.2 and compute the corresponding system reliability $R^0 = f(\mathbf{x}^0, \mathbf{r}^0)$.

- Step 1: Compute the value of sensitivity function S_j with respect to x_j as

$$S_j = \frac{\partial R_s / \partial x_j}{\min_{1 \le i \le m} \partial g_{ij} / \partial x_j},$$

for $j = 1, \dots, n$, where $\partial R_s / \partial x_j$ and $\partial g_{ij} / \partial x_j$ are the derivatives computed at point $(\mathbf{x}^0, \mathbf{r}^0)$. Find the descending order $(S_{v_1}, S_{v_2}, \dots, S_{v_n})$ of sensitivity function values $S_1, \dots, S_n$. Also, compute

$$\hat{S} = \max_{1 \le j \le n} \frac{\partial R_s / \partial r_j}{\min_{1 \le i \le m} \partial g_{ij} / \partial r_j}.$$

- Step 2: If $\hat{S} > S_{v_1}$ (=maximum S_j), go to step 5. Otherwise let $k = 1$.

- Step 3: Let $x_{v_k}^0 = x_{v_k}^0 + 1$. Find an optimal solution $\mathbf{r}'$ of Problem 8.2. If the problem is feasible and $R^0 < f(\mathbf{x}^0, \mathbf{r}')$, let $\mathbf{r}^0 = \mathbf{r}'$, $R^0 = f(\mathbf{x}^0, \mathbf{r}')$, and go to step 1.

- Step 4: Let $x_{v_k}^0 = x_{v_k}^0 - 1$ and $k = k + 1$. If $k \leq n$, go to step 3.
- Step 5: Take the pair of stages (v_1, v_n), and let $x_{v_1}^0 = x_{v_1}^0 + 1$ and $x_{v_n}^0 = x_{v_n}^0 - 1$. Find an optimal solution $\mathbf{r}'$ of Problem 8.2. If the problem is feasible and $R^0 < f(\mathbf{x}^0, \mathbf{r}')$, let $\mathbf{r}^0 = \mathbf{r}'$, $R^0 = f(\mathbf{x}^0, \mathbf{r}')$, and go to step 1. Otherwise, reset $x_{v_1}^0 = x_{v_1}^0 - 1$ and $x_{v_n}^0 = x_{v_n}^0 + 1$.
- Step 6: Repeat step 5 for pairs (v_1, v_{n-1}), (v_1, v_{n-2}), ... , (v_1, v_2), (v_2, v_n), ... , (v_2, v_3), ... , (v_{n-1}, v_n) until improvement in system reliability takes place. If the improvement takes place for some pair, go to step 1 (without repeating step 5 further). Otherwise, take $(\mathbf{x}^0, \mathbf{r}^0)$ as the optimal solution and stop.

Any standard method, such as the Lagrange multiplier method, can be used to solve problems of type $P(\mathbf{x}^0)$. When the functions are nondifferentiable, or it is inconvenient to analytically obtain the derivatives, numerical approximations to partial derivatives can be used. Xu et al. [326] demonstrated their method using three different examples. Performing extensive numerical experimentation, they compared its performance with that of the branch-and-bound method of Kuo et al. [178] and various involving combinations of search and heuristic redundancy allocation methods. Based on their experimental results, Xu et al. [326] reported that the XKL method is superior to all other heuristic methods with respect to performance criteria such as execution time, optimal value, etc.

Solution to Example 8-1 by the XKL method
Example 8-1 is now solved by the XKL method using a penalty method for optimizing nonlinear programming problems of the type indicated in Problem 8.2. It may be noted that any NLP method can be used for solving NLP problems that are encountered in the XKL method. Let $\mathbf{x}^0 = (2, 2, 2, 2, 2)$. For this $\mathbf{x}^0$, an optimal solution $\mathbf{r}^0$ of $P(\mathbf{x}^0)$ is

$$(r_1^0, \ldots, r_5^0) = (0.834\,502, 0.853\,789, 0.887\,422, 0.765\,242, 0.840\,939),$$

and the corresponding system reliability is $f(\mathbf{x}^0, \mathbf{r}^0) = 0.865\,498\,2$. Initially, R^0 is taken as $f(\mathbf{x}^0, \mathbf{r}^0)$. For the initial solution $(\mathbf{x}^0, \mathbf{r}^0)$, the values of the sensitivity function are

$$(S_1, \ldots, S_5) = (0.003\,242, 0.003\,528, 0.004\,146, 0.002\,816, 0.003\,420)$$

and

$$\hat{S} = \max(0.000\,849, 0.000\,874, 0.000\,922, 0.000\,875, 0.000\,879) = 0.000\,922.$$

Since $\hat{S} < S_3 = \max S_i$, increase x_3 by 1. The optimal solution with the new $\mathbf{x}^0$ is

$$(\mathbf{x}^0, \mathbf{r}') = (2, 2, 3, 2, 2, \ 0.834\,022, 0.860\,685, 0.822\,942, 0.768\,800, 0.844\,452),$$

and the corresponding system reliability is $f(\mathbf{x}^0, \mathbf{r}') = 0.875\,878$. Since $f(\mathbf{x}^0, \mathbf{r}') > R^0$, update $\mathbf{r}^0$ and R^0 as $\mathbf{r}^0 = \mathbf{r}'$ and $R^0 = 0.875\,878$.

Table 8.4. Sequence of solutions obtained by the XKL method

	$\mathbf{x}^0$	$\mathbf{r}^0$	R^0
1	(2, 2, 2, 2, 2)	(0.834 502, 0.853 789, 0.887 422, 0.765 242, 0.840 939)	0.865 498
2	(2, 2, 3, 2, 2)	(0.834 022, 0.860 685, 0.822 942, 0.768 800, 0.844 452)	0.875 878
3	(3, 2, 3, 2, 2)	(0.762 786, 0.866 451, 0.828 722, 0.776 427, 0.848 162)	0.894 873
4	(3, 3, 3, 2, 2)	(0.768 914, 0.798 171, 0.830 824, 0.784 645, 0.855 136)	0.910 080
5	(3, 2, 3, 2, 3)	(0.772 846, 0.866 571, 0.829 703, 0.786 934, 0.778 894)	0.912 075
6	(3, 3, 2, 2, 3)	(0.773 399, 0.796 138, 0.898 203, 0.788 263, 0.785 991)	0.917 275
7	(3, 3, 2, 3, 2)	(0.775 367, 0.802 277, 0.902 547, 0.710 485, 0.860 674)	0.929 721
8	(3, 2, 2, 3, 3)	(0.783 191, 0.872 255, 0.903 500, 0.709 306, 0.785 600)	0.931 622

For the current best solution $(\mathbf{x}^0, \mathbf{r}^0)$,

$$(S_1, \ldots, S_5) = (0.003\,311, 0.003\,066, 0.002\,783, 0.002\,717, 0.003\,226)$$

and

$$\hat{S} = \max(0.000\,868, 0.000\,744, 0.000\,870, 0.000\,837, 0.000\,821) = 0.000\,870.$$

Since $\hat{S} < S_1 = \max_{1 \le i \le 5} S_i$, increase x_1 by 1. The optimal solution with the new $\mathbf{x}^0$ is

$$(\mathbf{x}^0, \mathbf{r}') = (3, 2, 3, 2, 2, \ 0.762\,786, 0.866\,451, 0.828\,722, 0.776\,427, 0.848\,162),$$

and the corresponding system reliability is $f(\mathbf{x}^0, \mathbf{r}') = 0.894\,872\,8$. Since $f(\mathbf{x}^0, \mathbf{r}') > R^0$, update $\mathbf{r}^0$ and R^0 as $\mathbf{r}^0 = \mathbf{r}'$ and $R^0 = 0.894\,872\,8$.

The above solutions and those obtained in the subsequent iterations are given in Table 8.4. Each of the solutions 2, 3, and 4 in Table 8.4 is obtained from the preceding one by an increment in one of the x_js, whereas each of the solutions 5, 6, 7, and 8 is obtained from the preceding one by transferring a unit from one element to another in vector $\mathbf{x}^0$.

8.6   The surrogate constraints method

The surrogate constraints approach is effective in solving multi-constraint optimization problems when there are good techniques available to solve the same problems with a single constraint. Luenberger [203] showed that the surrogate constraints method gives an exact optimal solution if the problem is a quasi-convex programming problem. Nakagawa and Miyazaki [246] used this approach to solve a nonlinear integer programming formulation of a reliability optimization problem with two constraints. They demonstrated that, for multiple constraints, this approach is superior to the dynamic programming approach involving Lagrange multipliers. Later, Hikita et al. [126] adopted the surrogate constraints approach (HNNN method) to solve reliability–redundancy allocation problems. This approach involves solving a sequence of single-constraint mixed integer programming problems.

Problem 8.3

Consider the reliability–redundancy allocation problem of maximizing

$$f(\mathbf{x}, \mathbf{r}) = \prod_{j=1}^{n} R_j(x_j, r_j),$$

subject to

$$\sum_{j=1}^{n} g_{ij}(x_j, r_j) - b_i \leq 0, \qquad \text{for } i = 1, \ldots, m, \tag{8.4}$$

$$0 \leq r_j \leq 1, \qquad \text{for } j = 1, \ldots, n,$$

x_j being a nonnegative integer.

This is a special case of Problem 8.1 for series systems. A surrogate problem S($\mathbf{u}$) corresponding to the above problem is defined as to maximize $z_u = f(\mathbf{x}, \mathbf{r})$, subject to

$$\sum_{i=1}^{m} u_i \left[\sum_{j=1}^{n} g_{ij}(x_j, r_j) - b_i \right] \leq 0,$$

$$0 \leq r_j \leq 1, \qquad \text{for } j = 1, \ldots, n,$$

x_j being a nonnegative integer, where u_i is a fixed value satisfying

$$u_1 + \cdots + u_m = 1,$$

$$u_i \geq 0, \qquad \text{for } i = 1, \ldots, m.$$

The surrogate problem S($\mathbf{u}$) can be rewritten so as to maximize $z_u = f(\mathbf{x}, \mathbf{r})$, subject to

$$\sum_{j=1}^{n} \left\langle \sum_{i=1}^{m-1} u_i \left[g_{ij}(x_j, r_j) - g_{mj}(x_j, r_j) \right] + g_{mj}(x_j, r_j) \right\rangle - \left[\sum_{i=1}^{m-1} u_i(b_i - b_m) + b_m \right] \leq 0,$$

$$0 \leq r_j \leq 1, \qquad \text{for } j = 1, \ldots, n,$$

$$x_j \in \{1, 2, \ldots\},$$

where $u_1, \ldots, u_{m-1}$ are fixed values satisfying

$$u_1 + \cdots + u_{m-1} \leq 1,$$

$$u_i \geq 0, \qquad \text{for } i = 1, \ldots, m - 1.$$

Let z_u^* denote the optimal value of the objective function in the surrogate problem $S(\mathbf{u})$. The surrogate dual problem corresponding to optimization Problem 8.3 is to minimize z_u^* subject to $u \in U^1$, where

$$U^1 = \left\{ (u_1, \ldots, u_{m-1}): \sum_{i=1}^{m-1} u_i \leq 1, \quad \text{and } u_i \geq 0, \text{ for } i = 1, \ldots, m-1 \right\}.$$

In other words, the surrogate dual problem is to find

$$Z_u = \min_{u \in U^1} \max_{(\mathbf{x},\mathbf{r}) \in K_u} f(\mathbf{x}, \mathbf{r}),$$

where

$$K_u = \left\{ (\mathbf{x}, \mathbf{r}): \sum_{j=1}^{n} \left\langle \sum_{i=1}^{m-1} u_i \left[g_{ij}(x_j, r_j) - g_{mj}(x_j, r_j) \right] + g_{mj}(x_j, r_j) \right\rangle \right.$$
$$\left. - \left[\sum_{i=1}^{m-1} u_i(b_i - b_m) + b_m \right] \leq 0, x_j \in \{0, 1, 2, \ldots\}, \text{ and } 0 \leq r_j \leq 1 \right\}.$$

It can be seen that the optimal objective value of the surrogate dual problem is greater than or equal to that of the original Problem 8.3.

The procedure adopted by Hikita et al. [126] to solve the surrogate dual problem can be described as follows. First solve the surrogate problem $S(\mathbf{u}^1)$ by the dynamic programming approach (as explained later in this section), with $\mathbf{u}^1 = (1/m, \ldots, 1/m)$ as the vector of surrogate multipliers. The vector $\mathbf{u}^1$ is the centroid of the polytope U^1. Suppose $(\mathbf{x}^1, \mathbf{r}^1)$ is the optimal solution of this surrogate problem. Let $k = 1$ and

$$U^{k+1} = U^k \cap \left\{ \mathbf{u}: \sum_{i=1}^{m-1} h_i u_i > h_0 \right\},$$

where

$$h_i = \sum_{j=1}^{n} \left[g_{ij}(x_j^k, r_j^k) - g_{mj}(x_j^k, r_j^k) \right] - (b_i - b_m), \tag{8.5}$$

$$h_0 = b_m - \sum_{j=1}^{n} g_{mj}(x_j^k, r_j^k). \tag{8.6}$$

Hikita et al. [126] showed that $z_u^* \geq z_{u^1}^*$ for any $\mathbf{u}$ in U^1 satisfying $\sum h_i u_i \leq h_0$. Now determine the centroid $\mathbf{u}^{k+1}$ of the polytope U^{k+1} and solve the surrogate problem $S(\mathbf{u}^{k+1})$. Suppose $(\mathbf{x}^{k+1}, \mathbf{r}^{k+1})$ is the optimal solution of $S(\mathbf{u}^{k+1})$. Continue the procedure for $k = 2, 3, \ldots$, until the surrogate problem is infeasible or the optimal objective values are very close for two successive values of k. The method suggested by Nakagawa et al. [244] for finding U^{k+1} from U^k and $(\mathbf{x}^{k+1}, \mathbf{r}^{k+1})$ is described in Appendix 3. The following algorithm is a stepwise description of the above procedure.

The HNNN algorithm

- Step 0: Select a small positive ϵ. Take the polytope U^1 in R^{m-1} with a set of m vertices

$$V = \{(1, 0, \dots, 0), \ (0, 1, \dots, 0), \ (0, 0, \dots, 1), \ (0, 0, \dots, 0)\}.$$

Let $v_1 = (1, 0, \dots, 0)$, $v_2 = (0, 1, \dots, 0)$, $\dots$, $v_{m-1} = (0, 0, \dots, 1)$, and $v_m = (0, 0, \dots, 0)$. Consider the set of supporting hyperplanes $P = \{1, 2, \dots, m\}$ where element i in P represents the supporting hyperplane $u_i = 0$ for $i = 1, \dots, m-1$ and the element m represents the plane $\sum_{i=1}^{m-1} u_i = 1$. Define a matrix $W = (w_{ij})_{m \times m}$ as

$$w_{ij} = \begin{cases} 0, & \text{if } j = i, \\ 1, & \text{otherwise,} \end{cases}$$

for $i = 1, \dots, m$ and $j = 1, \dots, m$. (The rows of W correspond to hyperplanes in P and the columns correspond to vertices in V. If plane i passes through vertex j, then $w_{ij} = 1$.)

- Step 1: Obtain an optimal solution (x^1, r^1) of the surrogate problem $S(u^1)$ where $u^1 = (1/m, \dots, 1/m)$ is the centroid of the polytope U^1. Let $(x^*, r^*) = (x^1, r^1)$, $f^* = f(x^*, r^*)$, and $k = 1$.

- Step 2: Obtain the polytope U^{k+1} and its centroid u^{k+1} from U^k by using the solution (x^k, r^k) in the algorithm given in Appendix 3. Simultaneously update the set P of supporting hyperplanes, the set V of vertices, and the incidence matrix $W = (w_{ij})$ for U^{k+1}. If U^{k+1} is empty, go to step 5.

- Step 3: Obtain an optimal solution (x^{k+1}, r^{k+1}) of the surrogate problem $S(u^{k+1})$. If $f(x^{k+1}, r^{k+1}) < f(x^*, r^*)$, update $(x^*, r^*) = (x^{k+1}, r^{k+1})$ and $f^* = f(x^{k+1}, r^{k+1})$.

- Step 4: If $k \geq 2$ and

$$1 - \epsilon \leq \frac{f(x^{k-1}, r^{k-1})}{f(x^k, r^k)} \leq 1 + \epsilon,$$

go to step 5. Otherwise let $k = k + 1$, and go to step 2.

- Step 5: Take (x^*, r^*) as the required solution of the original Problem 8.3.

It is possible that the solution (x^*, r^*) yielded by the above algorithm is not feasible for Problem 8.3. Hikita et al. [126] proposed the two following counter measures for avoiding the infeasibility.

Counter measure 1

This measure modifies the original problem by tightening the constraints and solves its surrogate dual problem. The modification is continued until the optimal solution of the surrogate dual problem is feasible to the original problem. The measure is explained in detail in the following algorithm.

Algorithm CSG1

- Step 0: Select a value θ from the interval $(0, 1)$ and let $t = 1$.

- Step 1: Let $b_i = b_i(1 - \theta t \hat{u}_i)$, for $i = 1, \ldots, m - 1$, and $b_m = b_m[1 - \theta t(1 - \sum_{i=1}^{m-1} \hat{u}_i)]$, where $(\hat{u}_1, \ldots, \hat{u}_{m-1})$ is the vector of surrogate multipliers for which $(\mathbf{x}^*, \mathbf{r}^*)$ is derived as the optimal solution in the HNNN algorithm. Let $k = 1$ and

$$U^1 = \left\{ u \in R^{m-1}: \sum_{i=1}^{m-1} u_i \leq 1, u_i \geq 0 \right\}.$$

- Step 2: Obtain an optimal solution (x^k, r^k) of $S(u^k)$. If this solution is feasible for the original Problem 8.3, stop. Otherwise, go to step 3.

- Step 3: If

$$1 - \epsilon \leq \frac{f(\mathbf{x}^{k-1}, \mathbf{r}^{k-1})}{f(\mathbf{x}^k, \mathbf{r}^k)} \leq 1 + \epsilon,$$

then let $t = t + 1$, and go to step 1. Otherwise, go to step 4.

- Step 4: Generate $U^{k+1} = U^k \cap \{u \in R^{m-1}: \sum h_i u_i > h_0\}$, where h_i and h_0 are as described in eqs. (8.5) and (8.6). If $U^{k+1} = \phi$, let $t = t + 1$ and go to step 1. Otherwise, find the centroid $\mathbf{u}^{k+1}$ of U^{k+1}, let $k = k + 1$, and go to step 2.

Counter measure 2

In this measure, the component reliabilities in the infeasible solution given by the HNNN algorithm are decreased in the direction that is perpendicular to the tangential plane of the objective function at the solution $(\mathbf{x}^*, \mathbf{r}^*)$. The following algorithm is a stepwise description of the measure.

Algorithm CSG2

- Step 0: Compute $d_j = \partial f(\mathbf{x}, \mathbf{r})/\partial r_j$ for $j = 1, \ldots, n$, at $(\mathbf{x}^*, \mathbf{r}^*)$ given by the HNNN algorithm. Find $\theta^u = \min\{r_j^*/d_j: 1 \leq j \leq n\}$ and obtain $\mathbf{r}^R = \mathbf{r}^*$ and $\mathbf{r}^L = \mathbf{r}^* - \theta \mathbf{d}$, where $\mathbf{d} = (d_1, \ldots, d_n)^T$ and $0 < \theta < \theta^u$.

- Step 1: $\bar{\mathbf{y}} = \frac{1}{2}(\mathbf{r}^R + \mathbf{r}^L)$. If $(\mathbf{x}^*, \bar{\mathbf{y}})$ is not feasible for Problem 8.3, then let $\mathbf{r}^R = \bar{\mathbf{y}}$, and repeat step 1. Otherwise, go to step 2.

- Step 2: If

$$\left| 1 - \frac{f(\mathbf{x}^*, \mathbf{r}^L)}{f(\mathbf{x}^*, \bar{\mathbf{y}})} \right| > \epsilon,$$

let $\mathbf{r}^L = \bar{\mathbf{y}}$, and go to step 1. Otherwise, take $(\mathbf{x}^*, \bar{\mathbf{y}})$ as the optimal solution of Problem 8.3 and stop.

8.6.1   The DP approach to solve surrogate problem $S(\mathbf{u})$

The problem $S(\mathbf{u})$ can be rewritten so as to maximize

$$z_u = \prod_{j=1}^{n} R_j(x_j, r_j),$$

subject to

$$\sum_{j=1}^{n} \overline{g}_j(x_j, r_j) \le \sum_{i=1}^{m-1} u_i(b_i - b_m) + b_m,$$

$$0 \le r_j \le 1, \quad j = 1, \ldots, n,$$

x_j being a nonnegative integer, where

$$\overline{g}_j(x_j, r_j) = g_{mj}(x_j, r_j) + \sum_{i=1}^{m-1} u_i[g_{ij}(x_j, r_j) - g_{mj}(x_j, r_j)].$$

Select a large integer M and find

$$\Delta = \frac{\sum\limits_{i=1}^{m-1} u_i(b_i - b_m) + b_m}{M}.$$

Let

$$F_k(d) = \max\left\{ \prod_{j=1}^{k} R_j(x_j, r_j) : \sum_{j=1}^{k} \overline{g}_j(x_j, r_j) \le d\Delta, 0 \le r_j \le 1 \text{ and} \right.$$

$$\left. x_j \in \{0, 1, 2, \ldots\} \text{ for } 1 \le j \le k \right\}, \tag{8.7}$$

$$\phi_k(d) = \max\{R_k(x_k, r_k) : \overline{g}_k(x_k, r_k) \le d\Delta, x_k \in \{0, 1, 2, \ldots\} \text{ and } 0 \le r_k \le 1\}, \tag{8.8}$$

for $d = 0, 1, \ldots, M$.

When Δ is sufficiently small, we have the approximate relation

$$F_{k+1}(d) \approx \max_{0 \le i \le d} \{\phi_{k+1}(i) + F_k(d - i)\}, \tag{8.9}$$

for $k = 0, \ldots, n - 1$, where $F_0(d) = 0$ for any d.

Thus $\phi_j(d)$, and the values of x_j and r_j that give $\phi_j(d)$, can be obtained as follows. Since the functions R_j and $\overline{g}_j$ are monotonically nondecreasing in r_j for fixed x_j, the maximum value of r_j that satisfies $0 \le r_j \le 1$ and $\overline{g}_j(x_j, r_j) \le d$ maximizes $R_j(x_j, r_j)$. For each feasible nonnegative integer value of x_j, find $r'_j = \max\{r_j : \overline{g}_j(x_j, r_j) \le d, 0 \le r_j \le 1\}$ and find the value of x_j that gives the highest value of $R_j(x_j, r_j)$ that is equal to $\phi_j(d)$. Equation (8.9) can be used as a recursive relation in the DP approach to solve the problem $S(\mathbf{u})$.

Table 8.5. Initial parameters used in solving Problem 8.1 by the
surrogate constraints method

j	r^L	$\overline{y}$	r^R
1	0.611 411	0.697 693	0.783 974
2	0.563 872	0.719 261	0.874 649
3	0.659 155	0.780 382	0.901 608
4	0.433 272	0.577 696	0.722 120
5	0.649 961	0.724 492	0.799 023

Solution to Example 8-1

The surrogate constraints method is now applied to Example 8-1. The final solution
yielded by the HNNN algorithm is

$$(\mathbf{x}^*, \mathbf{r}^*) = (3, 2, 2, 3, 3, \ 0.783\,974, 0.874\,649, 0.901\,608, 0.722\,120, 0.799\,023),$$

which is an optimal solution of a surrogate problem with $(u_1, u_2) = (0.030, 0.488)$. In
the DP approach to solving the surrogate problem, M is fixed at 100. For the solution
$(\mathbf{x}^*, \mathbf{r}^*)$,

$$\sum_{j=1}^{5} g_{1j}(x_j^*, r_j^*) - b_1 = -27.0,$$

$$\sum_{j=1}^{5} g_{2j}(x_j^*, r_j^*) - b_2 = 9.073\,506,$$

$$\sum_{j=1}^{5} g_{3j}(x_j^*, r_j^*) - b_3 = -7.518\,918,$$

and the system reliability is 0.936 562. It may be noted that the solution violates the
second constraint. Let us use counter measure 2 to derive a feasible solution. The
partial derivatives at point $(\mathbf{x}^*, \mathbf{r}^*)$ are

$$d_1 = \frac{\partial f(\mathbf{x}, \mathbf{r})}{\partial r_1} = 0.132\,456, \qquad d_2 = \frac{\partial f(\mathbf{x}, \mathbf{r})}{\partial r_2} = 0.238\,546,$$

$$d_3 = \frac{\partial f(\mathbf{x}, \mathbf{r})}{\partial r_3} = 0.186\,102, \qquad d_4 = \frac{\partial f(\mathbf{x}, \mathbf{r})}{\partial r_4} = 0.221\,714,$$

$$d_5 = \frac{\partial f(\mathbf{x}, \mathbf{r})}{\partial r_5} = 0.114\,417, \quad \text{and} \quad \theta^u = 3.256\,994.$$

The value of θ is fixed to $0.4(\theta^u) = 1.302\,798$ and the value of ϵ is taken as 10^{-6}. The
initial r^L, $\overline{y}$, and r^R are given in Table 8.5.

The solution $(\mathbf{x}^*, \overline{\mathbf{y}})$ is feasible and

$$\left| 1 - \frac{f(\mathbf{x}^*, \mathbf{r}^L)}{f(\mathbf{x}^*, \overline{\mathbf{y}})} \right| = |1 - 0.683\,369| = 0.316\,631 > \epsilon.$$

Therefore, let $\mathbf{r}^L = \bar{\mathbf{y}}$ and subsequently update $\bar{\mathbf{y}}$ as $\bar{\mathbf{y}} = \frac{1}{2}(\mathbf{r}^L + \mathbf{r}^R)$. The final solution obtained through 19 iterations is

$$(\mathbf{x}^*, \mathbf{r}^*) = (3, 2, 2, 3, 3, \ 0.780\,040, 0.867\,565, 0.896\,081, 0.715\,536, 0.795\,625),$$

and the corresponding system reliability is $0.931\,357$. The residual resources for this solution are

$$b_1 - \sum_{j=1}^{5} g_{1j}(\mathbf{x}^*, \mathbf{r}^*) = 27.0,$$

$$b_2 - \sum_{j=1}^{5} g_{2j}(\mathbf{x}^*, \mathbf{r}^*) = 0.000\,416,$$

$$b_3 - \sum_{j=1}^{5} g_{3j}(\mathbf{x}^*, \mathbf{r}^*) = 7.518\,918.$$

The solution derived by Hikita et al. [126] using the same method is

$$(\mathbf{x}^*, \mathbf{r}^*) = (3, 2, 2, 3, 3, \ 0.774\,887, 0.870\,065, 0.898\,549, 0.716\,524, 0.791\,368),$$

and the corresponding system reliability is $0.931\,451$. The corresponding residual resources are

$$b_1 - \sum_{j=1}^{5} g_{1j}(\mathbf{x}^*, \mathbf{r}^*) = 27.0,$$

$$b_2 - \sum_{j=1}^{5} g_{2j}(\mathbf{x}^*, \mathbf{r}^*) = 0.108\,244,$$

$$b_3 - \sum_{j=1}^{5} g_{3j}(\mathbf{x}^*, \mathbf{r}^*) = 7.518\,918.$$

$\epsilon = 10^{-8}$ and $M = 100$ were selected in the application of the surrogate constraints method to Example 8-1.

8.7   Evolutionary algorithms

Evolutionary algorithms are quite useful for solving complex discrete optimization problems, see Chapter 7, and for solving the reliability–redundancy allocation problem, although the problem involves continuous variables. Such algorithms provide a good heuristic approach, and can play a significant role (see Böck [25] and Böck et al. [26]) in the development of a solution methodology for reliability–redundancy allocation problems in complex systems. A major advantage of these algorithms is that they simultaneously provide several good solutions. Hsieh et al. [131] have designed a genetic algorithm for this type of problem representing the component reliability by a 16-digit binary string and the redundancy level by an eight-digit binary string. Similarly, Prasad and Kuo [270] have developed a two-phase evolutionary algorithm, where each chromosome of the evolutionary algorithm consists of integers representing the redundancy levels and real values representing the component reliabilities. For various terms used in this section, one may refer to Chapter 7 on genetic algorithms.

8.7.1   The genetic algorithm for reliability–redundancy optimization

We now describe the genetic algorithm of Hsieh et al. [131] for the reliability–redundancy allocation problem.

A chromosome is a string of binary digits. The string consists of a substring for each of n stages of the reliability system. Each substring in turn contains two parts: (1) an eight-digit string for the redundancy level, and (2) a 16-digit string for component reliability. Thus the chromosome size becomes $24n$. The second part of each substring is similar to the binary-string representation of real values described by Michalewicz [217]. A string $\mathbf{r}$ of 16 binary digits represents an integer d between 0 and $2^{16} - 1$, and the ratio $d/(2^{16} - 1)$ gives a value in the interval $[0, 1]$. Suppose the lower and upper limits of the redundancy level at stage j are 1 and c, respectively. Let w be the integer represented (in binary form) by the first part of the substring corresponding to stage j. Note that the range of w is 0 to $2^8 - 1 (= 255)$. The redundancy level x_j at stage j is taken as the integer nearest to the value $1 + w(c - 1)/255$. The evolutionary process is similar to that described in Section 7.1. A single-cut crossover operation is performed on the chromosomes. The fitness value of a chromosome is the system reliability for the corresponding solution. If the solution is infeasible, then the fitness value is taken as zero.

Hsieh et al. [131] have demonstrated their genetic algorithm by solving Example 8-1 with the help of a MATLAB code. The parameters used in the demonstration are

population size	200
number of generations	500
crossover rate, p_c	0.85
mutation rate, p_m	0.03

The best solution obtained by the genetic algorithm is

$$(x_1, x_2, x_3, x_4, x_5) = (3, 2, 2, 3, 3),$$

$$(r_1, r_2, r_3, r_4, r_5) = (0.779\,427, 0.869\,482, 0.902\,674, 0.714\,038, 0.786\,896),$$

and the corresponding system reliability is 0.931 578, which is higher than the system reliability of 0.931 451 reported by Hikita et al. [126].

Hsieh et al. [131] have also used the genetic algorithm to solve Example 8-2 in the same way, except that the number of generations is taken as 100 instead of 500. The best solution obtained by them is

$$(x_1, x_2, x_3, x_4, x_5) = (3, 3, 3, 3, 1),$$

$$(r_1, r_2, r_3, r_4, r_5) = (0.814\,090, 0.864\,614, 0.890\,291, 0.701\,190, 0.734\,731),$$

and the corresponding system reliability is $0.999\,8792$, which is larger than the value $0.999\,7894$ obtained by the heuristic method of Hikita et al. [126] for the same problem.

8.7.2   The evolutionary algorithm for reliability–redundancy optimization

For large-scale problems, the genetic algorithm described above requires large computer memory since the length of each chromosome is 24 times the number of subsystems. This may reduce the scope of the algorithm for solving real-life problems. To overcome this difficulty, Prasad and Kuo [268] have developed the following two-phase evolutionary algorithm.

Let n be the number of stages in the reliability system. For component reliability r_j and redundancy level x_j at stage j, $1 \le j \le n$, the chromosome is defined as

$$(x_1, x_2, \ldots, x_n, r_1, r_2, \ldots, r_n).$$

It may be noted that the chromosome consists of two parts: one for redundancy levels and the other for component reliabilities. The length of either part is equal to the number of subsystems.

The first phase is very similar to the evolution process described in Section 7.1, except that the crossover operation is uniform instead of swapping with a single cut. First, an initial population of s chromosomes (solutions) is randomly generated. Let ℓ_j and u_j denote the lower and upper limits of the redundancy level at stage j. While generating each chromosome of the initial population, the redundancy level at stage j is generated by taking a random number from the set $\{\ell_j, \ell_j + 1, \ldots, u_j\}$ and reliability at any stage is generated by taking a five-decimal random number from the interval $(0.5, 1)$. The probability of selecting any particular chromosome as a parent in the crossover operation is p_c. The selected chromosomes are paired up for the crossover operation. A uniform crossover operation is performed for each pair of chromosomes. In the crossover operation on a pair of chromosomes $(x_1', x_2', \ldots, x_n', r_1', r_2', \ldots, r_n')$ and $(x_1'', x_2'', \ldots, x_n'', r_1'', r_2'', \ldots, r_n'')$, gene x_j' is equally likely to go to either offspring. If one of the offspring receives gene x_j', then the other one receives gene x_j''. The genes r_j' and r_j'' are also allocated to the offspring in the same random fashion. For example, a crossover operation on the chromosome pair

$$\{(3, 2, 3, 2, 2, \ 0.752\,34, 0.870\,69, 0.902\,645, 0.710\,15, 0.771\,09),$$

$$(2, 3, 2, 3, 1, 0.808\,27, 0.751\,07, 0.902\,645, 0.710\,15, 0.824\,56)\},$$

may yield the pair of offspring

$$\{(\mathbf{2}, 2, 3, \mathbf{3}, 2, \ 0.752\,34, \mathbf{0.751\,07}, 0.902\,645, 0.710\,15, \mathbf{0.824\,56}),$$

$$(\mathbf{3}, 3, 2, \mathbf{2}, 1, 0.808\,27, \mathbf{0.870\,69}, 0.902\,645, 0.710\,15, \mathbf{0.771\,09})\}.$$

The genes in bold font among the first offspring have come from the second parent and the rest have come from the first parent. Similarly, the genes in bold font in the second offspring have come from the first parent.

Mutation is performed on all of the chromosomes of the population and also on the newly created offspring. The probability that any particular gene in a chromosome is mutated is p_m. If gene x_j is to be mutated, then x_j is replaced with equal chance by one of the values in the set $\{\ell_j, \ell_j + 1, \ldots, u_j\} \backslash \{x_j\}$. The probability that a gene r_j is mutated is also p_m. If a gene r_j is to be mutated, it is replaced by a random number drawn from the interval $(0.5, 1.0)$. This interval is chosen on the basis of a realistic assumption that in any good solution, the component reliability is at least 0.5.

After performing crossover and mutation operations, all parents and children are evaluated and the best s individuals among them are selected, without duplication, to form the population of the next generation. To simplify the notation, let us denote the string $(x_1, x_2, \ldots, x_n)$ by $\mathbf{x}$ and the string $(r_1, r_2, \ldots, r_n)$ by $\mathbf{r}$. The fitness value of a chromosome $(\mathbf{x}, \mathbf{r})$ is

$$R(\mathbf{x}, \mathbf{r}) P(\mathbf{x}, \mathbf{r}),$$

where $P(\mathbf{x}, \mathbf{r})$ is a multiplier which depends on deviation of the chromosome from the feasible region. Let

$$d(\mathbf{x}, \mathbf{r}) = \max\left(0, \max_{1 \leq i \leq m}\left[\frac{1}{b_i}\sum_{j=1}^{n} g_{ij}(x_j, r_j) - 1\right]\right).$$

The penalty function $P(\mathbf{x}, \mathbf{r})$ is defined as

$$P(\mathbf{x}, \mathbf{r}) = \begin{cases} 1, & \text{if } d(\mathbf{x}, \mathbf{r}) = 0, \\ (0.8)[1 - d(\mathbf{x}, \mathbf{r})]^2, & \text{if } 0 < d(\mathbf{x}, \mathbf{r}) < 1, \\ 0, & \text{if } d(\mathbf{x}, \mathbf{r}) \geq 1. \end{cases}$$

The second phase is the same as the first one except for a change in the mutation operation. In the second phase, the mutation operation on r_j is replaced by a random perturbation of r_j. If a gene r_j is selected for mutation, then it is replaced by a value drawn randomly from the interval $[r_j - \delta, r_j + \delta]$, where δ is a small predetermined value. The mutation operation is done only on genes $r_1, r_2, \ldots, r_n$. Consider a chromosome

$$(2, 3, 2, 2, 3, 0.712, 0.805, 0.900, 0.751, 0.782).$$

If δ is chosen as 0.009, it is possible that this chromosome is transformed in the second phase as

$$(2, 3, 2, 2, 3, 0.712\,56, 0.805, 0.900, 0.7584, 0.782).$$

In this transformation, the values 0.712 and 0.751 are changed to 0.712 56 and 0.7584, respectively. The second phase is carried out for a specified number of generations.

Table 8.6. Frequency distribution for v and $G(v)$

v	$G(v)$	$(G(v)/300) \times 100$
0.924	273	91.0
0.925	243	81.0
0.926	240	80.0
0.927	233	77.7
0.928	217	72.3
0.929	183	61.0
0.930	137	45.7
0.931	105	35.0
0.9315	35	11.7
0.9316	11	3.7
0.931 66	3	1.0

The evolution algorithm is applied to solve Example 8-2. In this application, the population size $s = 100$, $p_c = 0.5$, $p_m = 0.1$, and $\delta = 0.009$. All the values generated for genes $r_1, \ldots, r_5$ contain five decimal places. The first and second phases are carried out for 50 generations each. In the second phase, p_m is taken as 0.25. It may be noted that the performance of any evolutionary algorithm is reflected in the probability distribution of the best objective value given. For this reason, the algorithm is run 300 times. Some important statistics of the best objective value given by the algorithm are

number of trials	300
minimum	0.919 26
maximum	0.931 672
average μ	0.928 796
median	0.929 547
standard deviation σ	0.002 945
coefficient of variation σ/μ	0.003 17

Let $G(v)$ denote the number of trials in which the best objective value is at least v. The number $G(v)$ and the corresponding percentage $[G(v)/300] \times 100$ are given in Table 8.6 for a few selected values of v.

The performance and the computational effort of the algorithm will increase with population size. However, the expected computational effort will increase only linearly with the population size for fixed values of genetic parameters p_c and p_m. The algorithm is run 20 times using a code in C on a Pentium 166 PC for the population size 350 (instead of 100). The average time required for each run is 13.4 s. The ten best solutions obtained in one of the runs are described in Table 8.7. It may be noted that the first solution in Table 8.7 is the best solution ever obtained for Example 8-1. In fact, the best objective value is more than 0.931 66 in 17 runs.

Table 8.7. Ten best solutions for Example 8-1 with $s = 350$

	$(x_1, x_2, x_3, x_4, x_5)$	$(r_1,$	$r_2,$	$r_3,$	$r_4,$	$r_5)$	R_s
1	(3, 2, 2, 3, 3)	(0.779 78, 0.872 32, 0.902 45, 0.710 81, 0.788 16)					0.931 678
2	(3, 2, 2, 3, 3)	(0.779 78, 0.872 32, 0.902 61, 0.710 60, 0.788 29)					0.931 673
3	(3, 2, 2, 3, 3)	(0.780 02, 0.872 32, 0.902 45, 0.710 60, 0.788 16)					0.931 660
4	(3, 2, 2, 3, 3)	(0.780 02, 0.872 32, 0.902 45, 0.710 60, 0.788 16)					0.931 660
5	(3, 2, 2, 3, 3)	(0.779 78, 0.872 32, 0.902 61, 0.710 60, 0.788 16)					0.931 657
6	(3, 2, 2, 3, 3)	(0.779 78, 0.872 32, 0.902 61, 0.710 60, 0.788 16)					0.931 657
7	(3, 2, 2, 3, 3)	(0.779 78, 0.872 32, 0.902 61, 0.710 60, 0.788 16)					0.931 657
8	(3, 2, 2, 3, 3)	(0.779 78, 0.872 32, 0.902 61, 0.710 60, 0.788 16)					0.931 657
9	(3, 2, 2, 3, 3)	(0.779 78, 0.872 32, 0.902 45, 0.710 60, 0.788 29)					0.931 644
10	(3, 2, 2, 3, 3)	(0.779 78, 0.872 32, 0.902 45, 0.710 60, 0.788 29)					0.931 644

Table 8.8. Output from evolutionary algorithm for Example 8-2

	$(x_1, x_2, x_3, x_4, x_5)$	$(r_1,$	$r_2,$	$r_3,$	$r_4,$	$r_5)$	R_s
1	(3, 3, 2, 3, 2)	(0.812 927, 0.865 209, 0.917 338, 0.699 618, 0.618 505)					0.999 861
2	(3, 3, 3, 3, 1)	(0.816 552, 0.863 207, 0.849 530, 0.725 504, 0.711 417)					0.999 887
3	(2, 3, 3, 3, 2)	(0.803 311, 0.852 606, 0.904 196, 0.712 127, 0.783 241)					0.999 846
4	(3, 3, 3, 2, 2)	(0.780 391, 0.890 740, 0.861 650, 0.729 695, 0.771 508)					0.999 836
5	(3, 3, 3, 3, 1)	(0.790 568, 0.870 927, 0.862 468, 0.714 822, 0.803 516)					0.999 883
6	(3, 3, 3, 3, 1)	(0.816 871, 0.874 474, 0.857 720, 0.697 376, 0.775 270)					0.999 888
7	(3, 3, 3, 2, 2)	(0.808 821, 0.889 717, 0.850 714, 0.709 907, 0.777 374)					0.999 835
8	(3, 3, 3, 3, 1)	(0.822 947, 0.858 445, 0.852 919, 0.733 247, 0.517 912)					0.999 874
9	(3, 3, 3, 3, 1)	(0.830 530, 0.884 357, 0.840 484, 0.678 928, 0.714 060)					0.999 882
10	(2, 3, 3, 3, 2)	(0.813 407, 0.853 004, 0.897 191, 0.716 796, 0.769 553)					0.999 848
11	(3, 3, 3, 2, 2)	(0.825 170, 0.890 624, 0.840 974, 0.710 697, 0.718 775)					0.999 833
12	(3, 3, 3, 3, 1)	(0.819 390, 0.860 914, 0.859 580, 0.719 267, 0.736 710)					0.999 888
13	(3, 3, 2, 3, 2)	(0.827 021, 0.871 494, 0.912 209, 0.678 127, 0.630 035)					0.999 864
14	(2, 3, 3, 3, 2)	(0.836 642, 0.850 744, 0.890 624, 0.717 221, 0.732 804)					0.999 847
15	(3, 3, 3, 3, 1)	(0.822 221, 0.886 687, 0.858 567, 0.643 569, 0.819 631)					0.999 872
16	(2, 3, 3, 3, 2)	(0.844 401, 0.877 577, 0.882 287, 0.652 621, 0.791 928)					0.999 840
17	(3, 3, 3, 3, 1)	(0.778 288, 0.874 613, 0.860 734, 0.716 315, 0.806 666)					0.999 877
18	(3, 3, 2, 3, 2)	(0.822 320, 0.877 053, 0.908 158, 0.666 197, 0.703 984)					0.999 864
19	(3, 3, 2, 3, 2)	(0.815 080, 0.860 326, 0.906 859, 0.702 932, 0.710 325)					0.999 858
20	(3, 3, 3, 3, 1)	(0.807 861, 0.865 399, 0.866 122, 0.719 674, 0.747 193)					0.999 889

The algorithm is also used to solve Example 8-2 in the same way as the previous example. The population size is $s = 100$ and other parameters are the same as described above. The algorithm is run 20 times to see the consistency of the output. The best solution of each trial is given in Table 8.8.

8.8   Discussion

When there is an option to select both component reliability (on a continuous scale) and redundancy level at all or some stages of a system, then the reliability optimization problem becomes a nonlinear mixed integer programming problem. Such a problem is called a reliability–redundancy allocation problem. Researchers in system-reliability optimization have proposed various heuristic methods to solve this problem. Misra and Ljubojevic [229] proposed a method in which a continuous version of the problem is solved by an NLP method and the variables $x_1, \ldots, x_n$, representing redundancy levels, are rounded off to the nearest integer values. The approach of Tillman et al. [304] is to maximize $T(r_1, \ldots, r_n)$ over all the feasible choices of component reliabilities $r_1, \ldots, r_n$, where $T(r_1, \ldots, r_n)$ is the maximum possible system reliability over the feasible choices of $x_1, \ldots, x_n$ for fixed $(r_1, \ldots, r_n)$. The determination of $T(r_1, \ldots, r_n)$ is a redundancy allocation problem which can be solved by any good heuristic method described in Chapter 3. Tillman et al. [304] used the Hooke–Jeeves search method (Appendix 2) to maximize $T(r_1, \ldots, r_n)$. In fact, one can use any good direct search method for maximization of $T(r_1, \ldots, r_n)$ and any good heuristic method for the subproblem of redundancy allocation. Gopal et al. [113] adopted a conceptually similar approach to solve the reliability–redundancy allocation problem. However, they developed a simple search method to maximize $T(r_1, \ldots, r_n)$ and used the heuristic method of Gopal et al. [112] for the determination of $T(r_1, \ldots, r_n)$ for fixed $(r_1, \ldots, r_n)$.

Xu et al. [326] numerically compared the performance of eight combinations of two search methods and four heuristic redundancy allocation methods. To solve reliability–redundancy allocation problems, they also proposed a heuristic method that passes through a sequence of redundancy allocation vectors $(x_1, \ldots, x_n)$ obtaining the best component reliability vector $(r_1, \ldots, r_n)$ by an NLP method for each $(x_1, \ldots, x_n)$. Kuo et al. [178] proposed a branch-and-bound method in which, at each node, an NLP problem is solved by the Lagrange multipliers method. Branching at any node is based on the noninteger value of a variable x_j in the corresponding NLP solution.

Hikita et al. [126] adopted a surrogate constraint method to solve problems in which the constraints are separable and the objective function is suitable for the multi-stage decision-making approach. This method involves solving a sequence of surrogate problems which are single-constraint NLP problems. They solve each surrogate problem by the DP approach. Hsieh et al. [131] and Prasad and Kuo [270] demonstrated that evolutionary algorithms can also be successfully adopted for reliability–redundancy allocation. Interested readers may refer to Floudas [94] for other methods, such as the Benders decomposition method, the outer approximation method, etc., for solving nonlinear mixed integer programming problems.

EXERCISES

8.1 What is a typical reliability–redundancy allocation problem? Which category of mathematical formulation does this problem belong to?

8.2 In Example 8-1, let $\mathbf{x}^0 = (3, 3, 2, 2, 2)$ and $\mathbf{r}^0 = (0.5, 0.5, 0.5, 0.5, 0.5)$. Perform one iteration of the GAG2 method. Find r and sensitivity factors $S_j(x^*, r)$ for $j = 1, 2, 3, 4, 5$.

8.3 In Example 8-1, let $\mathbf{x}^0 = (3, 2, 3, 2, 2)$ and $\mathbf{r}^0 = (0.762\,786, 0.866\,451, 0.828\,722, 0.776\,427, 0.848\,162)$. Perform one iteration of the XKL method. Find $\hat{S}, r$, and the value of the sensitivity function S_j.

8.4 In the KLXZ method, what kind of branching rule is used? Describe the case in which a node is fathomed. What is the main idea of the THK method?

8.5 Derive the constraint

$$\sum_{j=1}^{n} \left\{ \sum_{i=1}^{m-1} u_i \left[g_{ij}(x_j, r_j) - g_{mj}(x_j, r_j) + g_{mj}(x_j, r_j) \right] \right.$$
$$\left. - \left[\sum_{i=1}^{m-1} u_i (b_i - bm) + b_m \right] \right\} \leq 0$$

from $\sum_{i=1}^{m} u_i \left[\sum_{j=1}^{n} g_{ij}(x_j, r_j) - b_i \right] \leq 0$ (see pp. 181–182).

8.6 Show the surrogate problem formulation S(**u**) of Example 8-1 and find the recursive function of the dynamic programming approach.

8.7 In the surrogate constraints method, what methods are used to avoid infeasibility. Discuss how these methods avoid infeasibility.

8.8 When the usual genetic algorithm is applied to a large-scale MINLP, it has a major difficulty. Hence, Prasad and Kuo have developed the two-phase evolutionary algorithm. How do they overcome this difficulty?

8.9 In order to solve mixed integer nonlinear programming problems, what kind of methods are often used? Describe briefly each of them.

8.10 In Chapter 8, two methods are usually applied for each algorithm: one is for redundancy level **x**, the other is for reliability **r**. Discuss what kinds of methods are proper for each algorithm.

8.11 In solving Example 8-1, several methods are presented in Sections 8.2–8.5. For example, Tillman et al. [304] used the Hooke–Jeeves pattern search technique in conjunction with a suggested heuristic redundancy method proposed by Aggarwal et al. [7] to handle the mixed integer reliability optimization problem. The search method determines the optimal level of the component reliabilities, while the heuristic method ensures the integer number of redundancies. Other search techniques can be combined with the heuristic redundancy allocation methods as depicted in the figure below.

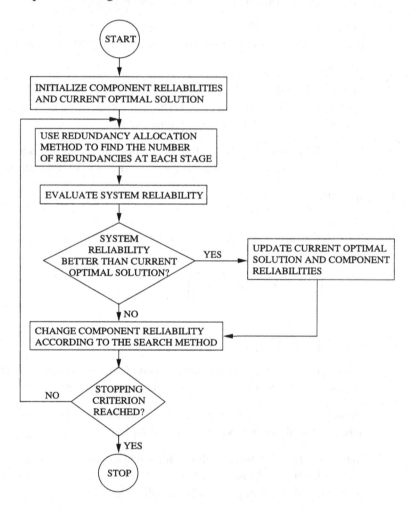

(a) Use the Nakagawa–Nakashima heuristic [247] to replace the heuristic of Aggarwal et al. [7].

(b) Use the Kohda–Inoue heuristic [163] to replace the heuristic of Aggarwal et al. [7].

(c) Use the genetic algorithm to replace the heuristic of Aggarwal et al. [7].

(d) Run the same problem 500 times using the above three methods with the different coefficients listed in Table 8.1. Compare the solutions and computational times of these methods.

9. Component assignment in reliability systems

9.1 Introduction

Consider a reliability system with components that can perform multiple functions: these components are sometimes interchangeable. The situation of interchangeability also arises when identical components of different ages are used in a system: the overall reliability will then depend on the optimal assignment of components to various positions in the system. Derman et al. [77], [78] initially considered the problem of optimal component assignment in reliability systems, with the primary objective of maximizing system reliability. Later, developments emerged for series-parallel and parallel-series systems, consecutive k-out-of-n systems, and general coherent systems. From a mathematical programming point of view, component assignment can be viewed as a nonlinear problem. The majority of the work on component assignment assumes that component reliabilities are position-invariant and is devoted to analytical derivation, which depends only on the increasing order of components. Heuristics and implicit enumeration methods have also been developed by some authors for optimal components assignment. Also, the theory of majorization and the nature of Schur-convex functions have been elegantly exploited to derive analytical solutions for series-parallel and parallel-series systems: a few heuristics have also been based on the majorization concept and mathematical programming methodology. Some work has been done on series-parallel and parallel-series systems with position-dependent component reliabilities.

Section 9.2 deals with the optimal assignment of components to series-parallel systems. It contains exact methods for some special cases and heuristics for the general problem. Section 9.2 also contains an algorithm based on a bicriteria approach for optimal component assignment in series-parallel systems with two path sets and general component reliabilities. Section 9.3 discusses the optimal component assignment in parallel-series systems. Heuristics as well as exact methods are presented for parallel-series systems in this section. Section 9.4 describes component assignment problems in coherent systems.

9.2 Optimal assignment of components in series-parallel systems

Consider a series-parallel reliability system with k series subsystems $P_1, \ldots, P_k$ connected in parallel. Each series subsystem is a minimal path set and vice versa. Let n_h be the number of positions in path set P_h for $h = 1, \ldots, k$, $N_h = \sum_{i=1}^{h} n_i$, for $h = 1, \ldots, k$, and $n = N_k$, that is, $n = n_1 + \cdots + n_k$. Number the n_1 positions of P_1 as $1, 2, \ldots, n_1$, the n_2 positions of P_2 as $N_1 + 1, \ldots, N_1 + n_2, \ldots$, and the n_k positions of P_k as $N_{k-1} + 1, \ldots, N_{k-1} + n_k$. There are m ($\geq n$) components to be fixed in n positions of the reliability system. The reliability of component j is r_{ij} when it is fixed in the ith position of the system.

Let $x_{ij} = 1$ if component j is assigned to position i and 0 otherwise. Then, we can represent component assignment by vector $\mathbf{x} = (x_{11}, x_{12}, \ldots, x_{1m}, \ldots, x_{n1}, \ldots, x_{nm})$. Note that any assignment $\mathbf{x}$ must satisfy the conditions

$$\sum_{j=1}^{m} x_{ij} = 1, \qquad \text{for } i = 1, \ldots, n, \tag{9.1}$$

$$\sum_{i=1}^{n} x_{ij} \leq 1, \qquad \text{for } j = 1, \ldots, m, \tag{9.2}$$

$$x_{ij} = 0 \text{ or } 1. \tag{9.3}$$

For an assignment $\mathbf{x}$, the system reliability is

$$R(\mathbf{x}) = 1 - \prod_{h=1}^{k} \left(1 - \prod_{i \in P_h} \prod_{j=1}^{m} r_{ij}^{x_{ij}} \right).$$

Problem 9.1

Mathematically, the problem of maximizing the reliability of a series-parallel system through component assignment is to maximize

$$R(\mathbf{x}) = 1 - \prod_{h=1}^{k} \left(1 - \prod_{i \in P_h} \prod_{j=1}^{m} r_{ij}^{x_{ij}} \right),$$

subject to

$$\sum_{j=1}^{m} x_{ij} = 1, \qquad \text{for } i = 1, \ldots, n,$$

$$\sum_{i=1}^{n} x_{ij} \leq 1, \qquad \text{for } j = 1, \ldots, m,$$

$$x_{ij} = 0 \text{ or } 1.$$

Let $t_{ij} = -\ln r_{ij}$, $z_h = \sum_{i \in P_h} \sum_{j=1}^{m} t_{ij} x_{ij}$, and $\mathbf{z} = (z_1, z_2, \ldots, z_k)$. Then, we can write the system reliability as

$$R(\mathbf{z}) = 1 - \prod_{h=1}^{k} (1 - e^{-z_h}). \tag{9.4}$$

z_h is called *a-hazard* of path set P_h and vector $\mathbf{z} = (z_1, \ldots, z_k)$ is called *a-hazard vector* of the system.

Let $\mathbf{u} = (u_1, \ldots, u_k)$ and $\mathbf{v} = (v_1, \ldots, v_k)$ be two vectors in R^k, and let $u_{[i]}(v_{[i]})$ be the ith least element among u_i (v_i). Then $\mathbf{u}$ is said to majorize $\mathbf{v}$ (denoted by $\mathbf{u} \succ \mathbf{v}$) if

$$u_{[1]} \le v_{[1]},$$

$$u_{[1]} + u_{[2]} \le v_{[1]} + v_{[2]},$$

$$\vdots$$

$$u_{[1]} + \cdots + u_{[k-1]} \le v_{[1]} + \cdots + v_{[k-1]},$$

$$u_{[1]} + \cdots + u_{[k]} = v_{[1]} + \cdots + v_{[k]}.$$

Let $f(\mathbf{u})$ be a real valued function defined on a nonempty subset S in R^k. The function $f(\mathbf{u})$ is said to be *Schur-convex* if $f(\mathbf{u}) \ge f(\mathbf{v})$ whenever $\mathbf{u} \succ \mathbf{v}$. Similarly, $f(\mathbf{u})$ is said to be *Schur-concave* if $f(\mathbf{u}) \le f(\mathbf{v})$ whenever $\mathbf{u} \succ \mathbf{v}$. El-Neweihi et al. [86] have observed that the reliability function $R(\mathbf{z})$ is a Schur-convex function. The reader may refer to Marshall and Olkin [209] for further details of majorization.

9.2.1   Optimal allocation of components in series-parallel systems

Let $m = n$ and $N = \{1, \ldots, n\}$. Assume that $r_{ij} = r_j$ for any i and j, that is, the component reliability is invariant of the position in which it is used. It can easily be seen that the reliability of any path set P_h depends on the set of components assigned to P_h but not on the specific assignment of these components to the positions of P_h. Therefore, the problem of maximizing the system reliability is reduced to the problem of finding an optimal allocation of components to the k path sets. Let A_h denote the set of components allocated to path set P_h for $h = 1, \ldots, k$. Then the partition $A = (A_1, \ldots, A_k)$ of set N into k subsets with $|A_h| = n_h$ represents the corresponding allocation. Now the problem is to maximize

$$R(A) = 1 - \prod_{h=1}^{k} \left(1 - \prod_{j \in A_h} r_j \right),$$

over the set of all allocations. The a-hazard of component j is $t_j = -\ln r_j$ for $j = 1, \ldots, n$. For any allocation $A = (A_1, \ldots, A_k)$, the a-hazard of path set P_h, is $z_h = \sum_{j \in A_h} t_j$ for $h = 1, \ldots, k$, and the system reliability in terms of a-hazard values is $R(\mathbf{z}) = 1 - \prod_{h=1}^{k} (1 - e^{-z_h})$. Now the problem is to maximize $R(\mathbf{z})$ over all allocations. Note that for every allocation $\sum_{h=1}^{k} z_h = \sum_{j=1}^{n} t_j$. Without loss of generality, it is assumed throughout this section that $n_1 \le n_2 \le \cdots \le n_k$ and $r_1 \ge r_2 \ge \cdots \ge r_n$.

A simple rule for optimal allocation

El-Neweihi et al. [86] have given a simple rule of component allocation in series-parallel systems. Using the majorization concept and the nature of the Schur-convex function, they have shown that the allocation obtained by the rule maximizes the system reliability. The rule is to allocate the n_1 most reliable components to the smallest path set P_1, the next n_2 most reliable components to the next smallest path set $P_2, \ldots$, and finally the n_k least reliable components to the largest path set P_k. The allocation $A^* = (A_1^*, \ldots, A_k^*)$ obtained by the rule is $A_1^* = \{1, 2, \ldots, n_1\}$ and $A_h^* = \{N_{h-1} + 1, \ldots, N_{h-1} + n_h\}$ for $h = 2, \ldots, k$. Let $\mathbf{z}^*$ be the a-hazard vector for allocation A^*. El-Neweihi et al. [86] have shown that $\mathbf{z}^*$ majorizes the a-hazard vector of any other allocation and consequently maximizes the function $R(\mathbf{z})$, which is Schur-convex. It may be noted that the optimal allocation A^* depends on the decreasing order of component reliabilities but not on their actual magnitudes. Such allocations are called invariant optimal allocations, and are discussed in more detail in Section 9.4.4.

Optimal allocation for a more general case

Prasad et al. [272] have derived some theoretical results for a more general case, and presented an exact algorithm and two greedy algorithms for obtaining an optimal allocation. The assumption $r_{ij} = g_i r_j$ holds when the positions of a system are subjected to shocks and the probability of the ith position being shock-free during the intended period is g_i. For an allocation $A = (A_1, A_2, \ldots, A_k)$, the a-hazard of path set P_h is

$$z_h = d_h + \sum_{j \in A_h} t_j,$$

where

$$d_h = -\sum_{i=N_{h-1}+1}^{N_h} \ln g_i, \quad \text{and } t_j = -\ln r_j.$$

Let

$$\alpha_h = \prod_{i=N_{h-1}+1}^{N_h} g_i, \quad \text{that is,} \quad \alpha_h = \exp(-d_h).$$

Note that d_h does not depend on the allocation. In the model considered by El-Neweihi et al. [86], $d_h = 0$ for all h. Even for this case, the reliability of any path set P_h depends on the set of components allocated to P_h but not on the specific assignment of

Table 9.1. Path set sizes and α_h values for Example 9-1

h	n_h	α_h
1	6	0.7524
2	4	0.5350
3	7	0.9781
4	9	0.8142

components to the positions of P_h. The nonzero values of d_h cause difficulty in deriving an optimal allocation analytically.

An allocation A is said to be *ordered* if a permutation $\mathbf{v} = (v_1, \ldots, v_k)$ of $1, \ldots, k$ exists, such that

$$A_{v_1} = \{1, \ldots, n_{v_1}\},$$

$$A_{v_2} = \{n_{v_1} + 1, \ldots, n_{v_1} + n_{v_2}\},$$

$$\vdots$$

$$A_{v_k} = \{[n_{v_1} + \cdots + n_{v_{(k-1)}}] + 1, \ldots, [n_{v_1} + \cdots + n_{v_{(k-1)}}] + n_{v_k}\}.$$

An ordered allocation can be represented by a permutation of $1, 2, \ldots, k$. The system reliability for an ordered allocation $\mathbf{v} = (v_1, \ldots, v_k)$ is denoted by $R(\mathbf{v})$. An ordered allocation is said to be *totally ordered* if $z_{v_1}(\mathbf{v}) \leq z_{v_2}(\mathbf{v}) \leq \cdots \leq z_{v_k}(\mathbf{v})$, where $z_{v_h}(\mathbf{v})$ is the a-hazard of path set P_{v_h} for allocation $\mathbf{v}$.

Example 9-1

Let $k = 4$, $n = 26$ and path set sizes and α_h values be as given in Table 9.1. Let the reliabilities of 26 components be 0.99, 0.99, 0.97, 0.90, 0.88, 0.87, 0.86, 0.81, 0.79, 0.76, 0.65, 0.65, 0.64, 0.62, 0.62, 0.62, 0.60, 0.59, 0.57, 0.52, 0.52, 0.52, 0.49, 0.48, 0.46, 0.45. Consider the allocation

$$A_2 = \{1, 2, 3, 4\},$$

$$A_1 = \{5, 6, 7, 8, 9, 10\},$$

$$A_3 = \{11, 12, 13, 14, 15, 16, 17\},$$

$$A_4 = \{18, 19, 20, 21, 22, 23, 24, 25, 26\}.$$

Note that this allocation is an ordered one, which can be represented by $(2, 1, 3, 4)$.

Now,

$$z_2 = -\ln(\alpha_2) - \sum_{j=1}^{4} \ln r_j = 0.7815,$$

$$z_1 = -\ln(\alpha_1) - \sum_{j=5}^{10} \ln r_j = 1.4233,$$

$$z_3 = -\ln(\alpha_3) - \sum_{j=11}^{17} \ln r_j = 3.2749,$$

$$z_4 = -\ln(\alpha_4) - \sum_{j=18}^{26} \ln r_j = 6.2659,$$

where $\alpha_h = \prod_{i=N_{h-1}+1}^{N_h} g_i$. Since $z_2 < z_1 < z_3 < z_4$, the allocation $(2, 1, 3, 4)$ is totally ordered. The system reliability for this allocation is

$$1 - \prod_{h=1}^{4}(1 - e^{-z_h}) = 1 - [1 - \exp(-1.4233)][1 - \exp(-0.7815)]$$
$$\times [1 - \exp(-3.2749)][1 - \exp(-6.2659)]$$
$$= 0.6047.$$

Prasad et al. [272] have shown that an optimal allocation which is totally ordered exists and have presented the following two greedy algorithms for obtaining a suboptimal totally ordered allocation.

Algorithm 1
- Step 0: Initialize $\ell = 0$, $I = \{1, \ldots, k\}$, and $h = 1$.
- Step 1: Choose $r \in I$ such that

$$\alpha_r \prod_{j=\ell+1}^{\ell+n_r} r_j = \max_{i \in I} \left(\alpha_i \prod_{j=\ell+1}^{\ell+n_i} r_j \right),$$

where $\alpha_i = \exp(-d_i)$ for $i = 1, 2, \ldots, k$.
- Step 2: Let $u_h = r$ and $h = h + 1$. If $h > k$, go to step 4. Otherwise, go to step 3.
- Step 3: Let $I = I - \{r\}$, $\ell = \ell + n_r$, and go to step 1.
- Step 4: Stop. Take the totally ordered allocation $\mathbf{u} = (u_1, \ldots, u_k)$ as the heuristic one.

Table 9.2. The values of $E_i(\ell)$

i	ℓ			
	0	7	11	17
1	0.4929	0.0989	**0.0448**	–
2	0.4577	**0.1691**	–	–
3	**0.5510**	–	–	–
4	0.2935	0.0255	0.0086	**0.0019**

Note that the value of r chosen from I in step 1 satisfies

$$d_r + \sum_{j=\ell+1}^{\ell+n_r} t_j = \min_{i \in I}\left(d_i + \sum_{j=\ell+1}^{\ell+n_i} t_j\right)$$

and, consequently, we have $z_{u_1} \le z_{u_2} \le \cdots \le z_{u_k}$. In the allocation $\mathbf{u}$, n_{u_1} most reliable components are allocated to P_{u_1}, n_{u_2} next most reliable components are allocated to $P_{u_2}, \ldots$, and finally n_{u_k} least reliable components are allocated to P_{u_k}.

Solution of Example 9-1 from algorithm 1

Let $E_i(\ell) = \alpha_i \prod_{j=\ell+1}^{\ell+n_i} r_j$. The values of $E_i(\ell)$ are given in Table 9.2.

- Iteration 1: Initially, $\ell = 0$, $I = \{1, 2, 3, 4\}$, and $h = 1$. The value of $E_i(0)$ is given in Table 9.2 for each $i \in I$. Since $E_3(0) = 0.5510$ is the maximum among them, u_1 is taken as 3. Now let $\ell = n_{u_1} = n_3 = 7$, $I = \{1, 2, 4\}$, and $h = 2$.

- Iteration 2: The value of $E_i(7)$ is given in Table 9.2 for each $i \in \{1, 2, 4\}$. Since $E_2(7) = 0.1691$ which is the maximum, u_2 is taken as 2. Now let $\ell = n_{u_1} + n_{u_2} = 11$, $I = \{1, 4\}$, and $h = 3$.

- Iteration 3: The value of $E_i(11)$ is given in Table 9.2 for $i = 1$ and 4. Now, $u_3 = 1$ as $E_1(11) > E_4(11)$. Now let $\ell = n_{u_1} + n_{u_2} + n_{u_3} = 17$ and $I = \{4\}$.

- Iteration 4: Since I contains element 4 only let $u_4 = 4$. The value of $E_4(17)$ is 0.0019. The ordered allocation yielded by the algorithm is $(3, 2, 1, 4)$.

Note that $E_1(11)$, $E_2(7)$, $E_3(0)$, and $E_4(17)$ give the reliabilities of path sets P_1, P_2, P_3, and P_4, respectively, for the allocation $(u_1, u_2, u_3, u_4) = (3, 2, 1, 4)$. Therefore, the system reliability for this allocation is

$$1 - [1 - E_1(11)][1 - E_2(7)][1 - E_3(0)][1 - E_4(17)]$$

$$= 1 - (1 - 0.0448)(1 - 0.1691)(1 - 0.5510)(1 - 0.0019)$$

$$= 0.6443.$$

The following greedy algorithm obtains an ordered allocation $\mathbf{v} = (v_1, \ldots, v_k)$ by determining first v_k, next $v_{k-1}, \ldots$, and finally v_1.

Table 9.3. The values of $F_i(\ell)$

i	ℓ			
	26	17	10	6
1	0.0099	0.0448	0.2409	**0.4929**
2	0.0260	0.0765	**0.2237**	–
3	0.0068	**0.0378**	–	–
4	**0.0019**	–	–	–

Algorithm 2
- Step 0: Initialize $\ell = n$, $I = \{1, \ldots, k\}$, and $h = k$.
- Step 1: Choose $s \in I$ such that

$$\alpha_s \prod_{j=\ell-n_s+1}^{\ell} r_j = \min_{i \in I}\left(\alpha_i \prod_{j=\ell-n_i+1}^{\ell} r_j \right).$$

- Step 2: Let $v_h = s$ and $h = h - 1$. If $h = 0$, go to step 4. Otherwise, go to step 3.
- Step 3: Let $I = I - \{s\}$, $\ell = \ell - n_s$ and go to step 1.
- Step 4: Stop. Take the totally ordered allocation $\mathbf{v} = (v_1, \ldots, v_k)$ as the suboptimal allocation.

In the allocation $\mathbf{v}$, the n_{v_k} least reliable components are allocated to P_{v_k}; the next $n_{v_{k-1}}$ least reliable components are allocated to $P_{v_{k-1}}, \ldots$, and finally the n_{v_1} most reliable components are allocated to P_{v_1}.

Solution of Example 9-1 from algorithm 2
Let $F_i(\ell) = \alpha_i \prod_{j=\ell-n_i+1}^{\ell} r_j$. The values of $F_i(\ell)$ are given in Table 9.3.

- Iteration 1: Initially, $\ell = 26$, $I = \{1, 2, 3, 4\}$, and $h = 4$. The value of $F_i(26)$ is given in Table 9.3 for each $i \in I$. Since $F_4(26) = 0.0019$ is the minimum among them, v_4 is taken as 4. Now let $\ell = 26 - n_{v_4} = 17$, $I = \{1, 2, 3\}$, and $h = 3$.
- Iteration 2: The value of $F_i(17)$ is given in Table 9.3 for each $i \in \{1, 2, 3\}$. Since $F_3(17) = 0.0378$ is the minimum among them, v_3 is taken as 3. Now let $\ell = 26 - n_{v_4} - n_{v_3} = 10$, $I = \{1, 2\}$, and $h = 2$.
- Iteration 3: The value of $F_i(10)$ is given in Table 9.3 for $i = 1$ and 2. Now, $v_2 = 2$ as $F_2(10) < F_1(10)$. Now let $\ell = 26 - n_{v_4} - n_{v_3} - n_{v_2} = 6$ and $I = \{1\}$.
- Iteration 4: Since set I contains the element 1 only, let $v_1 = 1$. The value of $F_1(6)$ is 0.4929. The ordered allocation yielded by the algorithm is $(1, 2, 3, 4)$.

For the allocation $(v_1, v_2, v_3, v_4) = (1, 2, 3, 4)$, the reliabilities of path sets P_1, P_2, P_3, and P_4 are $F_1(6), F_2(10), F_3(17)$, and $F_4(26)$, respectively. Therefore, the system reliability for this allocation is

$$R_s = 1 - [1 - F_1(6)][1 - F_2(10)][1 - F_3(17)][1 - F_4(26)]$$

$$= 1 - (1 - 0.4929)(1 - 0.2237)(1 - 0.0378)(1 - 0.0019)$$

$$= 0.6219.$$

It is proved by Prasad et al. [272] that if the two allocations yielded by algorithms 1 and 2 are the same, then they are optimal. It is found through numerical investigation that both algorithms, more often than not, yield the same allocation.

Dominance and admissibility conditions

A path set P_r is preferred to a path set P_s if $n_r \leq n_s$ and $d_r \leq d_s$. If $n_r = n_s$, $d_r = d_s$, and $r < s$, then P_r is preferred to P_s. Suppose an ordered allocation $\mathbf{v}$ is such that if P_r is preferred to P_s, then r precedes s in the permutation $\mathbf{v}$. The allocation $\mathbf{v}$ is called a *nondominated* ordered allocation. Prasad et al. [272] have shown that in order to obtain an optimal allocation, it is enough to consider only the allocations which are totally ordered and nondominated.

Let $(u_1, \dots, u_{r-1})$ be a partial sequence of $1, 2, \dots, k$ and $J = \{1, \dots, k\} \backslash \{u_1, \dots, u_{r-1}\}$. An element $c \in J$ is said to be admissible in stage r with respect to $(u_1, \dots, u_{r-1})$ if

$$d_{u_{r-1}} + \sum_{j=\ell+1}^{\ell + n_{u_{r-1}}} t_j \leq d_c + \sum_{j=\ell+n_{u_{r-1}}+1}^{\ell + n_{u_{r-1}} + n_c} t_j,$$

where $\ell = n_{u_1} + \cdots + n_{u_{r-2}}$. Any element in $\{1, \dots, k\}$ is admissible at stage 1. Note that if $(u_1, \dots, u_k)$ is a totally ordered allocation, then u_h is admissible in stage h with respect to $(u_1, \dots, u_{h-1})$ for $h = 2, \dots, k$ and the converse is also true.

In order to obtain an optimal allocation, Prasad et al. [272] have developed the following algorithm which implicitly enumerates all nondominated totally ordered (NDTO) allocations using admissibility and dominance conditions.

Algorithm 3

- Step 0: Take an NDTO allocation $\mathbf{u} = (u_1, \dots, u_k)$. Renumber the path sets such that the ordered allocation under consideration is $\mathbf{v} = (1, \dots, k)$ and obtain the preference relation set accordingly. Let $R^* = R(\mathbf{v})$ and $u_i = i$ for $i = 1, \dots, k$.

- Step 1: Let $s = k - 1$, $b = v_s$, $J = \{v_k\}$, and $I = \{h: h \in J, h > b\}$.

- Step 2: If $I = \phi$, let $J = J \cup \{b\}$ and go to step 3. Otherwise, take the least element c from I. If an element g exists in J, such that P_g is preferred to P_c, let $I = I \backslash \{c\}$, and repeat step 2. Otherwise, check whether c is admissible in stage s with respect to $(v_1, v_2, \dots, v_{s-1})$. If it is admissible, let $v_s = c$, $J = (J \cup \{b\}) \backslash \{c\}$ and go to step 4. Otherwise, let $I = I \backslash \{c\}$, and repeat step 2.

Table 9.4. Path set sizes and α_h values
for Example 9-1 by algorithm 3

h	n_h	α_h
1	7	0.9781
2	4	0.5350
3	6	0.7524
4	9	0.8142

Table 9.5. NDTO allocations

$(v_1, v_2, v_3, v_4)^a$	R_{v_1}	R_{v_2}	R_{v_3}	R_{v_4}	System reliability
(1, 2, 3, 4)	0.5510	0.1691	0.0448	0.0019	0.6443
(1, 3, 2, 4)	0.5510	0.0989	0.0765	0.0019	0.6271
(2, 1, 3, 4)	0.4577	0.2036	0.0448	0.0019	0.5882
(2, 3, 1, 4)	0.4577	0.2409	0.0378	0.0019	0.6047
(3, 1, 2, 4)	0.4929	0.1106	0.0765	0.0019	0.5843
(3, 2, 1, 4)	0.4929	0.2238	0.0378	0.0019	0.6220

^{a} The allocations are to be read in accordance with component renumbering.

- Step 3: Let $s = s - 1$. If s = 0, go to step 6. Otherwise, let $b = v_s$ and $I = \{h: h \in J, h > b\}$ and go to step 2.

- Step 4: Let $s = s + 1$, take the least element d in J which is admissible in stage s with respect to $(v_1, v_2, \ldots, v_{s-1})$, and let $v_s = d$; go to step 5. If d is not available, go to step 3.

- Step 5: If $s < k$, let $J = J \setminus \{d\}$ and go to step 4. Otherwise, evaluate $R(v_1, \ldots, v_k)$. If $R(v_1, \ldots, v_k) > R^*$, let $R^* = R(v_1, \ldots, v_k)$ and $u_i = v_i$ for $i = 1, \ldots, k$, and go to step 1.

- Step 6: Stop. The NDTO allocation $(u_1, \ldots, u_k)$ is optimal and the corresponding system reliability is R^*.

Solution of Example 9-1 from algorithm 3

Consider the NDTO allocation $(u_1, u_2, u_3, u_4) = (3, 2, 1, 4)$ obtained by algorithm 1 for Example 9-1. Renumber the path sets 1 and 3 as 3 and 1, respectively, such that this allocation becomes (1, 2, 3, 4). Consequently, the α_h and n_h values are as shown in Table 9.4. Note that the path set P_1 is preferred to the path set P_4, and there is no other preference relation among the other path sets. The sequence of NDTO allocations enumerated by algorithm 3 is given in Table 9.5.

The initial ordered allocation $(1, 2, 3, 4)$ is the required optimal allocation, since it is the best among all NDTO allocations. Incidentally, algorithm 1 has also yielded the same allocation.

9.2.2   A heuristic approach for optimal assignment of components

Problem 9.2

For general component reliabilities r_{ij}, the problem of maximizing the reliability of series-parallel systems through component optimal assignment is to maximize

$$R(\mathbf{x}) = 1 - \prod_{h=1}^{k}\left[1 - \exp\left(-\sum_{i \in P_h}\sum_{j=1}^{m} t_{ij}\, x_{ij}\right)\right],$$

subject to

$$\sum_{j=1}^{m} x_{ij} = 1, \qquad \text{for } i = 1, \ldots, n,$$

$$\sum_{i=1}^{n} x_{ij} \leq 1, \qquad \text{for } j = 1, \ldots, m,$$

$$x_{ij} = 0 \text{ or } 1,$$

where $t_{ij} = -\ln r_{ij}$.

This problem is a nonlinear 0–1 programming problem with assignment constraints. Prasad et al. [267] have presented an iterative heuristic algorithm which requires solving $k(k + 1)/2$ assignment problems. Let M be the set of all components available for assignment to the positions of the system. In iteration 1, the algorithm selects a path set that has the highest possible reliability when the components are selected from M and assigned to it. After eliminating the components from M that are to be assigned to the selected path set, the algorithm selects another path set from the remaining path sets following the same criterion. Repetition of this procedure gives an order of path sets and an assignment of components to all positions of each path set. The resulting assignment is taken as the heuristic optimal assignment: the algorithm for this is given below.

Algorithm 4
- Step 0: Let $J = M$ and $H = \{1, \ldots, k\}$.
- Step 1: If H is empty, go to step 4.
- Step 2: For each $h \in H$, minimize

$$z_h = \sum_{i \in P_h}\sum_{j \in J} t_{ij}\, x_{ij},$$

subject to

$$\sum_{j \in J} x_{ij} = 1, \qquad \text{for } i \in P_h,$$

$$\sum_{i \in P_h} x_{ij} \leq 1, \qquad \text{for } j \in J,$$

$$x_{ij} = 0 \text{ or } 1, \qquad \text{for } i \in P_h \text{ and } j \in J.$$

Let $x_{ij} = x_{ij}^0$ for $i \in I_h$, $j \in J$ be the corresponding optimal solution, and z_h^0 be the minimum value of z_h.

The optimization problem in step 2 of the above algorithm can be solved by any assignment technique.

- Step 3: Find an element r in H such that $z_r^0 = \min_{h \in H} z_h^0$. For each $i \in P_r$, let $x_{ij}^* = x_{ij}^0$ for $j \in J$ and $x_{ij}^* = 0$ for $j \in M \setminus J$. Also, let $z_r^* = z_r^0$, $H = H \setminus \{r\}$, and $J = J \setminus \{j \in J: \sum_{i \in I_r} x_{ij}^* = 1\}$; go to step 1.

- Step 4: Stop. The heuristic solution is $x_{ij} = x_{ij}^*$ for $i \in I$ and $j \in M$, and the corresponding system reliability is $1 - \prod_{h=1}^{k}[1 - \exp(-z_h^*)]$.

Example 9-2

Let $k = 3$, $m = n = 10$, and $(n_1, n_2, n_3) = (3, 3, 4)$; the component reliability matrix $[r_{ij}]$ and its corresponding t_{ij} matrix $[t_{ij}]$ can be expressed as

$$[r_{ij}] = \begin{pmatrix}
0.98 & 0.64 & 0.77 & 0.60 & 0.98 & 0.85 & 0.77 & 0.85 & 0.74 & 0.87 \\
0.63 & 0.90 & 0.66 & 0.96 & 0.95 & 0.73 & 0.69 & 0.69 & 0.90 & 0.89 \\
0.64 & 0.84 & 0.98 & 0.78 & 0.88 & 0.80 & 0.83 & 0.77 & 0.97 & 0.92 \\
0.77 & 0.98 & 0.91 & 0.63 & 0.64 & 0.97 & 0.69 & 0.67 & 0.72 & 0.68 \\
0.84 & 0.93 & 0.99 & 0.93 & 0.78 & 0.93 & 0.71 & 0.73 & 0.80 & 0.76 \\
0.67 & 0.72 & 0.81 & 0.80 & 0.95 & 0.82 & 0.91 & 0.67 & 0.75 & 0.82 \\
0.98 & 0.81 & 0.80 & 0.67 & 0.74 & 0.89 & 0.71 & 0.87 & 0.90 & 0.95 \\
0.85 & 0.90 & 0.73 & 0.88 & 0.86 & 0.75 & 0.84 & 0.63 & 0.68 & 0.61 \\
0.86 & 0.62 & 0.68 & 0.77 & 0.66 & 0.89 & 0.84 & 0.72 & 0.85 & 0.72 \\
0.81 & 0.62 & 0.84 & 0.94 & 0.71 & 0.88 & 0.68 & 0.99 & 0.81 & 0.88
\end{pmatrix},$$

$$[t_{ij}] = \begin{pmatrix} 0.020 & 0.446 & 0.261 & 0.511 & 0.020 & 0.163 & 0.261 & 0.163 & 0.301 & 0.139 \\ 0.462 & 0.105 & 0.416 & 0.041 & 0.051 & 0.315 & 0.371 & 0.371 & 0.105 & 0.117 \\ 0.446 & 0.174 & 0.020 & 0.248 & 0.128 & 0.223 & 0.186 & 0.261 & 0.030 & 0.083 \\ 0.261 & 0.020 & 0.094 & 0.462 & 0.446 & 0.030 & 0.371 & 0.400 & 0.329 & 0.386 \\ 0.174 & 0.073 & 0.010 & 0.073 & 0.248 & 0.073 & 0.342 & 0.315 & 0.223 & 0.274 \\ 0.400 & 0.329 & 0.211 & 0.223 & 0.051 & 0.198 & 0.094 & 0.400 & 0.288 & 0.198 \\ 0.020 & 0.211 & 0.223 & 0.400 & 0.301 & 0.117 & 0.342 & 0.139 & 0.105 & 0.051 \\ 0.163 & 0.105 & 0.315 & 0.128 & 0.151 & 0.288 & 0.174 & 0.462 & 0.386 & 0.494 \\ 0.151 & 0.478 & 0.386 & 0.261 & 0.416 & 0.117 & 0.174 & 0.329 & 0.163 & 0.329 \\ 0.211 & 0.478 & 0.174 & 0.062 & 0.342 & 0.128 & 0.386 & 0.010 & 0.211 & 0.128 \end{pmatrix},$$

respectively.

Solution of Example 9-2 from algorithm 4

- Iteration 1: $J = \{1, \dots, 10\}$ and $H = \{1, 2, 3\}$. z_1^0 is obtained by minimizing $\sum_{i=1}^{3} \sum_{j=1}^{10} t_{ij} x_{ij}$ subject to $\sum_{j=1}^{10} x_{ij} = 1$ for $i = 1, 2, 3$, $\sum_{i=1}^{3} x_{ij} \leq 1$ for $j = 1, \dots, 10$ and $x_{ij} = 0$ or 1. Assignment of components 1, 4, and 3 to positions 1, 2, and 3 gives $z_1^0 = 0.081$. Assignment of components 2, 3, and 5 to positions 4, 5, and 6, respectively, gives $z_2^0 = 0.081$; whereas assignment of components 1, 2, 6, and 8 to positions 7, 8, 9, and 10, respectively, gives $z_3^0 = 0.252$. Since z_1^0 is minimum, components 1, 4, and 3 are assigned to positions 1, 2, and 3 in path set P_1, respectively.

- Iteration 2: $J = \{2, 5, 6, 7, 8, 9, 10\}$ and $H = \{2, 3\}$. $z_2^0 = 0.144$ is obtained by assigning components 2, 6, and 5 in set J to positions 4, 5, and 6, respectively. Similarly, $z_3^0 = 0.283$ is obtained by assigning components 10, 2, 6, and 8 in set J to positions 7, 8, 9, and 10, respectively. Since $z_2^0 < z_3^0$, components 2, 6, and 5 in set J are assigned to positions 4, 5, and 6 in path set P_2, respectively.

- Iteration 3: $J = \{7, 8, 9, 10\}$ and $H = \{3\}$. The remaining components 7, 8, 9, and 10 are assigned to positions 8, 10, 9, and 7, respectively, so as to minimize $\sum_{i=7}^{10} \sum_{j \in J} t_{ij} x_{ij}$. This assignment gives $z_3^0 = 0.398$.

The complete assignment obtained by algorithm 4 is

$$(w_1, w_2, \dots, w_{10}) = (1, 4, 3, 2, 6, 5, 10, 7, 9, 8),$$

where w_j denotes the component assigned to position j. For this assignment, the reliability of path set P_1 is $r_{w_1} r_{w_2} r_{w_3} = (0.98)(0.96)(0.98) = 0.921\,98$.

Similarly, the reliabilities of path sets P_2 and P_3 are $0.865\,83$ and $0.671\,52$, respectively. Therefore, the system reliability for the above assignment $(w_1, w_2, \dots, w_{10})$ is

$$1 - (1 - 0.921\,98)(1 - 0.865\,83)(1 - 0.671\,52) = 0.996\,56.$$

9.2.3   Optimal assignment for two path sets: bicriteria approach

Consider a series-parallel system with two path sets, P_1 and P_2, and general component reliabilities. It is quite difficult to solve the optimization problem even for this simple system. Prasad et al. [266] have developed a methodology for maximization and minimization of a quasi-concave function of the form $g(c_1x, c_2x)$ subject to $Ax = b$, $x \geq 0$; where c_1x and c_2x are linear functions of x. They have shown that Problem 9.2 for $k = 2$ can be solved as a special case.

Define two vectors c_1 and c_2 such that

$$c_1x = \sum_{i=1}^{n_1} \sum_{j=1}^{m} t_{ij} x_{ij} \quad \text{and} \quad c_2x = \sum_{i=n_1+1}^{n} \sum_{j=1}^{m} t_{ij} x_{ij},$$

and define $g(z_1, z_2) = (1-e^{-z_1})(1-e^{-z_2})$. It is known that $g(z_1, z_2)$ is a quasi-concave function on R_+^2. Let K be the set of vectors x satisfying

$$\sum_{j=1}^{m} x_{ij} = 1, \quad \text{for } i = 1, \ldots, n, \tag{9.5}$$

$$\sum_{i=1}^{n} x_{ij} \leq 1, \quad \text{for } j = 1, \ldots, m, \tag{9.6}$$

$x_{ij} = 0$ or 1.

The optimization Problem 9.2 with $k = 2$ is the minimization of $g(c_1x, c_2x)$ subject to $x \in K$. Prasad et al.'s [266] algorithm 3 solves, as a special case, the problem of minimizing $g(c_1x, c_2x)$ subject to $x \in K$. The following algorithm is a minor modification of this algorithm, and it solves Problem 9.2 by minimizing $g(c_1x, c_2x)$ subject to $x \in K$.

Algorithm 5

- Step 0: Obtain $x^{(1)}$ and $x^{(2)}$, which minimizes c_1x and c_2x on K, and let $\lambda_1 = 0$. If $c_1x^{(1)} = c_1x^{(2)}$, take $x^{(2)}$ as the optimal solution and stop. If $c_2x^{(1)} = c_2x^{(2)}$, take $x^{(1)}$ as the optimal solution and stop. If neither of these two equalities holds, let $W = \{(1, 2)\}$, $r = 2$, $d_1 = c_1x^{(1)}$, and $d_2 = c_2x^{(2)}$.

- Step 1: If $g(c_1x^{(1)}, c_2x^{(1)}) \leq g(c_1x^{(2)}, c_2x^{(2)})$, let $q^* = g(c_1x^{(1)}, c_2x^{(1)})$ and $x^* = x^{(1)}$. Otherwise, let $q^* = g(c_1x^{(2)}, c_2x^{(2)})$ and $x^* = x^{(2)}$.

- Step 2: Choose any (i, j) from W. Let $W = W \setminus \{(i, j)\}$ and $r = r + 1$. Evaluate $\lambda_r = [c_1x^{(j)} - c_1x^{(i)}] / [c_2x^{(i)} - c_2x^{(j)}]$, let $d = (c_1 + \lambda_r c_2) x^{(i)}$ and find $x^{(r)}$, which minimizes $(c_1 + \lambda_r c_2)x$ on K. Let $d_r = (c_1 + \lambda_r c_2) x^{(r)}$. If $d_r = d$, go to step 6.

  Find the solution (h_1, h_2) of $z_1 + \lambda_i z_2 = d_i$ and $z_1 + \lambda_r z_2 = d_r$, and compute $q(i, r) = g(h_1, h_2)$. Find the solution (h'_1, h'_2) of $z_1 + \lambda_r z_2 = d_r$ and $z_1 + \lambda_j z_2 = d_j$ if $j \neq 2$. If $j = 2$, take $(h'_1, h'_2) = [d_r - \lambda_r c_2x^{(2)}, c_2x^{(2)}]$ and let $q(r, j) = g(h'_1, h'_2)$. If $g(c_1x^{(r)}, c_2x^{(r)}) < q^*$, let $x^* = x^{(r)}$ and $q^* = g(c_1x^{(r)}, c_2x^{(r)})$.

- Step 3: If $q(i, r) < q^*$, let $W = W \cup \{(i, r)\}$.
- Step 4: If $q(r, j) < q^*$, let $W = W \cup \{(r, j)\}$.
- Step 5: Delete from W each (u, v) for which $q(u, v) \geq q^*$.
- Step 6: If $W \neq \phi$, go to step 1: otherwise stop: $\mathbf{x}^*$ is the required optimal solution.

In the above algorithm, minimization of $(\mathbf{c}_1 + \lambda \mathbf{c}_2)\mathbf{x}$ on K can be done by any assignment technique.

Example 9-3

Let $m = n = 10$ and $n_1 = n_2 = 5$, and take r_{ij} of Example 9-2 as the component reliabilities.

Solution of Example 9-3 from algorithm 5

Since each element $\mathbf{x}$ in K is an assignment which can be represented by a permutation v, we denote each optimal solution derived in the algorithm by a permutation.

- Step 0: $\lambda_1 = 0$, $\mathbf{x}^{(1)} = (1, 4, 9, 2, 3, 6, 7, 8, 5, 10)$

  $\mathbf{c}_1\mathbf{x}^{(1)} = 0.121$, $\mathbf{c}_2\mathbf{x}^{(1)} = 1.546$, and $g(\mathbf{c}_1\mathbf{x}^{(1)}, \mathbf{c}_2\mathbf{x}^{(1)}) = 0.089\,68$.

  $\mathbf{x}^{(2)} = (6, 8, 3, 7, 9, 1, 4, 10, 5, 2)$, $\mathbf{c}_1\mathbf{x}^{(2)} = 1.083$, $\mathbf{c}_2\mathbf{x}^{(2)} = 0.303$,
  and $g(\mathbf{c}_1\mathbf{x}^{(2)}, \mathbf{c}_2\mathbf{x}^{(2)}) = 0.172\,90$.

  $W = \{(1, 2)\}$.

  $d_1 = 0.121$, $d_2 = 0.303$.

- Step 1: $q^* = 0.089\,68$ and $\mathbf{x}^* = \mathbf{x}^{(1)}$ since $g(\mathbf{c}_1\mathbf{x}^{(1)}, \mathbf{c}_2\mathbf{x}^{(1)}) \leq g(\mathbf{c}_1\mathbf{x}^{(2)}, \mathbf{c}_2\mathbf{x}^{(2)})$.
  At this stage, $W = \{(1, 2)\}$.

- Step 2: Choose $(1, 2)$ from W and let $i = 1$, $j = 2$, $W = W \setminus \{(1, 2)\} = \phi$, and $r = 3$. Then,

  $$\lambda_3 = \frac{\mathbf{c}_1\mathbf{x}^{(j)} - \mathbf{c}_1\mathbf{x}^{(i)}}{\mathbf{c}_2\mathbf{x}^{(i)} - \mathbf{c}_2\mathbf{x}^{(j)}} = \frac{1.083 - 0.121}{1.546 - 0.303} = 0.773\,93,$$

  $$d = \mathbf{c}_1\mathbf{x}^{(i)} + \lambda_3\,\mathbf{c}_2\mathbf{x}^{(i)} = 0.121 + (0.773\,93)(1.5460) = 1.317\,45,$$

  $$\mathbf{x}^{(3)} = (1, 4, 9, 2, 3, 5, 10, 7, 6, 8),$$

  $$[\mathbf{c}_1\mathbf{x}^{(3)}, \mathbf{c}_2\mathbf{x}^{(3)}] = (0.121, 0.403),$$

  $$g(\mathbf{c}_1\mathbf{x}^{(3)}, \mathbf{c}_2\mathbf{x}^{(3)}) = 0.037\,80,$$

  $$d_3 = \mathbf{c}_1\mathbf{x}^{(3)} + \lambda_3\,\mathbf{c}_2\mathbf{x}^{(3)} = 0.432\,88.$$

  Similarly,

  $$h_1 = \frac{\lambda_1 d_3 - \lambda_3 d_1}{\lambda_1 - \lambda_3} = \frac{0(0.432\,88) - (0.773\,93)(0.121)}{0 - 0.773\,93} = 0.121,$$

  $$h_2 = \frac{d_1 - d_3}{\lambda_1 - \lambda_3} = \frac{-0.311\,88}{-0.773\,93} = 0.403,$$

and

$$q(1, 3) = g(h_1, h_2) = 0.037\,80.$$

Since $j = 2$, take $h_1' = d_3 - \lambda_3 c_2 \mathbf{x}^{(2)}$. Now,

$$h_1' = 0.432\,88 - 0.773\,93(0.303) = 0.198\,29,$$

$$h_2' = 0.303.$$

Thus,

$$q(3, 2) = g(h_1', h_2') = 0.047\,04.$$

- Step 3: As $g(\mathbf{c}_1\mathbf{x}^{(3)}, \mathbf{c}_2\mathbf{x}^{(3)}) = 0.037\,80 < q^*$, let $\mathbf{x}^* = \mathbf{x}^{(3)}$ and $q^* = 0.037\,80$.
- Steps 4 and 5: Since $q(1, 3) \not< q^*$ and $q(3, 2) \not< q^*$, W remains as ϕ. Therefore,

$$\mathbf{x}^* = \mathbf{x}^{(3)} = (1, 4, 9, 2, 3, 5, 10, 7, 6, 8)$$

is the optimal solution, and the maximum system reliability is

$$R_s^* = 1 - q^* = 1 - 0.037\,80 = 0.9622.$$

9.3   Optimal assignment of components in parallel-series systems

Consider a parallel-series system with k minimal cut sets $C_1, \ldots, C_k$. Let n_h be the number of positions in cut set C_h for $h = 1, \ldots, k$, $N_h = \sum_{i=1}^{h} n_i$ for $h = 1, \ldots, k$, and $n = N_k$, that is, $n = n_1 + \cdots + n_k$. Number the n_1 positions of C_1 as $1, 2, \ldots, n_1$, the n_2 positions of C_2 as $N_1 + 1, \ldots, N_1 + n_2, \ldots$, and the n_k positions of C_k as $N_{k-1} + 1, \ldots, N_{k-1} + n_k$. There are m $(\geq n)$ components to be fixed in n positions of the system. Let r_{ij} denote the reliability of component j when it is assigned to position i and let $q_{ij} = 1 - r_{ij}$.

Let $x_{ij} = 1$ if component j is assigned to position i, and 0 otherwise. Then, the assignment of components can be represented by vector $\mathbf{x} = (x_{11}, x_{12}, \ldots, x_{1m}, \ldots, x_{n1}, \ldots, x_{nm})$. The system reliability for the assignment $\mathbf{x}$ is

$$R(\mathbf{x}) = \prod_{h=1}^{k} \left(1 - \prod_{i \in C_h} \prod_{j=1}^{m} q_{ij}^{x_{ij}} \right).$$

Problem 9.3

Maximize the reliability of a parallel-series system through assignment of interchangeable components

$$R(\mathbf{x}) = \prod_{h=1}^{k} \left(1 - \prod_{i \in C_h} \prod_{j=1}^{m} q_{ij}^{x_{ij}} \right),$$

subject to

$$\sum_{j=1}^{m} x_{ij} = 1, \qquad \text{for } i = 1, \ldots, n,$$

$$\sum_{i=1}^{n} x_{ij} \leq 1, \qquad \text{for } j = 1, \ldots, m,$$

$$x_{ij} = 0 \text{ or } 1.$$

Let

$$s_{ij} = -\ln q_{ij}, \qquad z_h = \sum_{i \in C_h} \sum_{j=1}^{m} s_{ij} x_{ij},$$

for $1 \leq h \leq k$ and $\mathbf{z} = (z_1, \ldots, z_k)$. The value of z_h is a-hazard of cut set C_h and vector $\mathbf{z}$ is a-hazard vector of the system. Then the system reliability is

$$R(\mathbf{z}) = \prod_{h=1}^{k} (1 - e^{-z_h}),$$

which is a Schur-concave function of $\mathbf{z}$. From a computational view point, Problem 9.3 is known to be NP-complete.

9.3.1   Optimal allocation of components in parallel-series systems

Let us assume that $m = n$ and $r_{ij} = r_j$. As argued in the case of a series-parallel system, it is enough to consider allocations of components to cut sets $C_1, C_2, \ldots, C_k$ in order to maximize the system reliability through assignment of interchangeable components. Let A_h denote the set of components allocated to C_h, $1 \leq h \leq k$. Then $A = (A_1, A_2, \ldots, A_k)$ represents an allocation. The system reliability for an allocation A is

$$R(A) = \prod_{h=1}^{k} \left(1 - \prod_{j \in A_h} q_j \right),$$

where $q_j = 1 - r_j$. In this case, a-hazard of cut set C_h is $z_h = \sum_{j \in A_h} s_j$ for $h = 1, \ldots, k$, where $s_j = -\ln q_j$. If the cut set hazard vector of allocation A majorizes that of allocation B, then allocation A cannot give higher reliability than allocation B. This is true because $R(\mathbf{z}) = \prod_{h=1}^{k} (1 - e^{-z_h})$ is a Schur-concave function of $\mathbf{z}$. El-Neweihi et al. [86] have shown that the allocation for $k = 2$ that minimizes $|z_1 - (z_1 + z_2)/2|$ is optimal. They have argued that if the cut set hazards $z_1, \ldots, z_k$ are all the same for an allocation A, then A maximizes the system reliability. The reliability of a parallel-series system increases with homogeneity among the cut set hazards whereas the reliability of a series-parallel system increases with heterogeneity among path set hazards. Derman et al. [77] have shown that if each cut set consists of only two positions, then $A_1^* = \{1, n\}$, $A_2^* = \{2, n - 1\}, \ldots, A_k^* = \{n/2, n/2 + 1\}$ is an optimal allocation.

Table 9.6. Reliabilities used in Example 9-4

j	r_j	$1-r_j$	j	r_j	$1-r_j$
1	0.990	0.010	7	0.656	0.344
2	0.934	0.066	8	0.593	0.407
3	0.898	0.102	9	0.549	0.451
4	0.863	0.137	10	0.487	0.513
5	0.781	0.219	11	0.479	0.521
6	0.728	0.272	12	0.439	0.561

The heuristic method of Baxter and Harche

Baxter and Harche [24] have proposed a heuristic method (B–H) for the optimal allocation of components in parallel-series systems. Assume that $r_1 \geq r_2 \geq \cdots \geq r_n$. The method sequentially allocates components $1, 2, \ldots, n$ to all the positions of the system. Let A_i denote the set of components allocated to cut set C_i for $i = 1, \ldots, k$. The following algorithm describes the method for $n_1 = n_2 = \cdots = n_k = s$.

Algorithm 6

- Step 1: Allocate component i to C_i for $i = 1, \ldots, k$.

- Step 2: Allocate component $k + i$ to C_{2k+1-i} for $i = 1, \ldots, k$.

- Step 3: Let $v = 2$.

- Step 4: Compute $R_i^{(v)} = 1 - \prod_{j \in A_i} (1 - r_j)$ for $i = 1, \ldots, k$. Allocate component $vk + 1$ to C_i, which has the smallest $R_i^{(v)}$; allocate component $vk + 2$ to C_i, which has the second smallest $R_i^{(v)}$, and so on; allocate component $vk + k$ to C_i with the largest $R_i^{(v)}$.

- Step 5: Let $v = v + 1$. If $v < s$, go to step 4. Otherwise, stop.

This method is called a *top–down heuristic* (TDH), since components are selected for allocation in decreasing order of reliability. As suggested by Baxter and Harche, the method can be directly extended to the case where $n_1, \ldots, n_k$ are not the same. Assume, without loss of generality, that $n_1 \leq n_2 \leq \cdots \leq n_k$. If the number of components allocated to a cut set C_i reaches n_i during the progress of the above algorithm, ignore C_i for further allocation.

Example 9-4

Let $n = 12$, $k = 3$, and $(n_1, n_2, n_3) = (4, 4, 4)$. Let $r_1, \ldots, r_{12}$ be as given in Table 9.6.

Solution of Example 9-4 from algorithm 6

The following is a stepwise description of the implementation of the B–H method for Example 9-4.

- Step 1: Allocating components 1, 2, and 3 to C_1, C_2, and C_3, respectively, we get $A_1 = \{1\}, A_2 = \{2\}$, and $A_3 = \{3\}$.

- Step 2: Allocating components 4, 5, and 6 to C_3, C_2, and C_1, respectively, we get $A_1 = \{1, 6\}, A_2 = \{2, 5\}$, and $A_3 = \{3, 4\}$.

- Step 3: Let $v = 2$.

- Step 4: We have

$$R_1^{(2)} = 1 - (1 - r_1)(1 - r_6) = 0.997\,280,$$

$$R_2^{(2)} = 1 - (1 - r_2)(1 - r_5) = 0.985\,546,$$

$$R_3^{(2)} = 1 - (1 - r_3)(1 - r_4) = 0.986\,026.$$

Since $R_2^{(2)} < R_3^{(2)} < R_1^{(2)}$, allocate components 7, 8, and 9 to C_2, C_3, C_1, respectively. This allocation gives $A_1 = \{1, 6, 9\}$, $A_2 = \{2, 5, 7\}$, and $A_3 = \{3, 4, 8\}$.

- Step 5: Let $v = 3$. Since $v < s\ (=4)$, go to step 4.

- Step 4: We have

$$R_1^{(3)} = 1 - (1 - r_1)(1 - r_6)(1 - r_9) = 0.998\,773,$$

$$R_2^{(3)} = 1 - (1 - r_2)(1 - r_5)(1 - r_7) = 0.995\,028,$$

$$R_3^{(3)} = 1 - (1 - r_3)(1 - r_4)(1 - r_8) = 0.994\,313.$$

Since $R_3^{(3)} < R_2^{(3)} < R_1^{(3)}$, allocate components 10, 11, and 12 to C_3, C_2, C_1, respectively. The resulting allocation is $A_1 = \{1, 6, 9, 12\}, A_2 = \{2, 5, 7, 11\}$, and $A_3 = \{3, 4, 8, 10\}$.

- Step 5: Since $v + 1 = 4$, stop. The system reliability for the final allocation is 0.993 815.

The heuristic method of Prasad and Raghavachari

Applying the majorization results, El-Neweihi et al. [86] have reduced the search for an optimal allocation to a subset D of allocations, such that the hazard vector of any allocation in D does not strictly majorize the hazard vector of any other allocation. Computationally, it is very difficult to enumerate all the allocations in D in order to find the best one among them.

Prasad and Raghavachari (P–R) [273] have developed an LP-based heuristic method for solving Problem 9.3 with $r_{ij} = r_j$. Let $y_{ij} = 1$ if component j is allocated to cut set C_i and 0 otherwise, and let $\mathbf{y} = (y_{11}, y_{12}, \ldots, y_{kn})$. For this case, the problem is as given below.

Problem 9.4

Maximize

$$R(\mathbf{y}) = \prod_{i=1}^{k} \left[1 - \exp\left(-\sum_{i=1}^{n} s_j y_{ij} \right) \right],$$

subject to

$$\sum_{i=1}^{k} y_{ij} = 1, \qquad \text{for } j = 1, \ldots, n, \tag{9.7}$$

$$\sum_{j=1}^{m} y_{ij} = n_i, \qquad \text{for } i = 1, \ldots, k, \tag{9.8}$$

$$y_{ij} = 0 \text{ or } 1, \qquad \text{for each } (i, j). \tag{9.9}$$

Let K be the set of all $\mathbf{y}$ satisfying the constraints (9.7)–(9.9). To derive a heuristic solution for Problem 9.4, Prasad and Raghavachari [273] have considered the problem of minimizing $\sum_{i=1}^{k} |\bar{z} - z_i|$, subject to $\mathbf{y} \in K$, where $z_i = \sum_{j=1}^{n} s_j y_{ij}$ and $\bar{z} = \frac{1}{k} \sum_{j=1}^{n} s_j$.

Introducing $2n$ nonnegative variables, this problem can now be written as Problem 9.5.

Problem 9.5

Minimize

$$f = \sum_{i=1}^{k} (d_i^+ + d_i^-),$$

subject to

$$\sum_{i=1}^{k} y_{ij} = 1, \qquad \qquad \text{for } j = 1, \ldots, n,$$

$$\sum_{j=1}^{n} y_{ij} = n_i, \qquad \qquad \text{for } i = 1, \ldots, k,$$

$$\sum_{j=1}^{n} s_j y_{ij} + d_i^+ - d_i^- = \bar{z}, \qquad \text{for } i = 1, \ldots, k,$$

$$y_{ij} = 0 \text{ or } 1, \qquad \qquad \text{for each } (i, j),$$

and d_i^+ and d_i^- being nonnegative variables.

Problem 9.5 is a mixed integer linear programming problem. The heuristic approach of Prasad and Raghavachari [273] involves

1. solving the relaxed version of Problem 9.5, that is, the linear programming problem resulting from the replacement of 0–1 constraints on y_{ij} by the constraints $0 \le y_{ij} \le 1$;

2. rounding off the values of y_{ij} in the LP solution; and

3. improving the solution iteratively by pairwise interchange of components among the cut sets.

Suppose $\mathbf{y} = (y_{11}, y_{12}, \dots, y_{kn})$ is the optimal solution of the relaxed version of Problem 9.5 derived using an LP technique.

The rounding off procedure

- Step 1: Take $m_i = n_i$ for $i = 1, \dots, k$. For each (i, j) for which $y_{ij} = 1$, update m_i as $m_i - 1$, let $J = \{j: y_{ij} = 1$ for some $i\}$, and $I = \{i: m_i = 0\}$.

- Step 2: Let $y_{rs} = \max\{y_{ij}: i \notin I, j \notin J\}$. A tie can be broken in any arbitrary manner. Take $y_{rs} = 1$, $y_{is} = 0$ for $i \ne r$, $J = J \cup \{s\}$, and $m_r = m_r - 1$. If $m_r = 0$, then let $I = I \cup \{r\}$,

- Step 3: If $I = \{1, 2, \dots, k\}$, stop. Otherwise, go to step 2.

Let $(A_1, A_2, \dots, A_k)$ be the allocation obtained by the above procedure. Since this allocation is not necessarily optimal, the following iterative algorithm can be used for improvement. In every iteration of the algorithm, two components are selected from two different cut sets and interchanged in order to improve the system reliability.

Algorithm 7

- Step 0: Determine the cut set hazard vector $(z_1, \dots, z_k)$ corresponding to the heuristic allocation $(A_1, \dots, A_k)$ obtained by the LP approach. Let INCR $= 1$.

- Step 1: If INCR $= 0$, take the current allocation $(A_1, \dots, A_k)$ as the optimal one and stop. Otherwise, let $g = 1, h = 2$, and INCR $= 0$.

- Step 2: If $g = k$, go to step 1. Otherwise, go to step 3.

- Step 3: If $z_h > z_g$, find $u \in A_g$ and $v \in A_h$, if they exist, such that $s_v > s_u$ and $\left|(s_v - s_u) - (z_h - z_g)/2\right| = \min\{\left|(s_j - s_i) - (z_h - z_g)/2\right|: i \in A_g, j \in A_h, 0 < (s_j - s_i) < (z_h - z_g)\}$, and let $A_g = \{A_g \cup \{v\}\} \setminus \{u\}$, $A_h = \{A_h \cup \{u\}\} \setminus \{v\}$, $z_g = z_g + s_v - s_u$, $z_h = z_h + s_u - s_v$, and INCR $= 1$. Similarly, if $z_g > z_h$, find $u \in A_g$ and $v \in A_h$, if they exist, such that $s_u > s_v$ and $\left|(s_u - s_v) - (z_g - z_h)/2\right| = \min\{\left|(s_i - s_j) - (z_g - z_h)/2\right|: i \in A_g, j \in A_h, 0 < (s_i - s_j) < (z_g - z_h)\}$, and let $A_g = A_g \cup \{v\} \setminus \{u\}$, $A_h = A_h \cup \{u\} \setminus \{v\}$, $z_g = z_g + s_v - s_u$, $z_h = z_h + s_u - s_v$, and INCR $= 1$. Let $h = h + 1$. If $h \le k$, repeat step 3. Otherwise, let $g = g + 1$ and $h = g + 1$, and go to step 2.

Example 9-5

Let $k = 3, n = 14$, and $(n_1, n_2, n_3) = (3, 5, 6)$. Suppose the reliabilities of the 14 components and the corresponding s_j values are as given in Table 9.7.

Table 9.7. Reliabilities and the corresponding s_j used in Example 9-5

j	r_j	s_j	j	r_j	s_j
1	0.415	0.5361	8	0.781	1.5187
2	0.549	0.7963	9	0.487	0.6675
3	0.479	0.6520	10	0.990	4.6052
4	0.656	1.0671	11	0.863	1.9878
5	0.934	2.7181	12	0.407	0.5226
6	0.439	0.5780	13	0.728	1.3020
7	0.898	2.2828	14	0.593	0.8989

These reliabilities are randomly generated from the uniform distribution over the interval [0.4, 1.0].

Solution of Example 9-5 from the Prasad–Raghavachari method
The allocation obtained from the LP solution and subsequent round off method is

$$A_1 = \{2, 4, 12\}, \ A_2 = \{1, 7, 9, 13, 14\}, \ A_3 = \{3, 5, 6, 8, 10, 11\},$$

and the corresponding system reliability is 0.904 917 40. The pairwise interchanges of components and the resulting allocations in the implementation of improvement algorithms are given in Table 9.8. In this table, (i, j) refers to the pair of cut sets whose components are interchanged and (r, s) represents the pair of components interchanged. For the first improvement iteration initially, $g = 1$ and $h = 2$. We have

$$z_1 = s_2 + s_4 + s_{12} = 0.7963 + 1.0671 + 0.5226 = 2.3860,$$

$$z_2 = s_1 + s_7 + s_9 + s_{13} + s_{14} = 5.6873,$$

and $z_2 - z_1 = 3.3013 > 0$. Since

$$s_7 - s_{12} = \max\{s_j - s_i : i \in A_1, j \in A_2, 0 < s_j - s_i < z_2 - z_1\},$$

the components $r = 12$ and $s = 7$ are taken from A_1 and A_2, respectively, and are interchanged to obtain the allocation $\{2, 4, 7\}$, $\{12, 9, 13, 14, 1\}$, $\{3, 6, 8, 10, 11, 5\}$, which is better than the previous one. All other iterations are also performed in the same fashion. The improvement algorithm has yielded the allocation

$$A_1 = \{2, 10, 13\}, \ A_2 = \{1, 3, 5, 7, 12\}, \ \text{and} \ A_3 = \{4, 6, 8, 9, 11, 14\},$$

with system reliability 0.996 352 08.

9.3.2 Optimal assignment in two cut sets: bicriterion approach

Consider a parallel-series system with two cut sets and general r_{ij}. This is a special case of Problem 9.3 with $k = 2$. Problem 9.6, the corresponding assignment problem can be written as shown.

Table 9.8. Pairwise interchanges of components

	(i, j)	(r, s)	Resulting allocation			System reliability
1	(1, 2)	(12, 7)	{2, 4, 7},	{12, 9, 13, 14, 1},	{3, 6, 8, 10, 11, 5}	0.964 780 36
2	(1, 3)	(2, 10)	{10, 4, 7},	{12, 9, 13, 14, 1},	{3, 6, 8, 2, 11, 5}	0.979 699 10
3	(2, 3)	(1, 5)	{10, 4, 7},	{12, 9, 13, 14, 5},	{3, 6, 8, 2, 11, 1}	0.995 119 51
4	(1, 2)	(7, 13)	{10, 4, 13},	{12, 9, 7, 14, 5},	{3, 6, 8, 2, 11, 1}	0.995 922 03
5	(1, 3)	(4, 6)	{10, 6, 13},	{12, 9, 7, 14, 5},	{3, 4, 8, 2, 11, 1}	0.996 226 48
6	(2, 3)	(14, 3)	{10, 6, 13},	{12, 9, 7, 3, 5},	{14, 4, 8, 2, 11, 1}	0.996 303 32
7	(1, 2)	(6, 9)	{10, 9, 13},	{12, 6, 7, 3, 5},	{14, 4, 8, 2, 11, 1}	0.996 334 02
8	(1, 3)	(9, 2)	{10, 2, 13},	{12, 6, 7, 3, 5},	{14, 4, 8, 9, 11, 1}	0.996 350 28
9	(2, 3)	(6, 1)	{10, 2, 13},	{12, 1, 7, 3, 5},	{14, 4, 8, 9, 11, 6}	0.996 352 08

Problem 9.6

Maximize

$$R(\mathbf{x}) = \left[1 - \exp\left(-\sum_{i=1}^{n_1}\sum_{j=1}^{m} s_{ij}x_{ij}\right)\right]\left[1 - \exp\left(-\sum_{i=n_1+1}^{n}\sum_{j=1}^{m} s_{ij}x_{ij}\right)\right],$$

subject to

$$\sum_{j=1}^{m} x_{ij} = 1, \qquad \text{for } i = 1, \ldots, n, \tag{9.10}$$

$$\sum_{i=1}^{n} x_{ij} \leq 1, \qquad \text{for } j = 1, \ldots, m, \tag{9.11}$$

$$x_{ij} = 0 \text{ or } 1. \tag{9.12}$$

As mentioned earlier, even this problem is NP-complete. Prasad et al. [266] have developed an algorithm for maximizing a quasi-concave function of the form $g(c_1\mathbf{x}, c_2\mathbf{x})$ subject to linear constraints $A\mathbf{x} = b$ and nonnegative restriction on $\mathbf{x}$. They have extended the algorithm for solving Problem 9.6 as a special case.

Define two vectors $\mathbf{c}_1$ and $\mathbf{c}_2$ such that

$$\mathbf{c}_1\mathbf{x} = \sum_{i=1}^{n_1}\sum_{j=1}^{m} s_{ij}x_{ij} \quad \text{and} \quad \mathbf{c}_2\mathbf{x} = \sum_{i=n_1+1}^{n}\sum_{j=1}^{m} s_{ij}x_{ij},$$

where $s_{ij} = -\ln(1 - r_{ij})$. Let $g(z_1, z_2) = (1 - e^{-z_1})(1 - e^{-z_2})$. Problem 9.6 is to maximize $g(\mathbf{c}_1\mathbf{x}, \mathbf{c}_2\mathbf{x})$ subject to the constraints (9.10)–(9.12). Algorithm 1 of Prasad et al. [266] gives an optimal solution of the relaxed version of Problem 9.6, that is, the problem with 0–1 constraints replaced by the constraints $0 \leq x_{ij} \leq 1$. If the optimal solution is an integral one, it is the required solution of Problem 9.6; otherwise, subsequent application of algorithm 2 of Prasad et al. [266] yields the optimal solution. Combining both algorithms with minor modifications, we now present an algorithm for obtaining an optimal solution of Problem 9.6.

Any assignment $\mathbf{x}$ can be represented by a permutation $\mathbf{v} = (v_1, \ldots, v_m)$ such that $x_{ij} = 1$ if $j = v_i$ and 0 otherwise. For a partial permutation $(u_1, \ldots, u_r)$ of $1, 2, \ldots, m$, let $h_1(u_1, \ldots, u_r)$ represent the maximum value of $\sum_{i=1}^{n_1} s_{iv_i}$ over the set of all permutations generated from $(u_1, u_2, \ldots, u_r)$. Similarly, let $h_2(u_1, \ldots, u_r)$ represent the maximum value of $\sum_{i=n_1+1}^{m} s_{iv_i}$ over the set of all permutations generated from $(u_1, \ldots, u_r)$. Every maximization problem in the following algorithm is required to be solved by an assignment technique.

Algorithm 8

- Step 0: Find $\mathbf{x}^{(1)}$ and $\mathbf{x}^{(2)}$ that maximizes $\mathbf{c}_1\mathbf{x}$ and $\mathbf{c}_2\mathbf{x}$ on K. If $g(\mathbf{c}_1\mathbf{x}^{(1)}, \mathbf{c}_2\mathbf{x}^{(1)}) \geq g(\mathbf{c}_1\mathbf{x}^{(2)}, \mathbf{c}_2\mathbf{x}^{(2)})$, let $\mathbf{x}^* = \mathbf{x}^{(1)}$ and $q^* = g(\mathbf{c}_1\mathbf{x}^{(1)}, \mathbf{c}_2\mathbf{x}^{(1)})$. Otherwise, let $\mathbf{x}^* = \mathbf{x}^{(2)}$ and $q^* = g(\mathbf{c}_1\mathbf{x}^{(2)}, \mathbf{c}_2\mathbf{x}^{(2)})$.

- Step 1: Find $\bar{\lambda} = [\mathbf{c}_1\mathbf{x}^{(1)} - \mathbf{c}_1\mathbf{x}^{(2)}] / [\mathbf{c}_2\mathbf{x}^{(2)} - \mathbf{c}_2\mathbf{x}^{(1)}]$ and let $F_1 = (\mathbf{c}_1 + \bar{\lambda}\,\mathbf{c}_2)\mathbf{x}^{(1)}$. Find $\bar{\mathbf{x}}$ that maximizes $(\mathbf{c}_1 + \bar{\lambda}\,\mathbf{c}_2)\mathbf{x}$ on K. If $(\mathbf{c}_1 + \bar{\lambda}\,\mathbf{c}_2)\bar{\mathbf{x}} = F_1$, go to step 4. If $g(\mathbf{c}_1\bar{\mathbf{x}}, \mathbf{c}_2\bar{\mathbf{x}}) > q^*$, go to step 3. Otherwise, go to step 2.

- Step 2: If $\mathbf{x}^* = \mathbf{x}^{(1)}$, let $\mathbf{x}^{(2)} = \bar{\mathbf{x}}$. If $\mathbf{x}^* = \mathbf{x}^{(2)}$, let $\mathbf{x}^{(1)} = \bar{\mathbf{x}}$. Go to step 1.

- Step 3: Let $\mathbf{x}^* = \bar{\mathbf{x}}$ and $q^* = g(\mathbf{c}_1\bar{\mathbf{x}}, \mathbf{c}_2\bar{\mathbf{x}})$. Compute $\alpha = [\exp(\mathbf{c}_1\bar{\mathbf{x}}) - 1] / [\exp(\mathbf{c}_2\bar{\mathbf{x}}) - 1]$. If $\alpha < \bar{\lambda}$, let $\mathbf{x}^{(2)} = \bar{\mathbf{x}}$ and go to step 1. If $\alpha > \bar{\lambda}$, let $\mathbf{x}^{(1)} = \bar{\mathbf{x}}$ and go to step 1. If $\alpha = \lambda$, take $\mathbf{x}^*$ as the optimal solution and stop. Otherwise, go to step 4.

- Step 4: Take the initial set of partial permutations as $G = \{(1), (2), \ldots, (m)\}$.

- Step 5: Select a partial permutation $(u_1, \ldots, u_r)$ from G and let $G = G \setminus \{(u_1, \ldots, u_r)\}$. If $r = m - 1$, go to step 7. Otherwise, evaluate $h_1(u_1, \ldots, u_r)$ and $h_2(u_1, \ldots, u_r)$. If $h_1(u_1, \ldots, u_r) > \mathbf{c}_1\mathbf{x}^{(2)}$ and $h_2(u_1, \ldots, u_r) > \mathbf{c}_2\mathbf{x}^{(1)}$, let

$$G = \bigcup_{y \in M'} \{(u_1, \ldots, u_r, y)\} \cup G,$$

where

$$M' = \{1, \ldots, m\} \setminus \{(u_1, \ldots, u_r)\}.$$

- Step 6: If $G = \phi$, take $\mathbf{x}^*$ as an optimal solution and stop. Otherwise, go to step 5.

- Step 7: If $(u_1, \ldots, u_m)$ is the permutation generated from $(u_1, \ldots, u_r)$ with $\sum_{i=1}^{n_1} s_{iu_i} < \mathbf{c}_1\mathbf{x}^{(2)}$ or $\sum_{i=n_1+1}^{m} s_{iu_i} < \mathbf{c}_2\mathbf{x}^{(1)}$, go to step 6. Otherwise, evaluate

$$\bar{q} = g\left(\sum_{i=1}^{n_1} s_{iu_i}, \sum_{i=n_1+1}^{m} s_{iu_i}\right).$$

If $\bar{q}$ greater than q^*, let $q^* = \bar{q}$ and $x_{ij}^* = 1$ if $j = u_i$ and 0 otherwise for $i = 1, \ldots, m$. Go to step 5.

Example 9-6

Let $m = n = 10$ and $n_1 = n_2 = 5$. Let r_{ij} be as given in Example 9-2.

Solution of Example 9-6 from algorithm 8

As explained earlier, each solution $\mathbf{x}$ in K is presented in the form of a permutation of $1, 2, \ldots, n$.

- Step 0:

$$\mathbf{x}^{(1)} = (4, 3, 1, 5, 7, 6, 2, 8, 9, 10),$$

$$\mathbf{c}_1\mathbf{x}^{(1)} = 2.161, \mathbf{c}_2\mathbf{x}^{(1)} = 1.162,$$

and

$$g\left(\mathbf{c}_1\mathbf{x}^{(1)}, \mathbf{c}_2\mathbf{x}^{(1)}\right) = 0.607\,97.$$

$$\mathbf{x}^{(2)} = (6, 8, 3, 7, 9, 1, 4, 10, 5, 2),$$

$$\mathbf{c}_1\mathbf{x}^{(2)} = 1.148, \mathbf{c}_2\mathbf{x}^{(2)} = 2.188,$$

and

$$g\left(\mathbf{c}_1\mathbf{x}^{(2)}, \mathbf{c}_2\mathbf{x}^{(2)}\right) = 0.606\,17.$$

Since $g\left(\mathbf{c}_1\mathbf{x}^{(1)}, \mathbf{c}_2\mathbf{x}^{(1)}\right) > g\left(\mathbf{c}_1\mathbf{x}^{(2)}, \mathbf{c}_2\mathbf{x}^{(2)}\right)$, let $\mathbf{x}^* = \mathbf{x}^{(1)}$ and $q^* = 0.607\,97$.

- Step 1:

$$\bar{\lambda} = \frac{\mathbf{c}_1\mathbf{x}^{(1)} - \mathbf{c}_1\mathbf{x}^{(2)}}{\mathbf{c}_2\mathbf{x}^{(2)} - \mathbf{c}_2\mathbf{x}^{(1)}} = \frac{2.161 - 1.148}{2.188 - 1.162} = 0.987\,33,$$

$$F_1 = \mathbf{c}_1\mathbf{x}^{(1)} + \bar{\lambda}\,\mathbf{c}_2\mathbf{x}^{(1)} = 2.161 + (0.987\,33)(1.162) = 3.308\,28,$$

$$\bar{\mathbf{x}} = (4, 6, 1, 5, 9, 8, 7, 10, 3, 2),$$

$$\mathbf{c}_1\bar{\mathbf{x}} = 1.941, \mathbf{c}_2\bar{\mathbf{x}} = 2.1,$$

and

$$g(\mathbf{c}_1\bar{\mathbf{x}}, \mathbf{c}_2\bar{\mathbf{x}}) = 0.751\,56.$$

Since $\mathbf{c}_1\bar{\mathbf{x}} + \bar{\lambda}\,\mathbf{c}_2\bar{\mathbf{x}} = 4.014\,39 \neq F_1$ and $g(\mathbf{c}_1\bar{\mathbf{x}}, \mathbf{c}_2\bar{\mathbf{x}}) > q^*$, go to step 3.

- Step 3: Let $\mathbf{x}^* = \bar{\mathbf{x}}$ and $q^* = 0.751\,56$.

$$\alpha = \frac{\exp(\mathbf{c}_1\bar{\mathbf{x}}) - 1}{\exp(\mathbf{c}_2\bar{\mathbf{x}}) - 1} = \frac{e^{1.941} - 1}{e^{2.1} - 1} = 0.832\,48.$$

Since $\alpha < \bar{\lambda}$, let $x^{(2)} = \bar{x}$ and go to step 1. At this stage

$\mathbf{x}^{(1)} = (4, 3, 1, 5, 7, 6, 2, 8, 9, 10)$,

$[\mathbf{c}_1\mathbf{x}^{(1)}, \mathbf{c}_2\mathbf{x}^{(1)}] = (2.161, 1.162)$,

$\mathbf{x}^{(2)} = (4, 6, 1, 5, 9, 8, 7, 10, 3, 2) = \mathbf{x}^*$,

$[\mathbf{c}_1\mathbf{x}^{(2)}, \mathbf{c}_2\mathbf{x}^{(2)}] = (1.941, 2.1)$,

and $q^* = 0.751\,56$.

- Step 1:

$$\bar{\lambda} = \frac{2.161 - 1.941}{2.1 - 1.162} = 0.234\,54,$$

$F_1 = \mathbf{c}_1\mathbf{x}^{(1)} + \bar{\lambda}\,\mathbf{c}_2\mathbf{x}^{(1)} = 2.161 + (0.234\,54)(1.162) = 2.433\,54$,

$\bar{\mathbf{x}} = (4, 3, 1, 5, 7, 8, 6, 9, 10, 2)$,

$\mathbf{c}_1\bar{\mathbf{x}} = 2.161, \mathbf{c}_2\bar{\mathbf{x}} = 1.71$,

and

$g(\mathbf{c}_1\bar{\mathbf{x}}, \mathbf{c}_2\bar{\mathbf{x}}) = 0.724\,76.$

Since $\mathbf{c}_1\bar{\mathbf{x}} + \bar{\lambda}\,\mathbf{c}_2\bar{\mathbf{x}} = 2.562\,06 \neq F_1$ and $g(\mathbf{c}_1\bar{\mathbf{x}}, \mathbf{c}_2\bar{\mathbf{x}}) \leq q^*$, go to step 2.

- Step 2: As $\mathbf{x}^* = \mathbf{x}^{(2)}$, let $\mathbf{x}^{(1)} = \bar{\mathbf{x}}$ and go to step 1. At this stage

$\mathbf{x}^{(1)} = (4, 3, 1, 5, 7, 8, 6, 9, 10, 2)$,

$[\mathbf{c}_1\mathbf{x}^{(1)}, \mathbf{c}_2\mathbf{x}^{(1)}] = (2.161, 1.71)$,

$\mathbf{x}^{(2)} = (4, 6, 1, 5, 9, 8, 7, 10, 3, 2) = \mathbf{x}^*$,

$[\mathbf{c}_1\mathbf{x}^{(2)}, \mathbf{c}_2\mathbf{x}^{(2)}] = (1.941, 2.1)$,

and $q^* = 0.751\,56$.

- Step 1:

$$\bar{\lambda} = \frac{2.161 - 1.941}{2.1 - 1.71} = 0.564\,10,$$

$F_1 = \mathbf{c}_1\mathbf{x}^{(1)} + \bar{\lambda}\,\mathbf{c}_2\mathbf{x}^{(1)} = 2.161 + (0.564\,10)(1.71) = 3.125\,61$,

$\bar{\mathbf{x}} = (4, 6, 1, 5, 8, 9, 7, 10, 3, 2)$,

$\mathbf{c}_1\bar{\mathbf{x}} = 2.033, \mathbf{c}_2\bar{\mathbf{x}} = 1.988$,

and

$g(\mathbf{c}_1\bar{\mathbf{x}}, \mathbf{c}_2\bar{\mathbf{x}}) = 0.750\,02.$

Since $\mathbf{c}_1\bar{\mathbf{x}} + \bar{\lambda}\,\mathbf{c}_2\bar{\mathbf{x}} = 3.154\,43 \neq F_1$ and $g(\mathbf{c}_1\bar{\mathbf{x}}, \mathbf{c}_2\bar{\mathbf{x}}) \leq q^*$, go to step 2.

- Step 2: As $x^* = x^{(2)}$, let $x^{(1)} = \bar{x}$ and go to step 1. At this stage

$$x^{(1)} = (4, 6, 1, 5, 8, 9, 7, 10, 3, 2),$$
$$\left[c_1 x^{(1)}, c_2 x^{(1)}\right] = (2.033, 1.988),$$
$$x^{(2)} = (4, 6, 1, 5, 9, 8, 7, 10, 3, 2) = x^*,$$
$$\left[c_1 x^{(2)}, c_2 x^{(2)}\right] = (1.941, 2.1),$$

and $q^* = 0.751\,56$.

- Step 1:

$$\bar{\lambda} = \frac{2.033 - 1.941}{2.1 - 1.988} = 0.821\,43,$$
$$F_1 = c_1 x^{(1)} + \bar{\lambda}\, c_2 x^{(1)} = 2.033 + (0.821\,43)(1.988) = 3.666\,00,$$
$$\bar{x} = (4, 6, 1, 5, 9, 8, 7, 10, 3, 2),$$
$$c_1 \bar{x} = 1.941, c_2 \bar{x} = 2.1,$$

and

$$g(c_1 \bar{x}, c_2 \bar{x}) = 0.751\,56.$$

As $c_1 \bar{x} + \bar{\lambda}\, c_2 \bar{x} = 3.666\,00 = F_1$, go to step 4.

- Step 4: Now, $\left[c_1 x^{(1)}, c_2 x^{(1)}\right] = (2.033, 1.988)$ and $\left[c_1 x^{(2)}, c_2 x^{(2)}\right] = (1.941, 2.1)$. We have

$$\psi(\theta) = g\left[2.033(1 - \theta) + 1.941\theta, 1.988(1 - \theta) + 2.1\theta\right].$$

$\psi(\theta)$ is maximum at $\theta = 1$ since $\psi(\theta)$ is concave over $(0, \infty)$ and $\psi'(\theta) = 0$ at $\theta = 1.0565$. This means that the solution $x^* = (4, 6, 1, 5, 9, 8, 7, 10, 3, 2)$ is optimal and the maximum system reliability is $g(c_1 x^*, c_2 x^*) = 0.751\,56$. Thus enumeration of permutations as explained in steps 5–7, of algorithm 8, is avoided.

9.4   Component assignment in coherent systems

System reliability depends not only on the component reliabilities but also on the structure of the system. Several methods have been presented in previous sections for deriving an optimal assignment of components to the positions of series-parallel and parallel-series systems. It is more difficult to derive such an assignment for a general reliability structure, even when the component reliabilities are invariant of the positions of the system. Methods for such optimal component assignment in coherent systems are discussed in Section 9.4.

Consider a reliability system with n components used in n positions $1, \ldots, n$ of the system. Let $r_1, \ldots, r_n$ be the reliabilities of the components $1, \ldots, n$, respectively. Let $x_j = 1$ if the component in position j works and 0 otherwise. Then the state of the reliability system can be described by a vector $(x_1, \ldots, x_n)$ with $x_j = 0$ or 1 for each j. Let S be the set of all binary vectors $(x_1, \ldots, x_n)$ with $x_j = 0$ or 1 for each j. Define a function $\phi(x_1, \ldots, x_n)$ on S as $\phi(x_1, \ldots, x_n) = 1$ if the system works when the components in all positions for which $x_j = 1$ work, and $\phi(x_1, \ldots, x_n) = 0$ otherwise. The system is said to have coherent structure if

1. $\phi(0, \ldots, 0) = 0,$

2. $\phi(1, \ldots, 1) = 1,$

3. $\phi(x_1, \ldots, x_n) \leq \phi(y_1, \ldots, y_n)$ when $(x_1, \ldots, x_n) \leq (y_1, \ldots, y_n)$.

The function $\phi(x_1, \ldots, x_n)$ is called the structure function. Let a_j denote the reliability of the component in position j for $j = 1, \ldots, n$ and $R(a_1, a_2, \ldots, a_n)$ denote the system reliability (Further details of coherent structures are given in Section 1.5.8.)

9.4.1   Optimal component assignment through pairwise interchanges

Boland et al. [40] have provided a comparison of criticality of two nodes in a coherent system. Let $(1_i, 0_j, \mathbf{x})$ denote the binary vector obtained from x by replacing x_i and x_j by 1 and 0, respectively. Similarly, let $(0_i, 1_j, \mathbf{x})$ denote the binary vector obtained from x by replacing x_i and x_j by 0 and 1, respectively. A node i is said to be more critical than node j (denoted by $i \overset{c}{>} j$), if $\phi(1_i, 0_j, \mathbf{x}) \geq \phi(0_i, 1_j, \mathbf{x})$ for every binary vector $\mathbf{x}$ and strict inequality holds for at least one x. For example, consider the system shown in Figure 9.1. It can easily be seen that $1 \overset{c}{>} j$ for any $j \geq 2$. A set of positions $i_1, \ldots, i_r$ are said to be *permutation equivalent* if the system reliability does not change when the components assigned to $i_1, \ldots, i_r$ are reassigned to the same set of positions in any manner. For example, when a subsystem has a series or parallel structure, all of the positions in the subsystem are permutation equivalent. A component assignment can be represented by a permutation of $1, \ldots, n$. A permutation $\pi = (\pi_1, \ldots, \pi_n)$ represents the assignment in which component π_i is assigned to position i for $i = 1, \ldots, n$. Two assignments are *equivalent* if one can be obtained from the other by reassigning the components among the equivalent positions. For example, consider the assignments

$(7, 5, 6, 1, 2, 3, 4), (7, 5, 6, 1, 2, 4, 3), (7, 5, 6, 2, 1, 3, 4),$ and $(7, 5, 6, 2, 1, 4, 3)$

for the system shown in Figure 9.1. All four assignments are equivalent and give the same system reliability.

Suppose components are renumbered such that $r_1 \leq r_2 \leq \cdots \leq r_n$. An assignment $\pi = (\pi_1, \ldots, \pi_n)$ of $1, \ldots, n$ is inadmissible if

$$i \overset{c}{>} j \text{ and } \pi_i < \pi_j \text{ for some pair } (i, j). \tag{9.13}$$

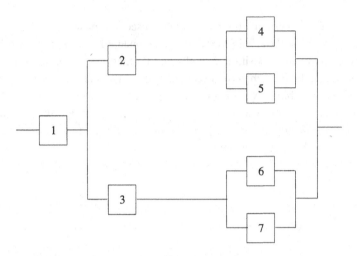

Figure 9.1. A seven-component coherent system

Suppose $i \overset{c}{>} j$ for some (i, j). If two components are to be assigned to positions i and j, it is always better to assign the component with the larger reliability to position i irrespective of the reliabilities of components in other positions. The procedure given by Boland and Proschan [39] for obtaining an optimal assignment is to:

1. collect only one assignment from each set of equivalent assignments,

2. eliminate all inadmissible assignments from the collection,

3. compute system reliability for each assignment in the remaining collection of inadmissible assignments and find the best one.

Example 9-7

Consider a seven-component system as shown in Figure 9.1. Here $1 \overset{c}{>} j$ for $j \geq 2$. Thus, let us first consider only the set A of 6! assignments that assign the largest reliable component 7 to position 1. Since $\{4, 5\}$ and $\{6, 7\}$ are sets of permutation equivalent positions, we can select from A only the assignments in which the index of the component assigned to position 4(6) is larger than that of the component assigned to 5(7). There are $6!/2^2$ such assignments in A.

Note that $2 \overset{c}{>} 4, 2 \overset{c}{>} 5, 3 \overset{c}{>} 6$ and $3 \overset{c}{>} 7$. This reduces the number of assignments to be searched to $6!/(2^2 3^2)$. It can easily be seen that the system reliability will not change if the components in positions 2, 4, and 5 are simultaneously interchanged with those in positions 3, 6, and 7, respectively. This further reduces the search to the following ten $[= 6!/(2^2 3^2 2)]$ assignments:

$(7, 6, 5, 4, 3, 2, 1)$
$(7, 6, 5, 4, 2, 3, 1)$
$(7, 6, 5, 4, 1, 3, 2)$
$(7, 6, 5, 3, 2, 4, 1)$
$(7, 6, 5, 3, 1, 4, 2)$
$(7, 6, 5, 2, 1, 4, 3)$
$(7, 6, 4, 5, 3, 2, 1)$
$(7, 6, 4, 5, 2, 3, 1)$
$(7, 6, 4, 5, 1, 3, 2)$
$(7, 6, 3, 5, 4, 2, 1)$

For any given component reliabilities, we have to renumber the components in increasing order of reliability and evaluate system reliability for the above ten assignments. The assignment that gives maximum system reliability is the optimal one.

9.4.2   The Malon greedy algorithm

Malon [208] proposed a greedy algorithm to maximize the reliability of a coherent system when the system is composed of several interchangeable modules, each of which consists of several positions, and interchangeable components are to be assigned to all modules. The greedy algorithm can be viewed as an extension of algorithm 1 (given for series-parallel structures) for general systems. It gives an optimal assignment when each module has a series structure and modules are to be optimally interchanged. The algorithm can briefly be described as follows.

First find the module that has the highest possible reliability when the most reliable components are assigned to it in an optimal manner. Assign the best components, as many as required in an optimal manner, to that module. Repeating this procedure with the remaining modules and the remaining components, select another module and assign the best available components to it. Repeat this procedure until components are assigned to all of the modules. Now rearrange all the modules of the system so as to maximize the reliability of the entire system. Malon [208] has proved that the greedy algorithm provides an optimal assignment irrespective of the system configuration when the modules have series structure.

9.4.3   The greedy algorithm of Lin and Kuo

Lin and Kuo [199] have presented the following heuristic algorithm for obtaining optimal component assignment for a general structure. Let $R[1_j, (a_1, \ldots, a_n)]$ denote the system reliability when the component reliability in the jth position is 1, and that in the ith position is a_i for $i \neq j$. Similarly, let $R[0_j, (a_1, \ldots, a_n)]$ denote the system reliability when a_i is the component reliability in the ith position for all $i \neq j$ and component reliability in the jth position is 0. The reliability importance of position j is

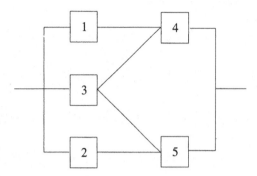

Figure 9.2. A complex coherent system

defined by Birnbaum [36] as:

$$I_j(a_1, \ldots, a_n) = \frac{\partial R(a_1, \ldots, a_n)}{\partial a_j}$$

$$= R[(1_j, (a_1, \ldots, a_n)] - R[(0_j, (a_1, \ldots, a_n)].$$

Let $I_j(p)$ denote the reliability importance of position j when the component reliability is p in every position. Let $I_j[(a_1, v_1), \ldots, (a_h, v_h); p]$ denote the reliability importance of position j when the component reliabilities in positions $v_1, \ldots, v_h$ are $a_1, \ldots, a_h$, respectively, and those in all other positions are equal to p.

Algorithm 9

- Step 0: Renumber the components such that $r_1 \geq r_2 \geq \cdots \geq r_n$. Let $J = \{1, \ldots, n\}$.

- Step 1: Find k such that

$$I_k(p_n) = \max_{j \in J} I_j(p_n).$$

Let $v_1 = k$, $J = J \setminus \{v_1\}$, and $h = 2$.

- Step 2: Find k in the set J such that

$$I_k[(r_1, v_1), (r_2, v_2), \ldots, (r_h, v_h); r_n] = \max_{j \in J} I_j[(r_1, v_1), (r_2, v_2), \ldots, (r_h, v_h); r_n],$$

and let $v_h = k$.

- Step 3: If $h = n$, assign component j to position v_j for $j = 1, \ldots, n$, and stop. Otherwise, let $h = h + 1$ and go to step 2.

Example 9-8

Consider the coherent system described in Figure 9.2. The system consists of five positions 1, 2, 3, 4, and 5. The system is neither series-parallel nor parallel-series.

Let the reliabilities of the components be $(r_1, r_2, r_3, r_4, r_5) = (0.5, 0.4, 0.3, 0.2, 0.1)$. The system reliability is given by

$$R(a_1, a_2, a_3, a_4, a_5) =$$
$$a_3[1 - (1 - a_4)(1 - a_5)] + (1 - a_3)[1 - (1 - a_1 a_4)(1 - a_2 a_5)],$$

where a_j is the reliability of the component in position j for $j = 1, 2, \ldots, 5$. The problem is to assign the five components to the five positions of the system so as to maximize the system reliability.

Solution of Example 9-8 from algorithm 9
- Iteration 1:

  - Step 0: The existing numbering of components ensures that $r_1 \geq r_2 \geq r_3 \geq r_4 \geq r_5$. Let $J = \{1, 2, 3, 4, 5\}$.
  - Step 1: $r_n = r_5 = 0.1$ and

  $$I_1(0.1) = R(1, 0.1, 0.1, 0.1, 0.1) - R(0, 0.1, 0.1, 0.1, 0.1)$$
  $$= 0.1171 - 0.0280$$
  $$= 0.0891.$$

  Similarly,

  $$I_2(0.1) = 0.1171 - 0.0280 = 0.0891,$$
  $$I_3(0.1) = 0.1900 - 0.0199 = 0.1701,$$
  $$I_4(0.1) = 0.1981 - 0.0190 = 0.1791,$$

  and

  $$I_5(0.1) = 0.1981 - 0.0190 = 0.1791.$$

  Since $I_4(0.1)$ is maximum among the values of $I_j(0.1)$ for $j \in J$, the value of k is taken as 4 and J is set to be $\{1, 2, 3, 5\}$. Let $h = 2$.

  $$I_1(0.1, 0.1, 0.1, 0.5, 0.1) = R(1, 0.1, 0.1, 0.5, 0.1) - R(0, 0.1, 0.1, 0.5, 0.1)$$
  $$= 0.5095 - 0.0640$$
  $$= 0.4455.$$

  Similarly,

  $$I_2(0.1, 0.1, 0.1, 0.5, 0.1) = 0.1855 - 0.1000 = 0.0855,$$
  $$I_3(0.1, 0.1, 0.1, 0.5, 0.1) = 0.5500 - 0.0595 = 0.4905,$$

  and

  $$I_5(0.1, 0.1, 0.1, 0.5, 0.1) = 0.2305 - 0.0950 = 0.1355.$$

Since $I_3(0.1, 0.1, 0.1, 0.5, 0.1)$ is maximum, $k = 3$, $v_h = v_2 = 3$, and $J = \{1, 2, 5\}$.

- Step 3: As $h = 2 < 4$, h is set to be 3. Go to step 2.

- Iteration 2:

  - Step 2:

    $$I_1(0.1, 0.1, 0.4, 0.5, 0.1) = R(1, 0.1, 0.4, 0.5, 0.1) - R(0, 0.1, 0.4, 0.5, 0.1)$$
    $$= 0.523 - 0.226$$
    $$= 0.297.$$

    Similarly,

    $$I_2(0.1, 0.1, 0.4, 0.5, 0.1) = 0.307 - 0.250 = 0.057$$

    and

    $$I_5(0.1, 0.1, 0.4, 0.5, 0.1) = 0.487 - 0.230 = 0.257.$$

    Since $I_1(0.1, 0.1, 0.4, 0.5, 0.1)$ is maximum, $k = 1$, $v_h = v_3 = 1$, and $J = \{2, 5\}$.
  - Step 3: As $h = 3 < 4$, h is set to be 4. Go to step 2.

- Iteration 3:

  - Step 2:

    $$I_2(0.3, 0.1, 0.4, 0.5, 0.1) = R(0.3, 1, 0.4, 0.5, 0.1) - R(0.3, 0, 0.4, 0.5, 0.1)$$
    $$= 0.361 - 0.310$$
    $$= 0.051.$$

    Similarly,

    $$I_5(0.3, 0.1, 0.4, 0.5, 0.1) = 0.541 - 0.290 = 0.251.$$

    Since $I_5(\mathbf{a}) > I_2(\mathbf{a})$, for $\mathbf{a} = (0.3, 0.1, 0.4, 0.5, 0.1)$, is maximum, $k = 5$, $v_h = v_4 = 5$ and $J = \{2\}$. Now take $v_5 = 2$ because J contains only the element 2.

Now assign components 1, 2, 3, 4, and 5 to positions 4, 3, 1, 5, and 2, respectively. The system reliability for this assignment is 0.3402. In fact, the assignment of components 1, 2, 3, 4, and 5 to the positions 3, 4, 5, 1, and 2, respectively, gives the maximum system reliability of 0.3438. The relative error of the heuristic solution is only about 1 percent.

9.4.4   Invariant optimal assignments

Suppose there exists a sequence $(v_1, \ldots, v_n)$ of n positions such that the assignment of the jth most reliable component to position v_j for $j = 1, \ldots, n$ maximizes the system reliability irrespective of the magnitudes of component reliabilities. This is called *invariant optimal assignment* and is denoted by the sequence $(v_1, \ldots, v_n)$. Some systems admit invariant optimal assignments. For example, the optimal component assignment derived by El-Neweihi et al. [86] for a series-parallel system and that derived by Derman et al. [77] for a parallel-series system with two components in each cut set are invariant optimal assignments. The advantage of invariant optimal assignment is that it is enough to know the increasing or decreasing order of component reliabilities without any knowledge of their actual magnitudes to find an optimal assignment.

El-Neweihi et al. [87] have considered the problem of assembling n systems from k types of components. Suppose system j, $1 \le j \le n$, is a series system consisting of m_{ij} components of type i, $1 \le i \le k$. El-Neweihi et al. have assumed that $m_{i1} \le m_{i2} \le \cdots \le m_{in}$ for each i. The total number of components of type i available for assembly is $M_i = \sum_{j=1}^{n} m_{ij}$. Let r_{ih} denote the reliability of the hth component of type i for $h = 1, \ldots, M_i$. Assume, without loss of generality, that $r_{i1} \le r_{i2} \le \cdots \le r_{iM_i}$. Consider the allocation A^* in which for each type i, the m_{i1} lowest reliable components are allocated to system 1, the next m_{i2} lowest reliable components are allocated to system 2, and so on. Let $R_j(A)$ denote the reliability of system j for allocation A and let $X_j(A) = \ln R_j(A)$. El-Neweihi et al. [87] have shown that vector $[X_1(A^*), \ldots, X_n(A^*)]$ majorizes $[X_1(A), \ldots, X_n(A)]$ for any allocation A.

If a reliability system G consists of all n systems as subsystems and the reliability of G is a Schur-convex function of $X_1(A), \ldots, X_n(A)$, then allocation A^* maximizes the reliability of G. This result generalizes the work of El-Neweihi et al. [86] on series-parallel systems. As a special case, allocation A^* maximizes the expected number of working components among the n systems. El-Neweihi et al. [87] have derived more general results for the case $m_{ij} = 1$ for all (i, j). In this case, a permutation $(a_{i1}, \ldots, a_{in})$ of $1, \ldots, n$ for each component type i determines an allocation. Let c_{ij} denote an attribute associated with component a_{ij}. The attribute c_{ij} may also be the actual reliability of component a_{ij}. Let $R_j(c_{1j}, \ldots, c_{kj})$ denote the corresponding reliability of system j. Let $x \vee y = \max(x, y)$ and $x \wedge y = \min(x, y)$. El-Neweihi et al. [87] have shown that if

1. $R_j(c_1, \ldots, c_k)$ is nondecreasing in each c_i,

2. $R_j(c_1, \ldots, c_k) + R_j(d_1, \ldots, d_k) \le R_j(c_1 \vee d_1, \ldots, c_k \vee d_k) + R_j(c_1 \wedge d_1, \ldots, c_k \wedge d_k)$,

then allocation A^* maximizes the reliability vector $[R_1(A), \ldots, R_n(A)]$ in a weak majorization sense. By taking $c_{ij} = \infty$ for the appropriate pairs (i, j), this result has been extended to the case where all n subsystems do not have the same number of components.

Table 9.9. Invariant optimal designs of linear consecutive k-out-of-n systems

k	F system	Reference	G system	Reference
$k = 1$	Any arrangement		Any arrangement	
$k = 2$	$(1, n, 3, n-2, \ldots, n-3, 4, n-1, 2)$	Malon [206], Du and Hwang [82]	Does not exist	Zuo and Kuo [337]
$2 < k < n/2$	Does not exist	Malon [207]	Does not exist	Zuo and Kuo [337]
$n/2 \leq k < n-2$	Does not exist	Malon [207]	Unknown	
$k = n-2$	$[1, 4, \text{(any arrangement)}, 3, 2]$	Malon [207]	$[1, 3, 5, \ldots, 2(n-k)-1,$ (any arrangement), $2(n-k), \ldots, 6, 4, 2]$	Kuo et al. [181]
$k = n-1$	$[1, \text{(any arrangement)}, 2]$	Malon [207]		
$k = n$	Any arrangement		Any arrangement	

Table 9.10. Invariant optimal designs of circular consecutive k-out-of-n systems

k	F system	Reference	G system	Reference
$k = 1$	Any arrangement		Any arrangement	
$k = 2$	$(1, n-1, 3, n-3, \ldots, n-4, 4, n-2, 2, n, 1)$	Du and Hwang [82]	Does not exist	Zuo and Kuo [337]
$2 < k < (n-1)/2$	Does not exist	Malon [207]	Does not exist	Zuo and Kuo [337]
$(n-1)/2 \leq k < n-2$	Does not exist	Malon [207]	Unknown	
$k = n-2$	$(1, n-1, 3, n-3, \ldots, n-4, 4, n-2, 2, n, 1)$	Zuo and Kuo [337]	$(1, 3, 5, \ldots, n, \ldots, 6, 4, 2, 1)$	Zuo and Kuo [337]
$k = n-1$	Any arrangement	Kuo et al. [181]	Any arrangement	Kuo et al. [181]
$k = n$	Any arrangement		Any arrangement	

Considerable work has been done on invariant optimal assignment in consecutive k-out-of-n systems. For consecutive 2-out-of-n: F linear systems, Derman et al. [79] conjectured that when $r_1 \leq r_2 \leq \cdots \leq r_n$, the assignment

$$\mathbf{u}^* = (1, n, 3, n - 2, \ldots, n - 3, 4, n - 1, 2)$$

maximizes system reliability. Du and Hwang [82] proved the conjecture of Hwang [139] that the assignment

$$\mathbf{v}^* = (n, 1, n - 1, 3, n - 3, \ldots, n - 4, 4, n - 2, 2, n)$$

maximizes the reliability of the consecutive 2-out-of-n: F cyclical system. Hwang [139] has shown that his conjecture for n components reduces to that of Derman et al. [79] for $(n - 1)$ components when the largest component reliability r_n is 1.0. Note that the assignments $\mathbf{u}^*$ and $\mathbf{v}^*$ are invariant optimal for linear and cyclical consecutive 2-out-of-n: F systems, respectively. Malon [206] has directly and independently proved the conjecture for the consecutive 2-out-of-n: F linear system. Later, Malon [207] has shown that the consecutive k-out-of-n: F linear system admits invariant optimal component assignment if, and only if, $k \in \{1, 2, n - 2, n - 1, n\}$. Kuo et al. [181] and Zuo and Kuo [337] have considered the component assignment problem in the consecutive k-out-of-n: G system, which consists of n components in a sequence such that it works, if and only if, at least k consecutive components work. Zuo and Kuo [337] have summarized the work on optimal invariant assignments in consecutive k -out-of-n systems as shown in Tables 9.9 and 9.10.

Zuo and Kuo [337] have developed two heuristics based on Birnbaum's component reliability importance and a randomization method for obtaining suboptimal assignments for linear consecutive k-out-of-n systems. They have also suggested a binary search method for deriving an exact optimal assignment for such systems. For a special type of consecutive k-out-of-n system, Papastavridis and Sfakianakis [259] presented an algorithm for optimal arrangement.

9.5   Discussion

The problem of assigning interchangeable components in a reliability system can be viewed as nonlinear. This is also described by several researchers as an optimal system assembly problem, which was initially considered by Derman et al. [77], [78]. Most of the work on this problem focuses on invariant optimal assignments for position-independent component reliabilities. Series-parallel and some consecutive k-out-of-n systems admit invariant optimal assignments. For reference, see Derman et al. [77], [78]; El-Neweihi et al. [86], [87]; Malon [206]–[208]; Hwang [139], [140]; Du and Hwang [82], [83]; Kuo et al. [181]; and Zuo and Kuo [337]. Heuristics are presented by Malon [208], Baxter and Harche [24], Lin and Kuo [199], Zuo and Kuo [337], and Prasad and Raghavachari [273]. Boland et al. [40] have developed an implicit enumeration method for a general coherent system based on comparison of the criticality of two positions. Prasad et al. [266], [267], [272] have developed exact algorithms and heuristics for series-parallel and parallel-series systems when the component reliabilities are not position-independent.

Regarding the solution methodology, El-Neweihi et al. [86], [87] have exploited the nature of Schur-convex functions in deriving invariant optimal assignments. The same approach is adopted by Prasad et al. [272]. Prasad and Raghavachari [267] have used mathematical programming techniques along with the concept of majorization. Prasad et al. [266] have used bicriteria linear programming methods to derive optimal assignments in series-parallel and parallel-series systems consisting of only two subsystems.

EXERCISES

9.1 Discuss the mathematical nature of the component assignment problem. How does the structure of the system affect the solution procedure for such a problem?

9.2 In Example 9-1, let $\alpha_h = (0.8010, 0.5410, 0.9100, 0.7942)$. Find the solutions using algorithms 1 and 2. Are these solutions optimal?

9.3 For Exercise 9.2, find all nondominated totally ordered (NDTO) allocations and the solution by algorithm 3. Is this optimal? Compare this solution with the solutions of Exercise 9.2.

9.4 Apply algorithm 4 to Exercise 9.2 to find the system reliability. Let $z_h = d_h + \sum_{j \in A_h}$. Is there any algorithm similar to algorithm 4?

9.5 In Example 9-4, let $(n_1, n_2, n_3) = (3, 4, 5)$. Find the allocation by the Baxter–Harche method.

9.6 A series-parallel system has two path sets, P_1 and P_2, and $n_1 = n_2 = 2$. Find the solution by algorithm 5. The corresponding t_{ij} matrix is given below.

$$\begin{pmatrix} 0.446 & 0.163 & 0.020 & 0.462 \\ 0.117 & 0.051 & 0.288 & 0.128 \\ 0.315 & 0.462 & 0.174 & 0.211 \\ 0.010 & 0.288 & 0.342 & 0.223 \end{pmatrix}$$

9.7 Apply the Baxter–Harche method to solve Example 9-5. Take this solution to be the initial solution to perform three iterations of the Prasad–Raghavachari method.

9.8 Given the following configuration,

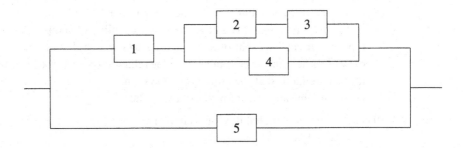

where $\mathbf{r} = (0.9, 0.8, 0.7, 0.6, 0.5)$, find the optimal assignment by the procedure of Boland et al. [40].

9.9 Find the allocation of Exercise 9.8 by the greedy algorithm of Lin and Kuo. Compare this allocation with the optimal allocation.

9.10 What is the invariant optimal assignment? Give an example of the use of an invariant optimal assignment and discuss its characteristics.

9.11 Consider a series-parallel system with k subsystems. Show that if components assigned to two subsystems r and s are reassigned in such a way that the deviation between a-hazard values z_r and z_s increases, then the system reliability increases.

9.12 Show that reassignment of components in two subsystems of a parallel-series system as described in Exercise 9.11 reduces system reliability.

9.13 Formulate the component allocation problem in series-parallel systems as a 0–1 programming problem.

9.14 Enumerate all ordered allocations in Example 9-1 and evaluate system reliability for each one of them. Identify all totally ordered allocations in Example 9-1.

9.15 Show that there exists an ordered allocation that maximizes the reliability of a series-parallel system over all allocations.

9.16 Consider a parallel-series system with k cut sets each consisting of two positions. Suppose the interchangeable components of this system are numbered such that $r_1 \leq \cdots \leq r_n$ or $r_1 \geq \cdots \geq r_n$ with $n = 2k$ and are paired up as $(1, n), (2, n - 1), \ldots, (n/2, n/2 + 1)$. Show that the allocation of these k pairs to k cut sets of the system maximizes system reliability.

9.17 Suppose the component reliabilities of a parallel-series system with four positions in each cut set are in geometric progression, that is, $r_j = c^j r$ for some $c > 0$ and $r > 0$. Show that the allocation: $A_1 = \{1, 2, n - 1, n\}$, $A_2 = \{3, 4, n - 3, n - 2\}, \ldots, A_k = \{n/2 - 1, n/2, n/2 + 1, n/2 + 2\}$ is optimal. Does any other type of optimal allocation exist for this system? If so, find it.

9.18 Show that if the reliability of component i in one type of consecutive k-out-of-n system (e.g. an F system) is equal to the unreliability of component i in the other type of consecutive k-out-of-n system (e.g. a G system) for all i, and if both types of systems have the same k and n, then the reliability of one type of system is equal to the unreliability of the other type of system. These consecutive k-out-of-n: G and F systems are mirror images of each other.

9.19 Given that the unreliability of a circular consecutive k-out-of-n: F system with $n \leq 2k + 1$ is

$$
Q_{CF}(n; k) = \sum_{i=1}^{n} \left(p_{i+k} \prod_{j=i}^{i+k-1} q_j \right) + \prod_{i=1}^{n} q_i
$$

$$
= \sum_{i=1}^{n} \left(\prod_{j=i}^{i+k-1} q_j \right) - \sum_{i=1}^{n} \left(\prod_{i=1}^{n} q_i \right) + \prod_{i=1}^{n} q_i,
$$

where $q_j = 1 - p_j = q_{j-n}$ if $j > n$, and p_j is the reliability of component i, prove that a necessary condition for the optimal design of a circular consecutive k-out-of-n: F system with $n = k + 2$ is

$(q_i - q_j)(q_{i-1} - q_{j+1}) < 0,$ for $j = 1 + 1, i + 2,$

$(q_i - q_j)[q_{i+1}q_{j-1}(q_{i-1} - q_{j+1}) + q_{i-1}q_{j+1}(q_{i-1} - q_{j+1})] < 0,$

for $j \geq i + 3,$

where i ranges from 1 to n. (Hint: First prove that it is true for $j = i + 1$, then for $j = i + 2$, and finally for $j \geq i + 3$.)

10 Reliability systems with multiple objectives

10.1 Introduction

Decision making is the process of selecting a possible course of action from all available alternatives. A major concern is that almost all decision problems have multiple, usually conflicting, criteria. Significant work has been done toward the development of solution methodologies for such problems. The methods for various multiple criteria decision-making (MCDM) problems are very diverse. However, even with such diversity, all of these methods share the following common characteristics: (1) a set of criteria for judgment, (2) a set of decision variables, and (3) a process of comparing the alternatives. MCDM problems can be broadly classified into two categories: (1) multiple attribute decision making (MADM), and (2) multiple objective decision making (MODM).

The distinguishing feature of MADM is that there are usually a limited (and countably small) number of predetermined alternatives. The alternatives are associated with the level of achievement of the attributes, which may not necessarily be quantifiable, based on which decisions are to be made. Final selection of the alternative is made with the help of inter- and intra-attribute comparisons with explicit or implicit tradeoffs. Criteria for job selection might include prestige, location, salary, advancement opportunities, work conditions, and others. When we want to buy a personal computer (PC), we consider its attributes: price, memory size, style, mechanical quality, processing speed, reliability, etc.

MODM, on the other hand, is not associated with problems where the alternatives are predetermined. The thrust of this model is to design the "best" alternative by considering the various interactions within the design constraints which best satisfy the decision maker by way of attaining acceptable-level objectives. Common characteristics of the MODM method are: (1) a set of quantifiable objectives; (2) a set of constraints; and (3) a process of obtaining some tradeoff information, implicit or explicit, between the stated objectives. Thus, MODM is associated with design problems (in contrast to selection problems for MADM). PC manufacturers want to design a model which maximizes memory size, reliability, mechanical quality, etc., and minimizes production cost, size, etc. A water resource development plan for a community could be evaluated in terms of costs, probability of water shortage, energy (reuse factor), recreation, flood protection, land and forest use, water quality, etc. The choice of missile systems for an air force would be based on speed, yield, accuracy, range, vulnerability, reliability, etc.

In dealing with MCDM, the four important keywords are: attributes, objectives, goals, and criteria. Some authors make distinctions in their usage of these keywords, while many use them interchangeably. Hwang and Yoon [138], and Hwang and Masud [135] define the keywords as follows:

- Criteria: are a measure of effectiveness and a basis for evaluation. Criteria emerge as a form of attributes, or objectives, in the actual problem setting.

- Goals: (synonymous with targets) are priority values or levels of aspiration. They are either to be achieved, suppressed or not exceeded. Often we refer to them as constraints, because they are designed to limit and restrict the alternative set. For example, the standard gas (petrol) mileage, say 20 miles/gallon, set up by the US Federal Government for 1998 models, is a constraint; whereas 30 miles/gallon may serve as a goal for the car manufacturer.

- Attributes: performance parameters, components, factors, characteristics, and properties are synonyms for attributes. An attribute should provide a means of evaluating the levels of an objective. Each alternative can be characterized by a number of attributes (chosen by a decision maker's conception of criteria, that is, gas mileage, purchasing cost, horsepower, etc. of a car).

- Objectives: an objective is something to be pursued to its fullest. For example, a car manufacturer may want to maximize gas mileage, minimize production cost, or minimize the level of air pollution. An objective generally indicates the direction of change desired.

Since the purpose of Chapter 10 is to provide state-of-the-art methods and applications for designing optimal reliability systems, the remainder of the chapter will focus on MODM problems and applications. In the last three decades, MODM techniques have been applied to solve practical problems such as academic planning, econometrics and development planning, financial planning, capital budgeting, portfolio selection, health care planning, land-use planning, manpower planning, production planning, public administration, systems reliability, transportation planning, traffic management, water resource management, forest management, etc. These problems can be mathematically represented as follows.

Problem 10.1

Maximize

$$\mathbf{z} = [f_1(\mathbf{x}), f_2(\mathbf{x}), \ldots, f_k(\mathbf{x})],$$

subject to

$$g_i(\mathbf{x}) \leq 0, \qquad \text{for } i = 1, \ldots, m, \tag{10.1}$$

where $\mathbf{x}$ is an n-dimensional vector of decision variables. The problem consists of n decision variables, m constraints, and k objectives. The functions $f_j(\mathbf{x})$ and $g_i(\mathbf{x})$ may be linear or nonlinear. In the literature, this problem is often referred to as a vector maximization problem (VMP).

If the objectives of the original problem are to maximize $f_l(\mathbf{x})$ for $l = 1, \ldots, h$ and minimize $f_l(\mathbf{x})$ for $l = h+1, \ldots, k$, then the objective in the mathematical formulation of Problem 10.1 is to maximize vector

$$\mathbf{z} = [f_1(\mathbf{x}), f_2(\mathbf{x}), \ldots, f_h(\mathbf{x}), -f_{h+1}(\mathbf{x}), -f_{h+2}(\mathbf{x}), \ldots, -f_k(\mathbf{x})].$$

Problem 10.2

MODM problems are, sometimes, also formulated so as to minimize

$$\mathbf{z} = [f_1(\mathbf{x}), f_2(\mathbf{x}), \ldots, f_k(\mathbf{x})],$$

subject to

$$g_i(\mathbf{x}) \leq 0, \qquad \text{for } i = 1, \ldots, m.$$

Traditionally, there are two approaches for solving the VMP. One approach is to optimize an objective while appending the other objectives to a constraint set, so that the optimal solution will satisfy these objectives at least up to a predetermined level. The problem is given so as to maximize

$$z = f_l(\mathbf{x}),$$

subject to

$$g_i(\mathbf{x}) \leq 0, \qquad \text{for } i = 1, \ldots, m,$$

$$f_h(\mathbf{x}) \geq a_h, \qquad \text{for } h = 1, \ldots, k \text{ and } h \neq l,$$

where a_h is an acceptable predetermined level for the hth objective.

The other approach is to optimize a super-objective function, created by multiplying each objective function with a suitable weight and then by adding them together. In the second approach, we maximize

$$z = \sum_{h=1}^{k} w_h f_h(\mathbf{x}),$$

subject to

$$g_i(\mathbf{x}) \leq 0, \qquad \text{for } i = 1, \ldots, m,$$

where w_h is the weight of the hth objective such that $\sum_{h=1}^{k} w_h = 1$.

Both of the above approaches are *ad hoc* at best. Often they lead to a solution which may not be the best or most satisfactory. Because of the incommensurability and the conflicting nature of the multiple criteria, the problem becomes complex, and it becomes difficult to choose acceptable levels of a_hs, which will result in a nonempty set in the first attempt at a solution. In the first approach, the implied value tradeoff between f_l and $f_i, i \neq l$, is

$$\text{Value tradeoff} = \begin{cases} 0, & f_l \geq a_l, \\ \infty, & f_l < a_l. \end{cases}$$

This may not be the actual value structure, and is sensitive to the level of a_l. For the second approach, the major problem is in determining the proper weight w_i, which is sensitive to the level of the particular objective as well as the levels of all other objectives.

MODM methods are the result of the desire to eliminate the above difficulties as well as to treat the objectives independently. Most of the progress in this area has taken place within the last two decades. One of the most thorough reviews of the existing literature and a systematic classification of methods for the guidance of future users is provided by Hwang and Masud [135].

10.2   Classification of multiple objective decision making

Conventional methodologies for solving MODM problems have been systematically classified and discussed by Hwang and Masud [135], as shown in Table 10.1. The first class requires no information from the decision maker once problem constraints and objectives have been defined. The analyst makes assumptions about the decision maker's preference and presents solution(s) to the decision maker. The second class assumes that the decision maker has, consciously or subconsciously, a set of goals to achieve and these are given to the analyst prior to the formulation of a mathematical model. The third class, also known as interactive methods, requires greater decision-making involvement in the solution process. The interaction takes place through a DM–analyst or DM–computer dialogue at each iteration. Tradeoff, or preference information, given by the decision maker after each iteration is used for determining a new solution. Hence, the decision maker actually gains an insight into the problem. Finally, the fourth class does just one thing: it determines a subset of the complete set of nondominated solutions to an MODM problem. In doing so, it deals strictly with physical constraints and makes no attempt to consider the decision maker's preference. The aim of this class, however, is to narrow down possible courses of action and to make the decision maker's selection of the preferred course of action easier.

In many cases, multiplicity of criteria for judging the alternatives is pervasive. Often the decision maker wants to attain more than one objective or goal in selecting a course of action, while satisfying constraints dictated by environment, processes, and resources.

Table 10.1. A taxonomy of methods for MODM (Hwang and Masud [135])

Stage at which information is needed	Type of information	Methods
Class 1 No articulation of preference information		Global criterion
Class 2 *A priori* articulation of preference information	Cardinal information	Utility function
	Ordinal and cardinal information	Bounded objective Lexicographic Goal programming Goal attainment
Class 3 Progressive articulation of preference information	Explicit tradeoff	Geoffrion Interactive goal programming Surrogate worth tradeoff Satisfactory goals Zionts–Wallenius
	Implicit tradeoff	STEM and related SEMOPS and SIGMOP Displaced ideal GPSTEM Steuer
Class 4 *A posteriori* articulation of preference information (nondominated solution generation method)	Implicit tradeoff	Parametric Constraint MOLP Adaptive search

10.3 MODM solutions

We need some basic terminology for discussing MODM problems, which are reviewed in this section. Consider Problem 10.1. Let $\mathbf{X}$ denote the set of all $\mathbf{x}$ that satisfies eq. (10.1) and S denote the objective function space. The MODM solutions to the problem (Figure 10.1) can be defined as:

1. optimal,

2. positive-ideal and negative-ideal,

3. nondominated (Pareto optimal),

4. preferred, and

5. satisfying.

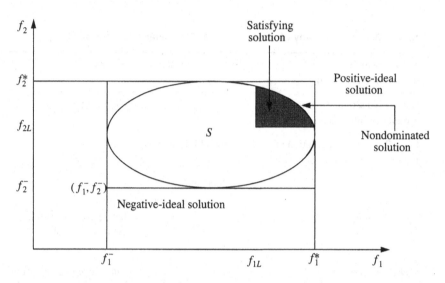

Figure 10.1. Solutions in the objective function space

An optimal solution

An optimal solution of the MODM is one which, simultaneously, results in the maximum value of each objective function. That is, $\mathbf{x}^*$ is an optimal solution to Problem 10.1, if $\mathbf{x}^* \in \mathbf{X}$ and $f_h(\mathbf{x}^*) \geq f_h(\mathbf{x}), \forall h$ for any $\mathbf{x} \in \mathbf{X}$. Figure 10.2 illustrates an optimal solution for the case of one decision variable and two objective functions. In the decision variable space representation, $\mathbf{X}$ is the set of all values between the upper limit x^U and the lower limit x^L. The two objectives, $f_1(x)$ and $f_2(x)$, simultaneously attain a maximum when $x = x^*$.

In objective function space representation, the optimal solution is located within the boundary of the feasible space S. Here, the optimal solution is also known as the superior solution or maximum solution. There is usually no optimal solution to an MODM problem because of conflicting objectives.

A positive-ideal solution

The positive-ideal solution is one that optimizes each objective function simultaneously. Let $f_h^* = \max_{\mathbf{x} \in \mathbf{X}} f_h(\mathbf{x})$ for maximization objectives, and $f_h^* = \min_{\mathbf{x} \in \mathbf{X}} f_h(\mathbf{x})$ for minimization objectives. A positive-ideal solution can then be defined as $A^* = \{f_1^*, f_2^*, \ldots, f_k^*\}$.

A negative-ideal solution

let $f_h^- = \min_{\mathbf{x} \in \mathbf{X}} f_h(\mathbf{x})$ for maximization objectives, and $f_h^- = \max_{\mathbf{x} \in \mathbf{X}} f_h(\mathbf{x})$ for minimization objectives. A negative-ideal solution can then be defined as $A^- = \{f_1^-, f_2^-, \ldots, f_k^-\}$, where f_h^- is the feasible and worst value for the hth objective function.

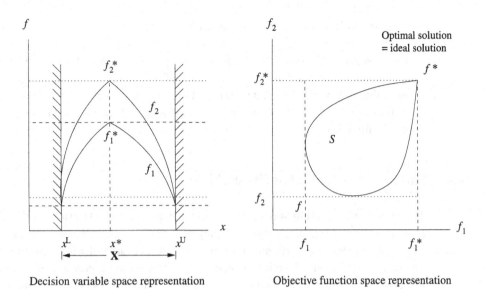

Decision variable space representation   Objective function space representation

Figure 10.2. An optimal solution of a maximization problem

A nondominated (Pareto optimal) solution

This solution is named differently in different doctrines: it is labeled a noninferior solution and an efficient solution in MODM; an admissible alternative in statistical decision theory; and a Pareto-optimal solution in economics.

Mathematically, a feasible solution in MODM is nondominated if no other feasible solution exists that will yield an improvement in one objective/attribute without causing a degradation in any other objective/attribute. It means that $\mathbf{x}^{ND}$ is a nondominated solution to the MODM problem if $\mathbf{x} \in \mathbf{X}$ does not exist such that $f_l(\mathbf{x}^{ND}) \leq f_l(\mathbf{x})$ for all l and $f_h(\mathbf{x}^{ND}) < f_h(\mathbf{x})$ for at least one h for maximization. In general, the number of nondominated solutions is quite large, and therefore, the decision maker must make a final selection of the most satisfactory solution by using other criteria.

A preferred solution

A preferred solution is a nondominated solution chosen by the decision maker through some additional criteria. As such, it lies in the region of acceptance of all criteria values for the problem. This preferred solution is also known as the best solution or final decision.

A satisfying solution

A satisfying solution when extended to an MODM problem, is a reduced subset of the feasible set, which exceeds all aspiration levels (goals) of each objective/attribute. However, satisfying solutions need not be nondominated. This method is credited for its simplicity, which matches the behavior processes of the decision maker whose knowledge and ability are limited. A satisfying solution may be taken as the final solution, though it is often utilized for screening out unacceptable solutions.

Traditionally, utility theory has been extensively applied to MODM to choose among a set of nondominated solutions. It assumes that the decision maker will choose the solution that will yield the greatest satisfaction. Thus, we can completely order the set of nondominated solutions, and the best nondominated solution is the one with the highest utility value.

10.4   Multiple objective reliability problems

The reliability of a multi-stage system can be improved by: (1) using more reliable components, and (2) adding redundant components in parallel. In many practical design situations, reliability apportionment is complicated because of the presence of several (mutually) conflicting objectives which cannot be combined into a single-objective function. For instance, a designer is required to minimize system cost and system weight, while simultaneously maximizing system reliability. Therefore, multiple objective functions become an important aspect in the reliability design of engineering systems.

Several techniques have been developed for solving constrained optimal-reliability allocation problems. In most of the work on reliability optimization, optimal redundancy is determined at each stage of the system assuming that component reliabilities are fixed. However, a more general problem is one where both the optimal component reliability and the optimal redundancy at each stage are determined to obtain the maximum system reliability. This results in a nonlinear mixed integer programming problem for which heuristic solution procedures have been described in Chapter 8.

Usually, two or more independent (conflicting) criteria are important in determining the replacement age of a critical item for a maintained system. In Hwang et al. [137], mathematical models have been developed for three such criteria: (1) minimum replacement cost-rate, (2) maximum availability, and (3) lower bound on mission reliability. The solutions are obtained using three methods for multiple criteria decision making: (1) strictest-selection, (2) lexicographic, and (3) the Waltz technique (SEMOPS).

A multiple objective formulation of a reliability allocation problem to maximize system reliability and minimize system cost has been considered by Sakawa [278] using surrogate worth tradeoff methods. Inagaki et al. [145] used interactive optimization to design a system with minimum cost and weight, and maximum reliability. Sakawa [279] solved a multiple objective reliability and redundancy allocation problem using an approximate technique based on the surrogate worth tradeoff method. Nakagawa [243] suggested a combined strategy of narrowing the feasible region, generating Pareto-optimal solutions, and, finally, selecting the best Pareto-optimal solution based upon the designer's experience. Later on, Sakawa [282] dealt with the mixed integer programming problem (reliability and redundancy allocation) and used a combination of the surrogate worth tradeoff method and the dual decomposition method while still treating the integer variables as continuous variables. Sakawa [283] conceptualized a technique called SPOT, which is basically an interactive decision-making procedure of choosing a preferred solution from a set of Pareto-optimum solutions. Misra and Sharma [231] provide an exact yet efficient search method to solve a wide variety of reliability design problems involving integer programming formulation. In [232], Misra and Sharma provided a mathematical formulation of a combined reliability and redundancy allocation problem with multiple objective optimization considering mixed types of redundancy. Dhingra [80] presented a multiple objective reliability apportionment problem for a series system. The problem he solved is a nonlinear mixed integer programming problem subjected to several design constraints. Sasaki et al. [287] use a fuzzy multiple objective 0–1 linear programming method which can adapt to changes in the utilization of system resources and the economic environment by making use of tolerances in the system control parameters. Their method considers the variability which arises from the subjective judgment of the decision maker by introducing fuzzy logic into the goal specification. They optimize planning unit selection and redundant distribution in order to maximize system availability.

10.5   Reliability–redundancy allocation with multiple objectives

We now provide an illustration of multiple objective optimization in reliability systems based on the methods presented by Hwang and coworkers [48], [134], [135], [138], [183], [331].

10.5.1   Problem description

Consider the overspeed protection system for the gas turbine depicted in Figure 10.3. Overspeed detection is continuously provided by the electrical and mechanical systems. When an overspeed occurs, it is necessary to cut off the fuel supply. For this purpose, four control valves (V1–V4) must close. The control system is modeled as a four-stage series system. Assume that all components have a constant failure rate.

Notation

b_i	upper limit on $g_i(\cdot)$
c_h	relative importance of objective h
d_h^+, d_h^-	over- and under-achievement from the goals for objective function f_h
F_h	$m_h(f_h)$, where $m_h = M/[f_h(\mathbf{x}^0)]$ for $h = 1, \ldots, k$ (a scaled value of f_h)
f_1	objective function for system reliability
f_2	objective function for system cost
f_3	objective function for system weight
f_h^0	goal set by the designer for objective h
$g_i(\cdot)$	ith constraint function

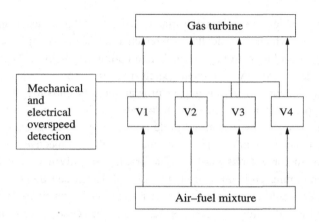

Figure 10.3. Schematic diagram for the overspeed protection system of a gas turbine

M	a constant
m	number of constraints
n	number of stages in the reliability system
$\mathbf{r}$	$(r_1, \ldots, r_n)$
R_s	system reliability
r_j	reliability of component in stage j
$\mathbf{x}$	$(x_1, \ldots, x_n)$
$\mathbf{x}^0$	starting design vector
x_j	number of redundant components in stage j

Problem 10.3

The multiple objective optimization problem considered by Dhingra [80] in this case is to minimize

$$\mathbf{z} = (-f_1, f_2, f_3),$$

subject to

$$g_i(\mathbf{r}, \mathbf{x}) \le b_i, \qquad \text{for } i = 1, \ldots, m.$$

The design constraints

1. A combination of weight and volume:

$$V = \sum_{j=1}^{n} v_j x_j^2 \le V_{\text{lim}}, \tag{10.2}$$

where v_j is the product of weight and volume per element at stage j.

2. The total system weight:

$$W = f_3 = \sum_{j=1}^{n} w_j x_j \exp\left(\frac{x_j}{4}\right) \leq W_{\lim}, \tag{10.3}$$

where w_j is the weight of each component at stage j. The term $\exp(x_i/4)$ accounts for the interconnecting hardware.

3. System reliability:

$$R_s = f_1 = \prod_{j=1}^{n} [1 - (1 - r_j)^{x_j}] \geq R_0. \tag{10.4}$$

4. The total cost:

$$C = f_2 = \sum_{j=1}^{n} c(r_j) \left[x_j + \exp\left(\frac{x_j}{4}\right) \right] \leq C_0, \tag{10.5}$$

where $c(r_j)$ is the cost of the component with reliability r_j at stage j.

The right-hand sides of eqs. (10.2)–(10.5) are constants.

Additional assumptions
1. The cost–reliability relation is

$$c(r_j) = \alpha_j / \lambda_j^{\beta_j}, \tag{10.6}$$

where α_j and β_j are constants representing the physical characteristics of each component at stage j, and λ_j is the constant component failure rate at stage j, that is, the component failure time at stage j follows an exponential distribution with mean $1/\lambda_j$. Since the component failure rate is constant, eq. (10.6) can be written as

$$c(r_j) = \alpha_j \left[\frac{-t}{\ln(r_j)} \right]^{\beta_j}, \tag{10.7}$$

where t is the mission time. Now, eq. (10.5) becomes

$$\sum_{j=1}^{n} \alpha_j \left[\frac{-t}{\ln(r_j)} \right]^{\beta_j} \left[x_j + \exp\left(\frac{x_j}{4}\right) \right] \leq C_0. \tag{10.8}$$

2. For all j

$$1 \leq x_j \leq 10, \, x_j \text{ being integer}, \tag{10.9}$$

$$0.5 \leq r_j \leq 1 - 10^{-6}, \, r_j \text{ being a real number}. \tag{10.10}$$

Problem 10.3 can now be stated as to select r_j and x_j at each stage j, such that system reliability R_s is maximized and system cost C and system weight W are minimized subject to eqs. (10.2)–(10.5), (10.8)–(10.10).

Table 10.2. Design data for reliability problem[a]

j	$10^5 \times \alpha_j$	β_j	v_j	w_j
1	1.0	1.5	1	6
2	2.3	1.5	2	6
3	0.3	1.5	3	8
4	2.3	1.5	2	7

[a] Number of stages $j = 4$;
$f_1^0 = R_0 =$ lower limit on $R_s = 0.75$;
$f_2^0 = C_0 =$ upper limit on cost $= 400.00$;
$f_3^0 = W_{\lim} =$ upper limit on weight $= 500.00$;
$f_3^0 = V_{\lim} =$ upper limit on volume $= 250.00$;
operating time $= 1000$ hours.

Table 10.3. Optimal solutions in single-objective optimization

Objective	j	r_i	x_i	Attributes
	1	0.816 04	6	$R_s = 0.999\,61$
max R_s	2	0.803 09	6	$C = 399.936$
	3	0.983 64	3	$W = 495.652$
	4	0.803 73	5	$V = 185.0$
	1	0.500 00	4	$R_s = 0.7604$
min C	2	0.500 00	4	$C = 20.7252$
	3	0.592 51	5	$W = 314.548$
	4	0.500 00	3	$V = 141.0$
	1	0.962 21	1	$R_s = 0.807\,86$
min W	2	0.923 15	1	$C = 399.509$
	3	0.987 87	1	$W = 34.6687$
	4	0.920 65	1	$V = 8.0$

Solutions of single-objective optimization problems

The optimal reliability allocation problem is a nonlinear mathematical programming problem with discrete as well as continuous design variables. A blending of sequential unconstrained minimization techniques and heuristic algorithms (Chapter 3) can be used to determine a heuristic optimum solution. Methods based on this approach are described in Chapter 8. Table 10.3 presents the results obtained by solving the three single-objective optimization problems for the data shown in Table 10.2. These solutions were obtained by Dhingra [80].

10.5.2 The multiple objective optimization approach

A general MODM problem with k objective functions can be transformed into Problem 10.2.

In general, no solution minimizes all objectives simultaneously. The concept of a Pareto-optimal solution [88] has been widely used to characterize optimal solutions for MODM programs, and in the context of minimizing a vector of objective functions, these can be defined as follows.

A solution $\mathbf{x}^*$ in the set $\mathbf{X}$ of feasible solutions is *Pareto optimal* if no $\mathbf{x} \in \mathbf{X}$ exists such that $f_l(\mathbf{x}) \leq f_l(\mathbf{x}^*)$ for $l = 1, \ldots, k$, and $f_h(\mathbf{x}) < f_h(\mathbf{x}^*)$ for at least one $h \in \{1, \ldots, k\}$.

A design vector $\mathbf{x}^*$ is Pareto optimal if no feasible vector $\mathbf{x}$ exists which would decrease some objective function f_h without causing a simultaneous increase in any other objective function f_l, $l \neq h$. Two techniques for generating Pareto-optimal solutions, namely, goal programming and goal attainment, were used to solve the design problem.

Goal programming

In the simplest version of goal programming, a designer sets goals and relative weights for each of the objective functions. An optimum solution $\mathbf{x}^*$ is then defined as one that minimizes the deviation from the set goals. Thus, the goal programming formulation leads to the problem of minimizing

$$z = \left[\sum_{h=1}^{k} c_h (d_h^+ + d_h^-)^p \right]^{1/p},$$

subject to

$$\mathbf{x} \in \mathbf{S},$$
$$F_h(\mathbf{x}) + d_h^+ - d_h^- = f_h^0, \qquad \text{for } h = 1, \ldots, k,$$
$$d_h^+ d_h^- = 0, \qquad \text{for } h = 1, \ldots, k,$$
$$d_h^+, d_h^- \geq 0, \qquad \text{for } h = 1, \ldots, k,$$

where p is a parameter not less than 1 and $\mathbf{S}$ is the set of all feasible solutions.

The scaling procedure $F_h(\mathbf{x}) = m_h f_h(\mathbf{x})$ for $h = 1, \ldots, k$ insures that all of the objective functions are equal at the starting vector $\mathbf{x}^0$. It also insures that the search procedure for optimization is not influenced by numerically different magnitudes of objective functions at $\mathbf{x}^0$. The value of p is chosen by the designer. Suppose

$$\text{goal } f_h^0 = \min_{\mathbf{x} \in \mathbf{X}} F_h(\mathbf{x}), \qquad \text{for } h = 1, \ldots, k.$$

Table 10.4. Pareto-optimal solutions for multiple objective optimization

Stage	r_i	x_i	Attributes	Goal vector
Goal programming				
1	0.943 27	3	$R_s = 0.984\,92$	
2	0.892 76	2	$C = 312.8$	(0.99, 300.0, 120.0)
3	0.953 54	2	$W = 128.7$	
4	0.891 32	3	$V = 47.0$	
Goal attainment				
1	0.941 29	2	$R_s = 0.977\,39$	
2	0.909 16	2	$C = 287.19$	(0.98, 300.0, 120.0)
3	0.940 80	2	$W = 89.031$	
4	0.912 86	2	$V = 32.0$	

Then $F_h(\mathbf{x}) \geq f_h^0$ for any $\mathbf{x} \in \mathbf{X}$ and, therefore, d_h^+ need not be defined and the goal programming formulation reduces to the problem of minimizing

$$
z = \left[\sum_{h=1}^{k} c_h (d_h^-)^p \right]^{1/p},
$$

subject to

$$
\mathbf{x} \in \mathbf{X},
$$
$$
d_h^- = F_h(\mathbf{x}) - f_h^0, \qquad \text{for } h = 1, \ldots, k,
$$
$$
d_h^- \geq 0, \qquad \text{for } h = 1, \ldots, k.
$$

Since deviation from the reliability goal is known to be twice as important as derivations from the cost and weight goals, the relative importance of the three objectives is taken as

$$
c_1 = 0.5, \, c_2 = 0.25, \text{ and } c_3 = 0.25.
$$

A Pareto-optimal solution derived from the goal programming formulation is given in Table 10.4.

Goal attainment method

This method requires setting up a goal f_h^0 and a nonnegative weight c_h for the objective function, $F_h, \, h = 1, \ldots, k$. The parameter c_h relates the relative under- or over-attainment for the goal f_h^0. To determine the optimum solution $\mathbf{x}^*$, minimize z,

subject to

$$g_i(\mathbf{x}) \leq 0, \qquad \text{for } i = 1, \ldots, m,$$

$$F_h(\mathbf{x}) - f_h^0 \leq c_h z, \qquad \text{for } h = 1, \ldots, k,$$

$$\sum_{h=1}^{k} c_h = 1.$$

For under-attainment (over-attainment) of the goals, a smaller c_h is associated with the more (less) important objectives.

The optimum solution using this formulation is fairly sensitive to the values of f_h^0 and c_h. Depending upon the values of f_h^0, it is possible that the weight c_h does not appreciably influence the optimum solution. Instead, the optimum solution can be determined by the nearest Pareto-optimal solution from $f_1^0, \ldots, f_k^0$. This might require that the weight be varied parametrically to generate a set of Pareto-optimal solutions.

For the reliability optimization problem, the goal attainment formulation is to minimize z, subject to

$$g_j(\mathbf{x}) \leq 0, \qquad j = 1, \ldots, m,$$

$$f_1(\mathbf{x}) + 0.2(z) \geq f_1^0,$$

$$f_2(\mathbf{x}) - 0.4(z) \leq f_2^0,$$

$$f_3(\mathbf{x}) - 0.4(z) \leq f_3^0,$$

$$z \geq 0.$$

The optimal solution derived from the goal attainment formulation is also shown in Table 10.4.

10.6 Fuzzy multiple objective optimization

When the objective and constraints of a practical problem are precisely known, one can formulate the problem in mathematical terms and adopt or develop an appropriate solution method for solving it. However, in most real-life situations, the objectives and the constraints are not precisely defined. Sometimes the resource constraints are not very rigid. For example, consider the constraints: (1) the total cost must be about US$ 10 000; and (2) the total volume must be preferably smaller than 20 m^3, but not significantly larger than 20 m^3. Under such imprecise conditions, the classical optimization approach does not serve much purpose. To solve optimization problems in the fuzzy environment, Bellman and Zadeh [32] first used fuzzy logic optimization. This approach is very useful in dealing with qualitative statements, vague objectives, and imprecise information. Park [260], Gupta and Al-Musawi [119], and Dhingra [80] made some interesting fuzzy applications in reliability optimization.

Table 10.5. Values of objective functions for individual optimal solutions

Objective	Optimal $\mathbf{x}$	f_1	f_2	$\cdots$	f_k
max f_1	$\mathbf{x}^1$	$f_1^* = f_1(\mathbf{x}^1)$	$f_2(\mathbf{x}^1)$	$\cdots$	$f_k(\mathbf{x}^1)$
max f_2	$\mathbf{x}^2$	$f_1(\mathbf{x}^2)$	$f_2^* = f_2(\mathbf{x}^2)$	$\cdots$	$f_k(\mathbf{x}^2)$
$\vdots$	$\vdots$	$\vdots$	$\vdots$		$\vdots$
max f_k	$\mathbf{x}^k$	$f_1(\mathbf{x}^k)$	$f_2(\mathbf{x}^k)$	$\cdots$	$f_k^* = f_k(\mathbf{x}^k)$
		$f_{1,\min} = \min\limits_{1 \le l \le k} f_1(\mathbf{x}^l)$	$f_{2,\min} = \min\limits_{1 \le l \le k} f_2(\mathbf{x}^l)$	$\cdots$	$f_{k,\min} = \min\limits_{1 \le l \le k} f_k(\mathbf{x}^l)$

Multiple objective optimization is itself fuzzy in the sense that we try to search for a solution that is satisfactory with respect to each objective. This is true even when the objectives are precisely defined. In fuzzy optimization for multiple objectives, we define a membership function for each fuzzy objective and fuzzy constraint. The value of the membership function measures the degree of satisfaction with respect to the objective/constraint. Suppose that k objective functions $f_1(\mathbf{x}), \ldots, f_k(\mathbf{x})$ are to be maximized simultaneously subject to $\mathbf{x} \in S$, where S is the set of all feasible solutions. Assume that there are no fuzzy constraints in the problem. Let $\mu_1(\mathbf{x}), \ldots, \mu_k(\mathbf{x})$ be the membership functions associated with the objective functions $f_1(\mathbf{x}), \ldots, f_k(\mathbf{x})$, respectively. By applying the max–min decision-making approach of Bellman and Dreyfus [31] and Zimmermann [336], we can formulate the corresponding fuzzy optimization problem of maximizing

$$\min\{\mu_1(\mathbf{x}), \ldots, \mu_k(\mathbf{x})\},$$

subject to $\mathbf{x} \in S$, which can be rewritten as to maximize α subject to

$$\mu_i(\mathbf{x}) \ge \alpha, \qquad \text{for } i = 1, \ldots, k, \text{ and } \mathbf{x} \in S.$$

A major difficulty in fuzzy multiple objective optimization is to define an appropriate membership function for each objective. There are several ways to define membership functions, see Lai and Hwang [183]. We now explain the procedure adopted by Zimmermann [336] in his max–min approach to fuzzy multiple objective optimization. Suppose a solution $\mathbf{x}^i$, $1 \le i \le k$, maximizes $f_i(\mathbf{x})$ subject to $\mathbf{x} \in S$. Let

$$f_i^* = f_i(\mathbf{x}^i), \qquad \text{for } i = 1, \ldots, k$$

and

$$f_{i,\min} = \min\limits_{1 \le l \le k} f_i(\mathbf{x}^l), \qquad \text{for } i = 1, \ldots, k.$$

These terms are systematically arranged in Table 10.5 for the sake of clarity.

A membership function of the ith objective, $1 \leq i \leq k$, is defined as

$$\mu_i(\mathbf{x}) = \begin{cases} 1, & \text{if } f_i(\mathbf{x}) \geq f_i^*, \\ [f_i(\mathbf{x}) - f_{i,\min}]/(f_i^* - f_{i,\min}), & \text{if } f_{i,\min} \leq f_i(\mathbf{x}) < f_i^*, \\ 0, & \text{if } f_i(\mathbf{x}) < f_{i,\min}. \end{cases}$$

Note that the membership function $\mu_i(\mathbf{x})$ is a piecewise linear function, and $0 \leq \mu_i(\mathbf{x}) \leq 1$ for all $\mathbf{x}$. The solutions that are satisfactory with respect to the ith objective give higher values of $\mu_i(\mathbf{x})$.

The fuzzy optimization approach can also be adopted for single-objective reliability optimization when the constraints are fuzzy. Consider the problem of maximizing $f(\mathbf{x})$ subject to $g_i(\mathbf{x}) \leq b_i$, $1 \leq i \leq m$, x_j being a nonnegative integer.

If $g_i(\mathbf{x})$ is preferred to be less than or equal to b_i and its values between b_i, and b_i' are tolerable; then, the membership function of $g_i(\mathbf{x})$ can be defined as

$$\mu_i(\mathbf{x}) = \begin{cases} 0, & \text{if } g_i(\mathbf{x}) \geq b_i', \\ [b_i' - g_i(\mathbf{x})]/(b_i' - b_i), & \text{if } b_i \leq g_i(\mathbf{x}) < b_i', \\ 1, & \text{if } g_i(\mathbf{x}) < b_i, \end{cases}$$

for $i = 1, \ldots, m$. Now, the corresponding fuzzy optimization problem can be formulated with two choices.

- Choice 1: maximize α, subject to $\mu_i(\mathbf{x}) \geq \alpha, i = 1, \ldots, m$, where $\mu_i(x)$ is constrained through fuzzy constraints.
- Choice 2: maximize $f(\mathbf{x})$, subject to $\mu_i(\mathbf{x}) \geq \alpha, i = 1, \ldots, m$.

Example 10-1

Consider a five-stage series system in which component redundancy is to be provided at each stage so as to, simultaneously, maximize the system reliability and minimize the total cost, subject to some constraints. When x_j components are used in parallel at stage j, $j = 1, \ldots, 5$, the system reliability and cost are

$$R_s(\mathbf{x}) = \prod_{j=1}^{5} [1 - (1 - r_j)^{x_j}],$$

$$c(\mathbf{x}) = \sum_{j=1}^{5} c_j[x_j + \exp(x_j/4)],$$

respectively, where $\mathbf{x} = (x_1, \ldots, x_5)$. The component reliabilities r_j and component costs c_j are as given in Table 10.6. Define

$$f_1(\mathbf{x}) = R_s(\mathbf{x})$$

Table 10.6. Component reliabilities and constants for Example 10-1

j	r_j	c_j	p_j	P	w_j	W
1	0.80	7	1		7	
2	0.85	7	2		8	
3	0.90	5	3	110	8	200
4	0.65	9	4		6	
5	0.75	4	2		9	

and

$$f_2(\mathbf{x}) = -c(\mathbf{x}).$$

The problem under consideration is to maximize $[f_1(\mathbf{x}), f_2(\mathbf{x})]$ subject to

$$\sum_{j=1}^{n} p_j x_j^2 \leq P,$$

$$\sum_{j=1}^{n} w_j x_j \exp(x_j/4) \leq W,$$

$$(1, 1, 1, 1, 1) \leq \mathbf{x} \leq (5, 5, 4, 4, 3),$$

x_j being an integer.

The constants in the above constraints are as given in Table 10.6.

The number of objectives in this example is $k = 2$. The solution that maximizes the first objective function $f_1(\mathbf{x})$ is

$$\mathbf{x}^1 = (3, 2, 2, 3, 3),$$

whereas the optimal solution for the second objective function is

$$\mathbf{x}^2 = (1, 1, 1, 1, 1).$$

The values $f_1(\mathbf{x}^i)$, $f_2(\mathbf{x}^i)$, $f_{1,\min}$, and $f_{2,\min}$ are given in Table 10.7.

A membership function of the first objective is defined as

$$\mu_1(\mathbf{x}) = \begin{cases} 1, & \text{if } R_s(\mathbf{x}) \geq 0.9045, \\ [R_s(\mathbf{x}) - 0.2984]/(0.9045 - 0.2984), & \text{if } 0.2984 \leq R_s(\mathbf{x}) < 0.9045, \\ 0, & \text{if } R_s(\mathbf{x}) < 0.2984. \end{cases}$$

Table 10.7. Values of objective functions for individual optimal solutions in Example 10-1

Objective	Optimal x	f_1	f_2
max f_1	$x^1 = (3, 2, 2, 3, 3)$	$f_1^* = f_1(x^1) = 0.9045$	$f_2(x^1) = -146.125$
max f_2	$x^2 = (1, 1, 1, 1, 1)$	$f_1(x^2) = 0.2984$	$f_2^* = f_2(x^2) = -73.089$
min$[f_i(x^1), f_i(x^2)]$		$f_{1,min} = 0.2984$	$f_{2,min} = -146.125$

Similarly, the membership function of the second objective is defined as

$$\mu_2(x) = \begin{cases} 1, & \text{if } c(x) \leq 73.089, \\ [146.125 - c(x)]/(146.125 - 73.089), & \text{if } 73.089 < c(x) \leq 146.125, \\ 0, & \text{if } c(x) > 146.125. \end{cases}$$

Now the fuzzy optimization formulation of the bicriteria problem is to maximize α, subject to

$$\mu_1(x) \geq \alpha,$$

$$\mu_2(x) \geq \alpha,$$

$$\sum_{j=1}^{n} p_j x_j^2 \leq P,$$

$$\sum_{j=1}^{n} w_j x_j \exp(x_j/4) \leq W,$$

$$(1, 1, 1, 1, 1) \leq x \leq (5, 5, 4, 4, 3),$$

x_j being an integer.

   The optimal solution of this problem is

$$x^* = (2, 1, 1, 2, 3)$$

and, correspondingly,

$$\alpha = 0.5488,$$

$$R_s(x^*) = 0.634\,37,$$

$$c(x^*) = 106.256.$$

10.7   Discussion

Multiple objective optimization simultaneously deals with several conflicting objectives. A single solution may not exist that optimizes every objective under consideration.

A solution which is optimal for one objective may be worse for some other objective. This can be seen in Table 10.3, where an optimal solution that maximizes system reliability gives high cost and large weight. Minimization of cost leads to low system reliability and large weight. Similarly, minimization of weight causes high cost and low system reliability.

The approach of multiple objective optimization either converts the problem into a single-objective optimization by associating weights/preferences with the objectives, or presents a set of nondominated solutions which help the decision maker in interactive decision making. Multiple objective optimization is useful in making tradeoffs among the conflicting objectives. However, Pareto-optimal solutions obtained using goal programming and goal attainment may be quite sensitive to the weights and relative importances associated with the objectives. Fuzzy optimization can be adopted when the parameters and goals of the problem cannot be stated precisely. For details on fuzzy multiple objective/attribute decision making refer to Chen and Hwang [48] and Lai and Hwang [183].

Sakawa [279] has adopted large-scale multiple objective optimization methods to solve large-scale reliability–redundancy allocation problems with multiple objectives. Sakawa [283] has applied a sequential proxy optimization technique (SPOT) to optimize system reliability, cost, weight, volume, and product of weight and volume for series-parallel systems subject to some constraints. SPOT is an interactive, multiple objective decision-making technique for selecting among a set of Pareto optimal solutions. To solve multiple objective redundancy allocation problems in reliability systems, Misra and Sharma [231] have adopted a multiple objective optimization method based on the min–max concept for obtaining Pareto optimal solutions. Misra and Sharma [232] have also presented a similar approach. A case study on multiple objective optimization in a maintenance system is described in Chapter 14.

EXERCISES

10.1   Describe the methods of multiple attribute decision making (MADM) and multiple objective decision making (MODM). What are the differences between them? Show the common characteristics of MODM methods.

10.2   There are two approaches to solving a vector maximization problem (VMP). Describe each approach and show the difficulties of these approaches. What kinds of methods can be used to overcome these difficulties?

10.3   If a feasible solution in MODM is nondominated, what are the conditions to be satisfied? Should a satisfying solution be nondominated?

10.4   There are two techniques for generating Pareto-optimal solutions. Describe each of them, and discuss the differences between them.

10.5   In MODM, what kind of solution simultaneously satisfies the maximum value of each objective function? Explain a positive-ideal solution and a negative-ideal solution?

10.6   A government decides to spend 2.3 billion dollars for military purposes to maximize defense ability. In addition, the government wants to minimize the number of soldiers. Discuss the corresponding attributes, objectives, goals, and criteria of this problem.

10.7   Why can fuzzy theory be applied to optimization problems? When is a fuzzy optimization approach useful? In fuzzy theory, which value is used to measure the degree of satisfaction with objectives or constraints?

10.8   In Example 10-1, let the solutions of the first and second objective functions be $\mathbf{x}^1 = (3, 2, 2, 3, 2)$ and $\mathbf{x}^2 = (1, 2, 1, 1, 1)$, respectively. Find the membership function of the first and the second objectives.

10.9   In Example 10-1, suppose that the objective is to maximize the system reliability and $P = 110$, $P' = 130$, $W = 200$, and $W' = 250$. Find the membership function of the first and second constraints, and formulate the corresponding fuzzy optimization problem.

10.10   Explain interactive methods for multiple objective decision making. Discuss the advantages and disadvantages of these methods.

11 Other methods for system-reliability optimization

11.1 Introduction

As mentioned in Chapter 6, the designer can improve system reliability by increasing the individual component reliabilities. However, this practice requires some effort in terms of resources such as cost, volume, weight, etc. When such effort can be quantified in terms of component reliabilities, it is possible to minimize the total effort required to improve system reliability to a desired level. We describe two methods for this kind of optimization.

11.2 Optimization of effort function

One of the standard approaches for enhancing system reliability is to increase the component reliabilities. However, an increase in component reliability requires some effort, which may be cost, volume, weight, power consumption, etc., and thus system-reliability enhancement also requires such effort. Assume that the effort to increase the reliability of any component from one level to another is measurable by a mathematical function. Such functions, called *effort functions*, are not necessarily explicit. Reliability engineers usually formulate effort functions based on knowledge of the development process. The problem under consideration is to minimize the total effort required to increase the reliability of a general coherent system from an existing level to a desired level through increments in component reliabilities. Albert [9] solved this problem for series systems when the effort functions are the same for all components. For a good description of this method refer to Lloyd and Lipow [202]. Dale and Winterbottom [68] provided a solution approach for a general coherent structure.

Notation

$G_j(a_j, r_j)$	effort required to increase reliability of component j from a_j to r_j
$G(R^0, R^1)$	effort required to increase system reliability from R^0 to R^1
$h(\mathbf{r})$	system reliability as a function of component reliabilities
n	number of components in the system
R^0	$h(\mathbf{r}^0)$, existing level of system reliability

R^1 $h(\mathbf{r}^1)$, desired level of system reliability
R_s system reliability
r_j reliability of component j
r_j^0 existing level of reliability of component j
$\mathbf{r}$ $(r_1, \ldots, r_n)$
$\mathbf{r}^0$ $(r_1^0, \ldots, r_n^0)$

Assume that

- A1: $G_j(r_j^0, r_j)$ is additive in the sense that $G_j(r_j^0, r_j) = G_j(r_j^0, v_j) + G_j(v_j, r_j)$, for $r_j^0 \le v_j \le r_j$.

- A2: $G_j(r_j^0, r_j)$ is strictly increasing in r_j $(> r_j^0)$, and $G_j(r_j, r_j) = 0$ for $0 \le r_j \le 1$.

- A3: $G_j(0, r_j)$ is differentiable with respect to r_j in the interval $(0, 1)$.

- A4: $\mathrm{d}^2 G_j(0, r_j)/\mathrm{d}r_j^2 > 0$, for $0 < r_j < 1$.

- A5: $G_j(r_j^0, r_j) \to \infty$, as $r_j \to 1$ for any fixed $r_j^0, 0 \le r_j^0 < 1$.

- A6: If component reliabilities are increased from $\mathbf{r}^0 = (r_1^0, \ldots, r_n^0)$ to $\mathbf{r} = (r_1, \ldots, r_n)$, in order to increase system reliability from $R^0 = h(\mathbf{r}^0)$ to $R_s = h(\mathbf{r})$, then

$$G(R^0, R_s) = \sum_{j=1}^{n} G_j(r_j^0, r_j). \tag{11.1}$$

Note that there are several choices for increments of component reliabilities to increase system reliability from one level to another. Suppose the system reliability is to be increased from R^0 to a desired level R^1. Then, the problem can be written so as to minimize

$$G(R^0, R^1) = \sum_{j=1}^{n} G_j(r_j^0, r_j),$$

subject to

$$r_j^0 \le r_j \le 1, \qquad \text{for } j = 1, \ldots, n,$$

$$h(\mathbf{r}) = R^1.$$

11.2.1 Albert method for a series system

The reliability of a series system which consists of n components with reliabilities $r_1^0, \ldots, r_n^0$ is

$$R^0 = \prod_{j=1}^{n} r_j^0.$$

Albert [9] presented a simple and elegant method for solving the effort minimization problem in a series system when the effort functions $G_j(0, r_j)$ associated with all the components are the same. The parameter $G_j(0, r_j)$ need not be explicitly known. For minimizing the total effort required to increase system reliability from R^0 to the desired level R^1, Albert's [9] method increases the lower component reliabilities to a common level which does not exceed the other component reliabilities. The method gives an exact optimal solution in every case, and is described as follows.

Albert algorithm

- Step 0: Renumber the components such that $r_1^0 \le r_2^0 \le \cdots \le r_n^0$.

- Step 1: Define $r_{n+1} = 1$ and find k such that

$$r_k \le \bar{r}_k \le r_{k+1},$$

where

$$\bar{r}_k = \left(\frac{R^1}{\displaystyle\prod_{j=k+1}^{n+1} r_j^0} \right)^{1/k}.$$

- Step 2: Increase the reliabilities of components $1, 2, \ldots, k$ to the level of $\bar{r}_k$ so that the system reliability increases to R^1. The optimal component reliabilities are $\bar{r}_k$ for the first k components and r_j^0 for component j, $j = k + 1, \ldots, n$.

Example 11-1

Let $n = 6$, $(r_1^0, \ldots, r_6^0) = (0.75, 0.80, 0.87, 0.90, 0.95, 0.99)$, and $r_7^0 = 1.00$. The corresponding system reliability is $R^0 = \prod_{j=1}^{6} r_j^0 = 0.4418$. Note that the components are indexed in increasing order of their reliabilities. Suppose the desired level of system reliability is $R^1 = 0.53$. We have

$$\bar{r}_1 = \frac{0.53}{(0.80)(0.87)(0.90)(0.95)(0.99)(1.00)} = 1.1996 > r_2,$$

$$\bar{r}_2 = \left[\frac{0.53}{(0.87)(0.90)(0.95)(0.99)(1.00)} \right]^{1/2} = 0.8484,$$

and $r_2 < \bar{r}_2 < r_3$. Therefore, $k = 2$ and the optimal reliabilities of the six components are

$$(0.8484, 0.8484, 0.8700, 0.9000, 0.9500, 0.9900).$$

11.2.2 The Dale–Winterbottom method for a general coherent system

We now describe the solution method of Dale and Winterbottom [68] for a general coherent system with n components. The rate of increase in total effort with respect to system reliability is

$$\frac{dG(0, R_s)}{dR_s} = \sum_{j=1}^{n} \frac{dG_j(0, r_j)}{dr_j} \left[\frac{\partial h(\mathbf{r})}{\partial r_j} \right]^{-1}.$$

Note that due to assumption A1,

$$\frac{dG(R^0, R_s)}{dR_s} = \frac{dG(0, R_s)}{dR_s}, \qquad \text{for } R_s > R^0,$$

$$\frac{dG_j(r_j^0, r_j)}{dr_j} = \frac{dG_j(0, r_j)}{dr_j}, \qquad \text{for } r_j > r_j^0.$$

Let

$$D_j = \frac{dG_j(0, r_j)}{dr_j} \left[\frac{\partial h(\mathbf{r})}{\partial r_j} \right]^{-1}, \qquad \text{for } j = 1, \ldots, n. \tag{11.2}$$

To find optimal increments in component reliabilities for increasing system reliability from R^0 to R^1 ($>R^0$), Dale and Winterbottom [68] developed an iterative method described by the following algorithm.

Dale–Winterbottom algorithm

- Step 0: Let $\mathbf{r} = (r_1^0, \ldots, r_n^0)$ and compute the corresponding $D_1, \ldots, D_n$. Renumber the components such that

$$D_1 \leq D_2 \leq \cdots \leq D_n.$$

- Step 1: If $D_1 < D_2$, increase the reliability r_1 until $D_1 = D_2$ or system reliability R_s reaches R^1, whichever happens first.
- Step 2: If $R_s = R^1$, go to step 4; otherwise, find the largest k such that

$$D_1 = D_2 = \cdots = D_k.$$

  If $k = n$, go to step 3; otherwise, increase $r_1, \ldots, r_k$ until $D_1 = D_2 = \cdots = D_k = D_{k+1}$ or system reliability R_s reaches R^1 with $D_1 = D_2 = \cdots = D_k$, whichever happens first, and repeat step 2.
- Step 3: Increase all r_i such that system reliability R_s reaches R^1 with $D_1 = D_2 = \cdots = D_n$.
- Step 4: Take $\mathbf{r}$ as the optimal solution, and stop.

The above procedure gives an exact optimal solution for a general coherent system. We illustrate this method using the numerical example of Dale and Winterbottom [68].

Example 11-2

Consider a parallel-series system in which components 1 and 2 and a parallel subsystem of components 3 and 4 are arranged in series. The reliability of this system for component reliabilities r_1, r_2, r_3, r_4 is

$$h(r_1, r_2, r_3, r_4) = r_1 r_2 (r_3 + r_4 - r_3 r_4).$$

Let $(r_1^0, r_2^0, r_3^0, r_4^0) = (0.90, 0.95, 0.60, 0.80)$. The corresponding system reliability is $R^0 = 0.7866$. Suppose the desired level of system reliability is $R^1 = 0.85$ and the effort functions are

$$G_j(r_j^0, r_j) = \alpha_j \ln\left(\frac{1 - r_j^0}{1 - r_j}\right), \qquad \text{for } j = 1, \dots, 4.$$

The parameter α_j provides a comparison of efforts for increments in component reliabilities. Let

$$(\alpha_1, \alpha_2, \alpha_3, \alpha_4) = (2, 1, 2, 3).$$

We have,

$$\frac{dG_j(0, r_j)}{dr_j} = \frac{\alpha_j}{(1 - r_j)}, \qquad \text{for } j = 1, \dots, 4,$$

and

$$D_1 = \frac{\alpha_1}{(1 - r_1)} \left[\frac{1}{r_2(r_3 + r_4 - r_3 r_4)}\right],$$

$$D_2 = \frac{\alpha_2}{(1 - r_2)} \left[\frac{1}{r_1(r_3 + r_4 - r_3 r_4)}\right],$$

$$D_3 = \frac{\alpha_3}{(1 - r_3)} \left[\frac{1}{r_1 r_2(1 - r_4)}\right], \qquad (11.3)$$

$$D_4 = \frac{\alpha_4}{(1 - r_4)} \left[\frac{1}{r_1 r_2(1 - r_3)}\right].$$

Using eq. (11.3) for $(r_1, r_2, r_3, r_4) = (r_1^0, r_2^0, r_3^0, r_4^0)$, we get

$$(D_1, D_2, D_3, D_4) = (22.8, 24.2, 29.2, 43.9).$$

There is no need to renumber the components as D_j is already in increasing order. We increase the reliability of component 1 from 0.90 to 0.904 76 following case 1. For $(r_1, r_2, r_3, r_4) = (0.904\,76, 0.95, 0.60, 0.80)$, we have system reliability $R_s = 0.790\,76$ and

$$(D_1, D_2, D_3, D_4) = (24.027, 24.027, 29.086, 43.629).$$

We continue the procedure since $R_s < 0.85$. Note that case 2 now holds for $k = 2$. Thus, we increase the reliabilities of components 1 and 2 to the levels

0.92 and 0.958 333, respectively, following case 2. For $(r_1, r_2, r_3, r_4) = (0.92, 0.958\,333, 0.60, 0.80)$, we have $R_s = 0.811\,133$ and

$$(D_1, D_2, D_3, D_4) = (28.355, 28.355, 28.355, 42.533).$$

Case 2 again holds for $k = 3$ with $R_s < 0.85$. The reliabilities of components 1, 2, and 3 are increased such that the new component reliability vector is $(r_1, r_2, r_3, r_4) = (0.937\,26, 0.967\,614, 0.686\,299, 0.80)$, $R_s = 0.850\,006$, and

$$(D_1, D_2, D_3, D_4) = (35.150, 35.150, 35.150, 52.725).$$

Therefore, the optimal solution for the desired reliability $R^1 = 0.85$ is

$$(r_1, r_2, r_3, r_4) = (0.937\,260, 0.967\,614, 0.686\,299, 0.800\,000).$$

11.3   Discussion

In this chapter we have discussed effort function minimization in reliability systems. It is relevant when the effort required to increase any component reliability is measurable by a mathematical function. The problem is to increase system reliability to a desired level by increasing the component reliabilities such that the total effort required is minimum. Albert [9] provided a mathematical basis for analyzing this problem. When the effort functions associated with components are all the same, Albert's method gives optimal increments in component reliabilities for a series system without explicit knowledge of the effort functions. The method of Dale and Winterbottom [68] gives optimal increments in component reliabilities for general coherent systems when the effort functions are explicitly known and differentiable.

For a series system with identical component effort functions, both of the methods yield the same optimal solution.

EXERCISES

11.1 Let $(r_1^0, \ldots, r_5^0) = (0.92, 0.88, 0.76, 0.95, 0.80)$, $r_6^0 = 1.0$, $R^0 = \prod_{j=1}^{5} r_j^0$, and $R^1 = 0.6$. Find the optimal reliabilities of the five components by Albert's method.

11.2 In Example 11-2, let $(r_1^0, r_2^0, r_3^0, r_4^0) = (0.93, 0.96, 0.6, 0.8)$ and $R^1 = 0.85$. Find the optimal solution using the Dale–Winterbottom algorithm.

12 Burn-in optimization under limited capacity

12.1 Introduction

The traditional view of reliability holds that reliability problems must be solved in order to eliminate customer complaints. This viewpoint and motivation influence the traditional manufacturing process, which includes inspection for detection of defects, statistical process control for process-stability, detection of special causes, and product assurance for failure prevention. From this perspective, it is through assessment and debugging that we grow reliability. The process is sequential, and typically takes place downstream. Methodologies developed from this approach have matured in recent years. We have gone from using a few gauges for inspection to an array of sophisticated tools and methods for assessing and tracking reliability.

The special issue of *IEEE Transactions on Reliability*, Kuo and Oh [179], focuses on reliability-related engineering design. The application areas of particular interest are advanced materials-processing and the design and manufacture of electro-mechanical devices. Among the many alternatives for designing products that are robust to manufacturing variations and customer use are concurrent engineering, computer-aided simulation, accelerated life testing (ALT), and physical experimentation. Combining these areas with a focus on reliability is a highly complex task, which requires the cooperation of engineers, statisticians, and designers working together.

Burn-in is a screening operation (in lieu of adequate process characterization and control) that combines appropriate electrical and thermal environments to approximate, in a shortened time span, the early life of a component or system. Burn-in has long been recognized as a useful method for detecting early failures of components or systems before customer delivery. Without burn-in, defective components are frequently delivered to customers. This results in costly field repairs and loss of confidence in the manufacturer. By using burn-in, the manufacturer delivers fewer defective components; consequently, the lower failure rates reduce field-repair costs. In today's market, burn-in is recognized by the microelectronics industry as essential for remaining competitive. The literature focusing on the use of burn-in has been reviewed in Kuo and Kuo [177] and Kuo et al. [174].

Burn-in is a particularly effective screening method for use in achieving and enhancing the field reliability of electronic devices. It is the most effective screen for detecting die-related faults, since time, bias, current, and temperature can be accelerated to the point of fault detection in a relatively short period of time. A typical burn-in would require devices to be subjected to a temperature stress of 125 °C for a minimum of 48 hours. Infant-mortalities are greatly influenced by a burn-in test, and a significant improvement in failure rates has been reported after burn-in. A reduction in the number of burn-in periods can result in transferring infant-mortalities from the manufacturing stage to the customer. There are several different types of burn-in used, each of which creates a variation of burn-in stresses. Among them, steady-state burn-in, static burn-in, dynamic burn-in, and test-during-burn-in are widely used for semiconductor devices. A detailed description of each method is presented in Kuo et al. [174].

Burn-in to minimize the life-cycle cost of the system has been investigated reasonably well in many applications, but the physical constraints during the decision process also need to be considered in searching for optimal burn-in time. Two types of constraints must be satisfied during decision making: (1) the minimum system-reliability requirement, and (2) the maximum capacity available for burn-in. Detailed descriptions are available in Chi and Kuo [53]. Guidelines are suggested for making burn-in decisions in the following text.

12.2 Statement of the problem

The Weibull failure distribution is commonly used to model the infant-mortality of microelectronic components. Its probability density function is

$$g(t) = ab(at)^{b-1} \exp[-(at)^b], \quad t \geq 0, \ a > 0, \ 0 < b < 1.$$

The hazard rate for Weibull can be written as

$$h(t) = ab(at)^{b-1}.$$

Failure rates with and without burn-in are depicted in Figure 12.1.

Considering a microelectronics system that consists of a number of boards (units), each of which contains electronic components, we adopt the notation below.

Notation

a_{ij}, b_{ij}	Weibull scale and shape parameters for component i in unit j
c_{ij}	cost of component i in unit j
$c_{ij,b}$	device burn-in cost for component i in unit j
$c_{ij,f}$	field-repair cost for component i in unit j
$c_{ij,s}$	shop-repair cost for component i in unit j
$e_{ij,b}$	expected fraction of failure during burn-in for component i in unit j
$e_{ij,f}$	expected fraction of failure at time t after $t_{ij,b}$ hours of burn-in
$h_{ij}(t)$	hazard rate of component i in unit j at field operation time t [173]
ℓ	a ratio used to calculate the cost of the loss of credibility

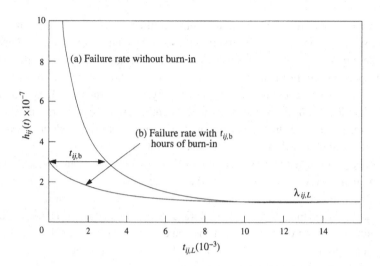

Figure 12.1. Failure rate of components with and without burn-in

n_{ij}	number of ith type of components used in jth unit
$R_{ij}(t \mid t_{ij,b})$	reliability of component i in unit j at time t after a burn-in period of $t_{ij,b}$
$R_{ij,\min}(t)$	minimal reliability requirements for component i of unit j at time t
$R_s(t)$	system reliability at time t
$R_s(t \mid t_{ij,b})$	system reliability at time t given component i in unit j burned-in for $t_{ij,b}$ hours
$R_{s,\min}(t)$	minimal system-reliability requirement at time t
t	field operation time after the component (with or without burn-in) is assembled
$t_{ij,L}$	time at which $\lambda_{ij,L}$ is reached
V_t	batch capacity of burn-in facility in terms of columns or racks of ovens
$\lambda_{ij,L}$	steady-state hazard rate for component i in unit j

Components are burned-in for various periods $t_{ij,b}$ before they are placed together in a system. Therefore, when the components are assembled, each one is at its own distinct place on the infant-mortality curve [173]. The question of how long the components should be burned-in has been partially answered through experimentation and unconstrained cost-minimization procedures. A more restricted burn-in optimization problem subject to reliability requirements was addressed by Kuo [173]. The burn-in problem under both the reliability and the capacity restrictions for a general system structure is formulated below.

12.2.1 Objective function and reliability constraints

Assume that all components are burned-in for time $t_{ij,b}$ hours under a stressed environment before they are assembled into a unit. A schematic operation is shown

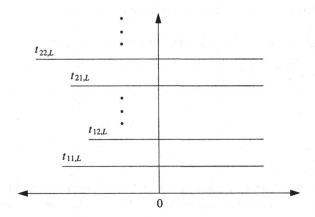

Figure 12.2. A schematic diagram

in Figure 12.2. Kuo [173] has obtained the optimal burn-in strategy by solving the problem of minimizing

$$C_s = f(t_{11,b}, t_{12,b}, \dots),$$

subject to

$$R_s(t|t_{ij,b}) \geq R_{s,\min}(t), \tag{12.1}$$

$$R_{ij}(t|t_{ij,b}) \geq R_{ij,\min}(t), \qquad \text{for all } i \text{ and } j, \tag{12.2}$$

where C_s is a function of the costs of burn-in, shop-repair, field repair, and loss of product-reputation. From a system viewpoint, this cost also depends on the system configuration and the component burn-in period $t_{ij,b}$.

System reliability is a function of the component reliabilities: this function depends on the system configuration. For details of the system cost and the reliability functions refer to Kuo [173] and Kuo et al. [174].

12.2.2 Burn-in capacity

In addition to determining an optimal burn-in time with only reliability constraints, there are other limitations, that is, capacity constraints to consider. Do we have a large enough burn-in facility? Are we allowed enough time to perform burn-in? What is the tradeoff cost?

Consider a company that produces various types of electronic systems. The capacity of the burn-in facility F_t available for a planning production period T_p is

$$F_t = V_t \times T_p, \tag{12.3}$$

where V_t is the batch capacity of the burn-in facility in terms of columns or racks of ovens.

The amount of capacity (space) required for component i of unit j is F_{ij}. The capacity required for component i of unit j for burn-in period $t_{ij,b}$ is $F_{ij}t_{ij,b}$. Therefore, the capacity constraint becomes:

$$\sum_i \sum_j F_{ij}t_{ij,b} \le F_t. \tag{12.4}$$

12.2.3 Problem formulation

If there is no minimum reliability constraint and if there is infinite capacity, then the burn-in optimization problem becomes trivial. Unfortunately, in most situations physical constraints do exist and make the optimization problem difficult to solve. For example, when the reliability constraint is violated, we can increase the burn-in time to satisfy this constraint up to the maximum system reliability obtainable by burn-in. However, the capacity constraint may be violated at some point because of the increased burn-in time.

The maximum system reliability obtainable by burn-in is the system reliability at a field operation time t after all components are burned-in for up to $t_{ij,L}$ (see Figure 12.1). In other words, the optimal, cost-effective burn-in policy might not be consistent with the minimum reliability requirements; this is particularly true when the system-reliability requirement is very tight. In contrast, the capacity constraint can be satisfied by decreasing burn-in time, but the reliability constraint may be violated by doing so.

Hence the objective is to minimize the system cost while satisfying both the reliability and capacity constraints. The problem to be solved is to minimize

$$C_s = \sum_i \sum_j [c_{ij} + c_{ij,b} + e_{ij,b}c_{ij,b} + (1 + \ell)e_{ij,f}c_{ij,f}], \tag{12.5}$$

subject to

$$R_s(t|t_{ij,b}) \ge R_{s,min}(t),$$

$$R_{ij}(t|t_{ij,b}) \ge R_{ij,min}(t), \qquad \text{for all } i \text{ and } j,$$

$$\sum_i \sum_j F_{ij}t_{ij,b} \le F_t,$$

$$0 < t_{ij,b} < t_{ij,L}, \qquad \text{for all } i \text{ and } j.$$

12.3   Optimization and the decision tree

Kuo [173] has suggested first solving the problem that has only the constraint $0 < t_{ij,b} < t_{ij,L}$. The optimal burn-in period (assuming that the minimal system reliability

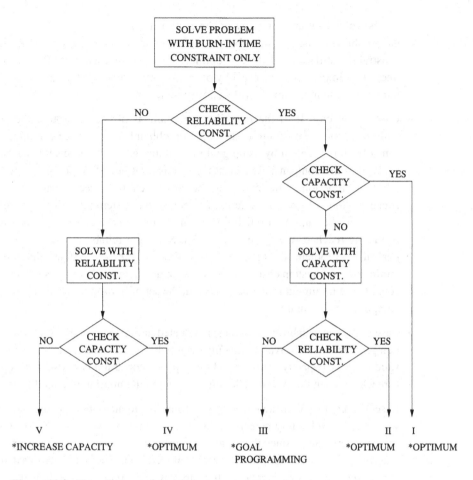

Figure 12.3. Flow diagram of decisions

is not required and burn-in capacity is enough) is

$$t_{ij,b}^* = \left(\frac{a_{ij}[(1 + \ell)c_{ij,f} - c_{ij,s}]}{(1 + \ell)c_{ij,f}\lambda_{ij,L} + Br} \right)^{1/b_{ij}}, \qquad \text{for all } i \text{ and } j, \qquad (12.6)$$

where B and r are cost coefficients associated with $c_{ij,b}$.

For eq. (12.5), the optimal burn-in time obtained from (12.6) is used to check the validity of the reliability constraints. If the reliability constraints are satisfied, then check the capacity constraint. If the reliability constraints are not satisfied, solve the cost optimization problem with the reliability constraints. The procedure is described in the form of a decision tree as shown in Figure 12.3, and each case is described below.

- Case I: If the constraints are not violated, the optimum is reached (Case I in the decision tree). Otherwise, a longer or shorter burn-in period can be used.

- Case II: If the reliability constraint is satisfied but the capacity constraint is violated, then the problem can be solved by using only the capacity constraint. This problem can be solved by applying any of the nonlinear programming techniques. Check the reliability constraint with the new optimal burn-in times $(t_{ij,b})$. If the reliability constraint is still satisfied, then the optimal solution is reached (Case II in the decision tree). This implies that the capacity of the facility is fully used or that it satisfies the minimum system reliability with capacity remaining.

- Case III: In case II, if the reliability constraint is not satisfied after the problem with the capacity constraint is solved, the problem becomes more complicated. We can solve this problem by using goal programming. The goal level depends on the strategy of the company. The most rigid constraint becomes the first goal. Let the reliability function be the first goal, the cost function the second, and the capacity constraint the last. When a solution is reached, the system reliability is greater than or equal to the minimum reliability required. In the case of equality, the optimum is already reached (the same as Case V). Since the reliability function is the first goal of the problem, the optimal burn-in time satisfies the system reliability while minimizing the system cost. Therefore, when the system reliability is greater than or equal to the minimum reliability required, the set of burn-in times obtained for each component is optimum.

- Case IV: If the reliability constraint is violated at the beginning, the cost minimization problem can be solved using only the reliability constraint and then checking the validity of the capacity constraint. If the capacity constraint is satisfied, the optimum is reached. Then the system reliability is equal to the minimum reliability required.

- Case V: In Case IV, if the capacity constraint is violated after solving the problem with only the reliability constraint, an increase in capacity is needed to satisfy the capacity constraint. Since the system reliability is equal to the minimum reliability required, the optimal burn-in time cannot be reduced. The only way to satisfy both of the constraints is by increasing the capacity. This type of solution is the same as in Case IV, except that greater capacity is needed.

If the burn-in temperature in the Arrhenius equation [173] is less than the maximum allowable burn-in temperature, the capacity constraint can be partially or totally satisfied by increasing the burn-in temperature to the maximum temperature allowed for burn-in, namely, the temperature where the component suffers no physical damage and the Arrhenius equation is still valid. Rationally, however, the maximum allowable burn-in temperature should be used at the beginning.

If the capacity cannot be increased for any reason (such as lack of funds or lack of time), then a subcontract with independent, outside laboratories might be made for burning-in some components. The price would be the purchase price of the burned-in component from the other company minus the component cost.

The component to be subcontracted is the one that has the minimum marginal cost. The marginal cost is the difference between the burn-in cost and the subcontract cost. However, the marginal cost is usually the same when the burn-in cost and the subcontract cost linearly depend on the burn-in time. In this case, any component can be sent out for further improvement. Otherwise, the differences among marginal costs need to be considered because of the size of the components and the size of the bulk order.

If the capacity is still not large enough to burn-in all remaining components, we can subcontract another component which has the next largest minimum marginal cost. This procedure continues until the system capacity requirement is met.

In Case V, the system reliability is equal to the minimum system-reliability requirement, and a set of optimal burn-in times for each component has been obtained to minimize the system cost. The total burn-in time and the burn-in times for individual components have been calculated to minimize the system cost. The total burn-in time and component burn-in times cannot be decreased or increased unless there is flexibility in the minimum system-reliability requirement or the system capacity. A subcontract has been considered because the company does not have enough capacity to burn-in all components in a given time. In other words, there is no burn-in time increment and no marginal system-reliability gain because of the subcontract.

Finally, we need to consider the global constraints before a decision is made:

- Condition 1: $c_{ij,b} > c_{ij,f}$, no burn-in,

- Condition 2: $c_{ij,f} \gg c_{ij,b}$, burn-in until $t_{ij,L}$.

Condition 1 means that the burn-in cost is more expensive than the field-repair cost, so burn-in would not pay off. Condition 2 means that the field-repair cost is very high or the components are not repairable, so that a much longer burn-in period would be beneficial. If any components belong to these categories, it is better to fix the burn-in times of these components to eliminate unnecessary components involved in the optimization problem.

12.4   Application to an electronic product

12.4.1   Assumptions

1. The system has four different units as described in Figure 6.1. The failure times of the units are mutually independent and identically distributed. Integrated circuits (ICs) are used in four different units. A unit fails if any of its ICs fail. Data for all of the components are known. For convenience, the index j is dropped in the following discussion.

2. Non-IC components have negligible failure rates.

3. $t_{i,L} = 10^4$ hours, for all i. No component is burned-in for longer than $t_{i,L}$. For justification, see [128] and [173].

4. The burn-in capacity already exists and many ICs of each type are burned-in together.

5. The starting point for system operation is time zero, at which time each component is located differently on its distinct infant-mortality curve [179].

Table 12.1. Number of various ICs used, their Weibull parameters, and cost factors[a]

i	n_{i1}	n_{i2}	n_{i3}	n_{i4}	$\sum_j n_{ij}$	b_i	a_i (10^3FIT)	$\lambda_{i,L}$ (FIT)	$c_{i,f}$	$c_{i,s}$
1	3	20	30	1	54	0.80	160	100	27.00	21.00
2	27	9	9	12	57	0.75	300	300	38.00	24.35
3	22	0	2	0	24	0.75	50	50	18.00	9.00
4	0	24	0	18	42	0.70	63	100	25.50	14.35
5	29	41	22	21	113	0.80	240	151	29.25	19.35
Total	81	94	63	52	290					

[a] i, component serial number; $t_{i,L}$, 10^4 for each i; $Br = 0$; j, unit serial number; FIT, failures/10^9 device hours.

Table 12.2. Optimal burn-in periods under usual operating temperature with and without constraints

Component	Unconstrained $t_{i,b}^*$			Constrained $t_{i,b}^*$
	$\ell = 0$	$\ell = 1$	$\ell = 3$	$\ell = 0$
1	1554	5467	7722	5774
2	2553	5975	7923	4714
3	3969	6814	8369	7694
4	3016	6224	8035	7239
5	2591	6074	8072	5476
R_s	0.977 06	0.985 59	0.986 87	0.9850[a]

[a] Value is for $R_{s,min}$.

6. The field operation time of interest is $t = 2 \times 10^4$ hours for calculation of reliabilities.

7. The terms "series" and "parallel" refer to the reliability logic diagrams, and have nothing to do with the schematic or layout diagrams.

12.4.2 Unconstrained minimization

Based on the assumption that $t_{i,L} = 10^4$ hours, the optimal burn-in period for $t > 10^4$ hours (assuming that the minimal system reliability is not required and burn-in capacity is large enough) is calculated using eq. (12.6). On the basis of the cost factors $c_{i,f}$ and $c_{i,s}$ given in Table 12.1, the $t_{i,b}^*$ for $\ell = 0$ (low penalty), $\ell = 1$ (medium penalty), and $\ell = 3$ (high penalty) are calculated under the usual operating temperature of 25 °C and are listed in Table 12.2.

12.4.3 System reliability

A system with a complex configuration is considerably more difficult to optimize than a "series" or "parallel" system. The reliability of the system shown in Figure 6.1 is

$$R_s = 1 - \left[(1 - R_1')(1 - R_4')\right]^2 R_3' - \left\{1 - R_2'\left[1 - (1 - R_1')(1 - R_4')\right]\right\}^2 (1 - R_3'),$$

(12.7)

where R_j' is the reliability of unit j, $j = 1, 2, 3, 4$. Each individual component reliability with $t_{i,b}$ hours of burn-in is calculated from the equations:

$$R_i(t|t_{i,b}) = \begin{cases} \exp\left\langle -\dfrac{a_i}{b_i'}\left[(t + t_{i,b})^{b_i'} - (t_{i,b})^{b_i'}\right]\right\rangle, & \text{if } 0 \le t \le t_{i,L} - t_{i,b}, \\ \exp\left\langle -\dfrac{a_i}{b_i'}\left[(t_{i,L})^{b_i'} - (t_{i,b})^{b_i'}\right] - \lambda_{i,L}(t - t_{i,L} + t_{i,b})\right\rangle, & \text{otherwise,} \end{cases}$$

(12.8)

where $b_i' = 1 - b_i$.

Since all components in each unit are in "series," the reliability of unit j is

$$R_j' = \prod_{i=1}^{5}(R_i)^{n_{ij}},$$

(12.9)

where five types of components are used in each unit.

Unconstrained optimal system reliability is obtained for various penalties using eq. (12.7) and listed at the end of Table 12.2.

12.4.4 Constrained minimization

Suppose the minimum required system reliability is 0.9850; low penalty ($\ell = 0$) is considered; and the capacity (V_t) available for burn-in is 2000 racks.

As long as the unconstrained optimal R_s for $\ell = 0$ is at least $R_{s,\min}$ and $\sum_i F_i t_{i,b} \le F_t$ (for $\ell = 0$), the optimal solution is reached. However, the values of $t_{i,b}^*$ given in Table 12.2 are not feasible for situations in which either $R_{s,\min} > 0.977\,06$ or $\sum_i F_i t_{i,b} > F_t$, or both. Note that the unconstrained optimal R_s (=0.977 06) for $\ell = 0$ is less than the minimum required system reliability of 0.9850. Therefore, the cost minimization problem with the reliability constraint needs to be solved. The randomized Hooke–Jeeves search technique [187] is applied to solve this optimization problem.

Using reliability constraints

Since maximum reliability is achieved when burn-in is performed to reach a steady-state failure rate, system reliability cannot be greater than $R_s\left(t = 20\,000 \mid t_{i,b} = t_{i,L}\right) = 0.9872$ by burn-in alone (see Figure 12.1). Therefore, in order to have a feasible solution to this burn-in optimization problem, $R_{s,\min}$ must be less than or equal to 0.9872.

Suppose that $\ell = 0$, $R_{s,\min} = 0.9850$, and capacity is always sufficient. Then the objective function to be minimized is, as given by Kuo [173],

$$C_s = \sum_{i=1}^{5} n_i \left[c_i + c_{i,b} + e_{i,s} c_{i,s} + (1 + \ell) e_{i,f} c_{i,f} \right], \tag{12.10}$$

where $e_{i,s}$ and $e_{i,f}$ are the expected fraction of failures in the shop and in the field, respectively. These are given as

$$e_{i,s} = \int_0^{t_{i,b}} a_i w^{-b_i} \, dw = \frac{a_i}{b_i'} (t_{i,b})^{b_i'}, \text{ and}$$

$$e_{i,f} = \begin{cases} \dfrac{a_i}{b_i'} \left[(t + t_{i,b})^{b_i'} - (t_{i,b})^{b_i'} \right], & \text{if } 0 \le t \le t_{i,L} - t_{i,b}, \\[2mm] \dfrac{a_i}{b_i'} \left[(t_{i,L})^{b_i'} - (t_{i,b})^{b_i'} \right] - \lambda_{i,L} (t - t_{i,L} + t_{i,b}), & \text{otherwise,} \end{cases}$$

where $b_i' = 1 - b_i$. After substituting the constants and factoring out the common terms, one can minimize

$$Y = 54 \left(-0.0048 t_{1,b}^{0.20} + 2.70 \times 10^{-5} t_{1,b} \right) + 57 \left(-0.0164 t_{2,b}^{0.25} + 1.14 \times 10^{-5} t_{2,b} \right)$$

$$+ 24 \left(-0.0018 t_{3,b}^{0.25} + 9.00 \times 10^{-7} t_{3,b} \right) + 42 \left(-0.0023 t_{4,b}^{0.30} + 2.55 \times 10^{-6} t_{4,b} \right)$$

$$+ 113 \left(-0.0120 t_{5,b}^{0.20} + 4.43 \times 10^{-6} t_{5,b} \right), \tag{12.11}$$

subject to

$$R_s \ge 0.9850,$$

where R_s is calculated from eqs. (12.7)–(12.9). Now, application of the randomized Hooke–Jeeves method yields the optimal burn-in times and system reliability as shown in the right-hand column of Table 12.2.

Using capacity constraints

Suppose a laboratory has a facility where 2000 board assemblies (V_t) can be used for burn-in simultaneously. Each different component board assembly can be put in a rack of the facility, that is, $F_i = 1$ for all i. Suppose 100 systems with data as shown in Table 12.1 are ordered for burn-in within two months ($T_p = 1440$ hours).

The capacity F_t available within the burn-in facility during the production planning period is $F_t = V_t \times T_p = 2\,880\,000$ rack-hours. Burn-in at an elevated temperature of T_2 can appreciably reduce the burn-in period at normal operating temperature. If t_{i,T_2}^* is the optimal burn-in period at temperature T_2, then according to the Arrhenius equation,

$$t_{i,T_2}^* = t_{i,b}^* \exp \left[-\frac{E_a}{k} \left(\frac{1}{T_1} - \frac{1}{T_2} \right) \right], \tag{12.12}$$

where E_a is the activation energy (0.4 eV); k is the Boltzmann constant (8.60 $\times$ 10^{-5} eV/K); and T_1, T_2 are the absolute temperatures.

Table 12.3. Burn-in time at 125 °C and the required capacity

i	t^*_{i,T_2} (hours)	Number of units for burn-in ($\times 100$)	Capacity required	$c_{i,b}$	P	Marginal cost
1	114.4	5 400	617 652	2.860	2.96	0.100
2	93.4	5 700	532 266	2.335	2.38	0.045[a]
3	152.4	2 400	365 784	3.810	3.91	0.100
4	143.4	4 200	602 280	3.585	3.67	0.085
5	108.5	11 300	1 225 824	2.712	2.84	0.128
Total capacity requirement (F_{req})			3 343 806			

[a] Component with minimum marginal cost.

Thus, t^*_{i,T_2} and the capacity requirement based on $T_1 = 25\,°C$ and $T_2 = 125\,°C$ are given in Table 12.3. In order to achieve the minimum cost and the required reliability level, a capacity of 3 343 806 rack-hours is necessary.

As discussed in Case V of Figure 12.3, the burn-in times for each component that minimize the system cost cannot be changed. Since the manufacturer's laboratory does not have a large enough capacity to burn-in all components at 125 °C, some components should be burned-in at a temperature higher than 125 °C or burned-in outside laboratories with the purchasing price of P.

Assume that burn-in at a temperature higher than 125 °C is not feasible. If the marginal cost for each component is not the same due to differences in component sizes, then the next step is to find the minimum marginal cost. The values $c_{i,b}$ and P of each component are listed in Table 12.3 for comparison. According to Table 12.3, component 2 has the minimum marginal cost. As long as $P < c_{i,b}$ for $i = 2$, all or part of component 2 needs to be sent outside for burn-in instead of reducing the burn-in time of component 2. Since they need only an extra 463 806 $(=F_{req} - F_t)$ unit hours capacity, only 86 percent of component 2 must be sent to outside laboratories and the remaining 14 percent can be burned-in at the manufacturer's laboratory.

12.5 Discussion

Burn-in screening is critical to electronic device manufacturers to maintain competitiveness in international markets. Reliability, as perceived by the customers, appears to be the most important factor driving the competition. However, burn-in is very costly and time consuming. Traditionally, manufacturers conducted burn-in in an *ad hoc* way. Using a reliability requirement and limited resources available in terms of both burn-in time and facilities for burn-in, in this chapter we have introduced a systematic approach to optimizing sound decision making for reliability burn-in.

Given the well-defined burn-in optimization problem described in Section 12.2.3, we can use many of the techniques presented earlier to provide a solution. However, instead of using a designated optimization technique to optimize burn-in cost subject to multiple constraints, we adopt an interactive decision diagram to handle one constraint at a time. The diagram in Figure 12.3 can be best understood by engineers. Since, it is often found that not all constraints are active in an optimization procedure, solving a full-scale optimization problem is not needed all of the time.

This case study shows that not only optimization methodologies, but also a conceptual approach, are important in handling a real problem.

EXERCISES

12.1 What is the burn-in process? Why do we need a burn-in operation?

12.2 Can burn-in reduce the total cost of products? If yes, describe the reasons. When, during the three periods of the bathtub curve, should we apply a burn-in test and process?

12.3 If the capacity of a burn-in facility is inadequate but cannot be increased, a subcontract is a reasonable solution for burning-in some components. Which components will you select to burn-in? Describe how many components should be selected for burn-in.

12.4 To reduce early failures, burn-in is necessary and commonly used by the electronics industry. Who performs burn-in and what are the advantages and disadvantages?

12.5 In Table 12.2, show the calculation of $t_{i,b}$ and R_s when $l = 0$. Show $R_s(20\,000|t_{i,b} = t_{i,L}) = 0.9872$.

12.6 Find $e_{i,s}$ and $e_{i,f}$ by using Table 12.1 when $l = 0$. Find C_s when $l = 1$. (Ignore the constant terms.)

12.7 Burn-in is a technique used to weed out infant-mortalities by applying higher-than-usual levels of stress to speed up the deterioration of electronic devices. To select a realistic burn-in method for integrated circuits, we must study the basic conditions related to integrated circuit burn-in, such as the internal construction and fabrication of the chip, circuit function, circuit layout, number of actually activated and stressed circuit nodes, fault coverage, possible failure modes and mechanisms, and acceleration factors. There are three burn-in types related to the levels of product: package-level burn-in (PLBI), die-level burn-in (DLBI), and wafer-level burn-in (WLBI). Consider only WLBI and PLBI, since

DLBI is usually applied for the known-good dies (KGD) and similar optimization procedures can be applied to DLBI. Burn-in decision making with WLBI and PLBI includes responding to the following two problems: (1) which burn-in policy gives a cost-effective solution, and (2) how much burn-in time is needed for WLBI and PLBI? Burn-in time is strongly limited by the cost factor and reliability requirements. In order to reduce the cost incurred during burn-in, a short test time and small test samples are generally recommended.

The specifications of WLBI and PLBI are determined by several factors, such as the physics of the devices, process maturity, market requirements, and resource constraints. The conditions for burn-in are dependent upon device physics. The process progress includes performance of burn-in, WLBI failure rate, yield loss, and failures due to processes after WLBI. Resources, such as cost and capacity, also contribute to the determination of an optimal burn-in policy.

To find the optimal burn-in time for WLBI and PLBI, we propose a three-stage approach:

- Stage 1: Calculate the cost of no burn-in.
- Stage 2: Calculate the optimal WLBI and PLBI times respectively, and compare them with the cost of no burn-in. If no burn-in is more profitable, then stop; otherwise, proceed.
- Stage 3: Calculate an optimal burn-in time for WLBI prior to PLBI in order to assign each burn-in time to WLBI and PLBI.

In stage 1, the optimal WLBI and PLBI times are obtained to guarantee the required level of mission reliability, R_{req}, at t_r. That is,

$$R(t_r \mid t_{\text{bw}}^*) \geq R_{\text{req}} \quad \text{and} \quad R(t_r \mid t_{\text{bp}}^*) \geq R_{\text{req}},$$

where $R(t)$ is the reliability; R_{req} is the required level of mission reliability at time t_r; t_b, t_e are the total burn-in time and equivalent total burn-in time, respectively; $t_{\text{bp}}, t_{\text{bp}}^*$ are the PLBI time and optimal PLBI time, respectively; $t_{\text{bw}}, t_{\text{bw}}^*$ are the WLBI time and optimal WLBI time, respectively; and t_{ew} is the equivalent WLBI time.

Stage 2 assigns an optimal burn-in time allocation between WLBI and PLBI. Thus the results of stage 2 may be $t_{\text{bw}}^* > 0, t_{\text{bp}}^* = 0$, or $t_{\text{bw}}^* > 0, t_{\text{bp}}^* > 0$, or $t_{\text{bw}}^* = 0, t_{\text{bp}}^* > 0$.

When $t_{\text{bw}}^* = 0$ or $t_{\text{bp}}^* = 0$ is obtained, only one burn-in type yields an optimal solution.

(a) A general burn-in cost model includes burn-in fixed cost, burn-in variable cost, shop-repair cost, and field-repair cost (see [172], [174]). Set up the cost model by considering either the WLBI or PLBI.

(b) Assume that the failure rate at the infant-mortality stage follows the Weibull distribution given in Figure 12.1. Find the reliability with WLBI only, with PLBI only, and with WLBI prior to PLBI.

(c) Set up optimization problems with WLBI only, with PLBI only, and with WLBI prior to PLBI.

(d) If DLBI is considered along with WLBI and PLBI, how would you formulate the optimization problem to minimize the total system cost?

13 Case study on design for software reliability optimization

13.1 Introduction

Software technology has been criticized for its high cost, low reliability, and frequent delays. Forty percent of software development cost is spent on testing to remove errors and to ensure high quality. But high cost and delays are still cited in Simmons et al. [298] as the result of low reliability. Guaranteeing and quality reliability in software systems is one of the most important tasks we are facing today. Unfortunately, software reliability is also a difficult subject to address. By focusing on the overall system, one can improve low software reliability by:

1. debugging the program for longer periods of time,

2. using better testing strategies, and

3. adding redundant software.

Debugging programs for longer periods of time requires more debuggers, more computer time, and more failure-correction personnel, by which one hopes to identify more failures and use less CPU hours. All these activities will increase software reliability but at the same time add significant software development costs.

To efficiently remove bugs from software, one needs to design sampling techniques which mimic realistic operational profiles with the hope that critical and frequently occurring bugs can be identified earlier. This is a difficult task unless we can correctly simulate operational profiles. Testing strategies need to be selected to maximize the coverage of potential failures.

The techniques of using more reliable components and adding redundancies to improve system reliability have been widely used in hardware systems. Nevertheless, software differs from hardware in terms of failure causes and reliability modeling measures. Unless redundant components of software are statistically independent, the conventional approach for hardware redundancy without modification will not apply to software redundancy. Using N-version programming, Ashrafi et al. [16] demonstrated that software reliability can be enhanced.

13.2 The basic execution time model

Software reliability can be modeled by various stochastic processes. To illustrate optimal decision making in software development, we now choose a very promising model for further elaboration. Realizing that failures occur as a random process, we would like to use an inhomogeneous Poisson process to exhibit the variation of failure intensity over time. In the basic execution time model of Musa et al. [241], we assume that the failure intensity function λ is linearly related to the mean failures experienced μ by

$$\lambda(\mu) = \lambda_0 \left(1 - \frac{\mu}{v_0} \right), \tag{13.1}$$

where λ_0 is the initial failure intensity at the start of execution, and v_0 is the total number of failures that would occur in infinite time. The basic model implies a uniform operational profile. For a highly nonuniform operational profile, a modification on the basic model is needed. The basic model makes no assumptions about the quality of the repair process, namely, it allows for possible introduction of new faults during repair. However, it assumes that the correction of faults following a failure is immediate. When the mean failures experienced μ is related to the execution time τ, μ will be an increasing function of τ, but it will be finite at an infinite time

$$\mu(\tau) = v_0 \left[1 - \exp\left(-\frac{\lambda_0}{v_0} \tau \right) \right]. \tag{13.2}$$

Combining (13.1) and (13.2), we can express the failure intensity function in terms of execution time

$$\lambda(\tau) = \lambda_0 \exp\left(-\frac{\lambda_0}{v_0} \right) \tau. \tag{13.3}$$

From eq. (13.1), we can use the objective and the present value of failure intensity to determine the additional expected number of failures that must be experienced to reach that objective. This is

$$\Delta\mu = \frac{v_0}{\lambda_0} (\lambda_p - \lambda_F), \tag{13.4}$$

where $\Delta\mu$ is the expected number of failures to reach the failure intensity objective, λ_p is the present failure intensity, and λ_F is the failure intensity objective.

Similarly, we can use (13.3) to determine the additional execution time $\Delta\tau$ required to reach the failure intensity objective

$$\Delta\tau = \frac{v_0}{\lambda_0} \ln \frac{\lambda_p}{\lambda_F}. \tag{13.5}$$

13.3   Resource usage

Improving software reliability requires a substantial amount of resources. According to Musa et al. [241], resource usage is linearly proportional to execution time and mean failures experienced. Assume that three kinds of resources are needed for software reliability enhancement: failure-identification personnel (indexed by I), failure-correction personnel (indexed by F), and computer time (indexed by C). Let χ_r be the usage of resource r (I, F, or C). Then

$$\chi_r = \theta_r \tau + \mu_r \mu, \qquad (13.6)$$

where θ_r is the usage of resource r per CPU hour and μ_r is the usage of resource r per failure.

For example, a test team runs test cases for 10 CPU hours to identify 12 failures. It takes four person hours of effort for each hour of execution time, and it takes three hours on the average to verify and identify each failure. Therefore, the total identification effort required is

$$\chi_I = \theta_I \tau + \mu_I \mu$$
$$= (4)(10) + (3)(12)$$
$$= 76 \text{ person hours.}$$

The reliability-related cost function represents the resources required to improve reliability of the software. For bug-counting models, software reliability is a function of the number of initial faults and debugging time. Thus, the cost of improving software from one reliability level to another can be related to the number of faults removed during the debugging period, as well as to the debugging time.

By associating the resources of failure-identification personnel and computer time with $\Delta \tau$, and the resources of failure-identification personnel, failure-correction personnel, and computer time with $\Delta \mu$, we can formulate a software reliability-related cost function

$$f(\lambda_P, \lambda_F) = C_1 \theta_r \Delta \tau + C_2 \mu_r \Delta \mu, \qquad (13.7)$$

where C_1 is the cost per unit time of the failure-identification personnel and computer usage, and C_2 is the cost per failure of the failure-correction personnel. The formulations of the extra debugging time $\Delta \tau$ and that of the extra faults removed $\Delta \mu$ to reach λ_F from λ_P depend on the choice of the reliability model. Users can specify their own software reliability models. In order to illustrate our procedure, one software reliability model is employed to formulate $\Delta \tau$ and $\Delta \mu$. In this study, the final formulations of $\Delta \tau$ and $\Delta \mu$ are introduced as given in (13.4) and (13.5).

13.4   Reliability modeling

In software development, redundancies are programs developed by different groups of people or different companies based on the same specifications. These programs are designed to perform the same function. In order to make the failures of redundant copies as independent as possible, we can apply different computer languages, development tools, development methodologies, and testing strategies to different redundant programs.

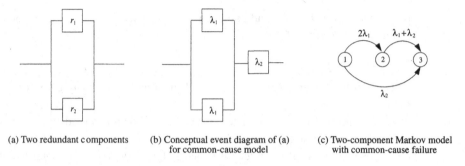

(a) Two redundant components   (b) Conceptual event diagram of (a)   (c) Two-component Markov model
                                 for common-cause model   with common-cause failure

Figure 13.1. Configurations of two-component software models

It has been shown that software redundancies are not totally independent (Echhardt and Lee [84], Knight and Leveson [162]). Some input data will fail more in one redundancy because of common errors made by different development teams. This partial independence of software redundancies can be represented by a common-cause model as shown in Chi and Kuo [55] and Chi et al. [56]. Some specific common-cause models have been proposed, especially in the area of nuclear safety. A general description of the multi-cause failure model is given in Section 1.5.11. The common-cause model for software redundancy is developed below.

13.4.1 The two-component model

Because of common-cause failure, a system with two partially independent software components in parallel, as shown in Figure 13.1(a), can be transformed into a parallel-series system in which there is a subsystem with two independent components in parallel and a common-cause component, Figure 13.1(b).

To compute the system reliability, the theory of Markov processes is used. Consider the states:

1. All components are good.

2. One of the two parallel components has failed and the common-cause component is good.

3. The system has failed, that is, either both the parallel components have failed or the common-cause component has failed.

Suppose the failure time of either of the two parallel components, shown in Figure 13.1(b), follows an exponential distribution with rate λ_1 and that of the common-cause component follows an exponential distribution with rate λ_2. Then, the state of the system at any time can be described by a Markov process and depicted in Figure 13.1(c). The states of the Markov process represent the number of failed components. Let $P_i(t)$ denote the probability that the system is in state i at time t and $P_i'(t) = dP_i(t)/dt$ for $i = 1, 2, 3$. Let

$$\mathbf{P}(t) = \begin{bmatrix} P_1(t) \\ P_2(t) \\ P_3(t) \end{bmatrix} \quad \text{and} \quad \mathbf{P}'(t) = \begin{bmatrix} P_1'(t) \\ P_2'(t) \\ P_3'(t) \end{bmatrix}.$$

Then we can write

$$\mathbf{P}'(t) = A\mathbf{P}(t),$$

where

$$A = \begin{bmatrix} -(2\lambda_1 + \lambda_2) & 2\lambda_1 & \lambda_2 \\ 0 & -(\lambda_1 + \lambda_2) & \lambda_1 + \lambda_2 \\ 0 & 0 & 0 \end{bmatrix}.$$

The solution of these differential equations is

$$\mathbf{P}(t) = e^{At} P(0),$$

that is,

$$\mathbf{P}(t) = \left[I + \sum_{n=1}^{\infty} (t^n/n!) A^n \right] P(0), \tag{13.8}$$

where $P(0)^T = (1\ 0\ 0)$. Suppose v_1, v_2, and v_3 are the eigen values of the matrix A, and $\mathbf{V}_i$ is the eigen vector of A corresponding to v_i for $i = 1, 2, 3$. Let V denote the matrix with columns $\mathbf{V}_1, \mathbf{V}_2$, and $\mathbf{V}_3$, that is,

$$V = (\mathbf{V}_1 \mathbf{V}_2 \mathbf{V}_3),$$

and let

$$D = \begin{pmatrix} v_1 & 0 & 0 \\ 0 & v_2 & 0 \\ 0 & 0 & v_3 \end{pmatrix}.$$

If the eigen vectors are linearly independent, then we have

$$A = VDV^{-1},$$
$$A^n = VD^nV^{-1}, \tag{13.9}$$

for any $n \geq 1$, where

$$D^n = \begin{pmatrix} v_1^n & 0 & 0 \\ 0 & v_2^n & 0 \\ 0 & 0 & v_3^n \end{pmatrix}.$$

Substituting eq. (13.9) into (13.8) we get

$$\mathbf{P}(t) = V \left[I + \sum_{n=1}^{\infty} (t^n/n!) D^n \right] V^{-1} P(0)$$

$$= V \begin{bmatrix} \exp(v_1 t) & 0 & 0 \\ 0 & \exp(v_2 t) & 0 \\ 0 & 0 & \exp(v_3 t) \end{bmatrix} V^{-1} P(0). \tag{13.10}$$

The eigen values of matrix A are

$$v_1 = -(2\lambda_1 + \lambda_2),$$

$$v_2 = -(\lambda_1 + \lambda_2),$$

$$v_3 = 0,$$

and the corresponding eigen vectors are

$$\mathbf{V}_1 = \begin{bmatrix} 1 \\ -2 \\ 1 \end{bmatrix}, \quad \mathbf{V}_2 = \begin{bmatrix} 0 \\ -1 \\ 1 \end{bmatrix}, \quad \text{and} \quad \mathbf{V}_3 = \begin{bmatrix} 0 \\ 0 \\ 1 \end{bmatrix}.$$

Now,

$$V = \begin{bmatrix} 1 & 0 & 0 \\ -2 & -1 & 0 \\ 1 & 1 & 1 \end{bmatrix}, \quad D = \begin{bmatrix} -(2\lambda_1 + \lambda_2) & 0 & 0 \\ 0 & -(\lambda_1 + \lambda_2) & 0 \\ 0 & 0 & 0 \end{bmatrix},$$

and

$$V^{-1} = \begin{pmatrix} 1 & 0 & 0 \\ -2 & -1 & 0 \\ 1 & 1 & 1 \end{pmatrix}.$$

It can be numerically verified that $A = VDV^{-1}$. Using (13.10), we get

$$\mathbf{P}(t) = V \left\{ \begin{array}{ccc} \exp[-(2\lambda_1 + \lambda_2)t] & 0 & 0 \\ 0 & \exp[-(\lambda_1 + \lambda_2)t] & 0 \\ 0 & 0 & 1 \end{array} \right\} V^{-1} \begin{bmatrix} 1 \\ 0 \\ 0 \end{bmatrix}$$

$$= \left\{ \begin{array}{c} \exp[-(2\lambda_1 + \lambda_2)t] \\ 2\exp[-(\lambda_1 + \lambda_2)t] - 2\exp[-(2\lambda_1 + \lambda_2)t] \\ \exp[-(2\lambda_1 + \lambda_2)t] - 2\exp[-(\lambda_1 + \lambda_2)t] + 1 \end{array} \right\}.$$

Therefore, the system reliability is

$$R_s(t) = 1 - P_3(t) = 2\exp[-(\lambda_1 + \lambda_2)t] - \exp[-(2\lambda_1 + \lambda_2)t]. \tag{13.11}$$

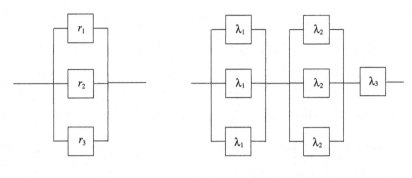

(a) Three redundant components   (b) Conceptual event diagram of (a)
                                            for common-cause model

Figure 13.2. Configurations of three-component software system

13.4.2 The three-component model

A system with three partially independent software components in parallel is shown in Figure 13.2(a).

Since some input data can cause failure in one component, or two or three components at the same time, such a system can be transformed into a parallel-series system of three independent subsystems as shown in Figure 13.2(b) where

- the first subsystem has three independent components in parallel, each having a failure rate λ_1;

- the second subsystem has three independent components in parallel such that the first one fails when software components 1 and 2 fail simultaneously; the second fails when software components 2 and 3 fail simultaneously; and the third fails when software components 1 and 3 fail simultaneously, with the failure rate of the three independent parallel components being λ_2; and

- the third subsystem is a single component which fails when all three software components fail simultaneously with failure rate λ_3.

A system with n partially independent software components in parallel, such as the simplified one shown in Figure 13.3, can be analyzed accordingly as explained in Chi and Kuo [55]. The Markov model of this system with common-cause failures is shown in Figure 13.4, where the failure rates of all independent components are assumed to be the same, λ_c. In this simplified model, the only common-cause failures considered are the ones that cause all of the redundancies to fail and the transition rate from one state to another depends on the number of survival components, namely, the transition rate from state m to $m + 1$ is $(n - m)\lambda_1$, where λ_1 is the transition rate from state $n - 1$ to state n.

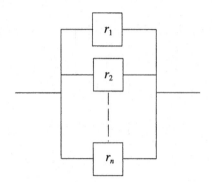

Figure 13.3. Configuration of an n-component redundant system

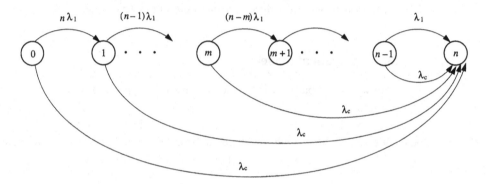

Figure 13.4. Generic n-component Markov model

13.5   Formulation of the software reliability optimization problem

To optimize the reliability of a software system, the reliability–redundancy allocation approach is applied. A general formulation is given below.

Problem 13.1

Maximize

$$R_s = f(r_1, \dots, r_n; x_1, \dots, x_n),$$

subject to

$$\sum_{j=1}^{n} g_{ij}(r_j, x_j) \le b_i, \qquad \text{for } i = 1, \dots, m,$$

$$0 \le r_j \le 1 \text{ and } x_j \in \{1, 2, \dots\}, \qquad \text{for } j = 1, \dots, n.$$

where r_j and x_j denote the component reliability and the component redundancy level at stage j.

13.5.1 A pure software system

Software is always accompanied by hardware. However, when the reliability of the hardware component in the system is known, one can optimize the system reliability by including only software components. When only n software components of a system are considered for reliability optimization, the above problem can be transformed into the following form.

Problem 13.2
Maximize

$$R_s = f(r_1, \ldots, r_n; x_1, \ldots, x_n),$$

subject to

$$\sum_{j=1}^{n} g_{ij}[f_{ij}(r_j, r_j^*), h_{ij}(x_j)] \leq b_i, \qquad \text{for } i = 1, \ldots, m,$$

where $f_{ij}(r_j, r_j^*)$ is the ith reliability-related cost function to increase the reliability of the jth module from r_j to r_j^* and $h_{ij}(x_j)$ is the redundancy cost function of x_j components consumed for resource i.

The objective function of the above formulation is represented in terms of the module reliabilities that are, in turn, functions of the number of redundant modules and the module reliabilities. The constraint function of the above formulation is a reliability-related cost function. The reliability-related cost function for a specific i and j is

$$f(\lambda_P, \lambda_F) = C_1 \theta_r \Delta\tau + C_2 \mu_r \Delta\mu,$$

where the formulation of extra debugging time $\Delta\tau$ and that of the extra faults removed $\Delta\mu$ to reach λ_F from λ_P, depend on the choice of the software reliability model. C_1 and C_2 are the unit cost coefficients. When the basic model is used, the formulations of the extra debugging time ($\Delta\tau$) and the extra faults removed $\Delta\mu$ become (13.4) and (13.5), respectively.

The redundancy cost function $h_{ij}(x_j)$ depends upon the type of constraints involved. A constant function, an increasing function, or a decreasing function can be used as needed and should be described in a generic form reflecting the software development life cycle.

13.5.2 A hardware and software mixed system

When the reliabilities of hardware and software components are not fixed, they should be optimized along with component redundancy levels. In a pure software system, each stage represents an independent functional module or subsystem. However, this model can be extended to optimize the reliability of a hardware and software system by adding the constraint function of the hardware part. Suppose the hardware components are indexed as $1, 2, \ldots, d$ and the software components as $d + 1, \ldots, n$. Here, the objective function to be maximized can be written as

$$R_{\mathrm{s}} = f(r_1, \ldots, r_n; x_1, \ldots, x_n).$$

The new constraints become

$$\sum_{j=1}^{d} g_{ij}[f_{ij}(r_j), h_{ij}(x_j)] + \sum_{j=d+1}^{n} g_{ij}[f_{ij}(r_j, r_j^*), h_{ij}(x_j)] \le b_i, \qquad \text{for } i = 1, \ldots, m.$$

The objective function of the above formulation is represented in terms of stage reliabilities. Each stage can be a pure software component, a pure hardware component, or a hardware and software mixed component. The constraint function is represented as the product of a reliability-related cost function and a redundancy cost function. For hardware components (Lin and Kuo [200]), the reliability-related cost function is

$$r_j = \exp(-\lambda_j t),$$

$$f(r_j) = v_j \left(\frac{-t}{\ln r_j} \right)^{u_j}, \tag{13.12}$$

where t, u_j, and v_j are given constants. For software components, the reliability-related cost function is

$$f(\lambda_{\mathrm{P}}, \lambda_{\mathrm{F}}) = C_1 \theta_r \Delta\tau + C_2 \mu_r \Delta\mu, \tag{13.13}$$

where the formations of the extra debugging time $\Delta\tau$ and the extra faults removed $\Delta\mu$ to reach λ_{F} from λ_{P} again depend on the choice of the software reliability model. Constraints on typical resources, Musa et al. [241], can be expressed as:

1. Personpower and computer time (total cost)

$$f_{1j} = f(r_j, r_j^*) = C_1 \theta_r \Delta\tau + C_2 \mu_r \Delta\mu \le b_1.$$

2. Project duration

$$f_{2j} = t_r \Delta\tau \le b_2,$$

where $t_r \Delta\tau$ is the failure-identification time. It is assumed that the failure-correction time is small enough to be ignored.

3. Memory space

$$f_{3j} = M \le b_3.$$

Table 13.1. Data for hardware in Example 13-1

Module i	$\lambda_{H,i}$	v_i	u_i
A	0.000 016 79	0.040 302	3.5
B	0.000 022 45	0.055 045	3.5

The redundancy cost function can be expressed as

$$h_{ij}(x_j) = kx_j. \tag{13.14}$$

The redundant components share with the original component: (1) the cost of specification, (2) some of the design cost, (3) most of the testing cost, and (4) most of the documentation cost. All of these determine the k value. Because of this, the cost of increasing one redundant component is usually less than 1.5 times the original unit development cost. The analysis of these costs should be investigated further.

Example 13-1

A NASA system is to be designed for launching space shuttles. The system has two modules in series. Module A consists of one hardware and one software component and cannot have any redundant components. The second module B can have redundant components. Each hardware component in module B has a corresponding software in series. Software for both modules has not yet been developed. The software failures of modules A and B are independent. However, there are common-cause failures among the software redundancies of module B. NASA wants to decide the redundancy level of module B and the software reliabilities for both modules to maximize system reliability for mission time $t = 900$ hours. The total hardware development cost must not exceed US\$ 2 000 000. The failure-identification personnel cost is not more than US\$ 1 000 000.

Define

$r_{H,A}$	hardware reliability of module A, which is $\exp(-\lambda_{H,A}t)$
$r_{H,B}$	hardware reliability of module B, which is $\exp(-\lambda_{H,B}t)$
$r_{S,A}$	software reliability of module A, which is $\exp(-\lambda_{S,A}t)$
$r_{S,B}$	software reliability of module B, which is $\exp(-\lambda_{S,B}t)$
x_1	redundancy of module A, which is fixed at 1
x_2	redundancy of module B

The reliability and development costs of hardware components for both modules are presented in Table 13.1. From Table 13.1 and using eqs. (13.12) and (13.13), we can calculate reliabilities and the development cost of hardware for modules A and B. For module A we have

$$r_{H,A} = \exp(-\lambda_{H,A}t) = \exp[(-0.000 016 79)(900)] = 0.985,$$

Table 13.2. Parameters for software used in Example 13-1

Module i	θ_{1i}	μ_{1i}	v_{0i}	λ_{0i}	C_{1i}	C_{2i}	k	δ	b_0(US$)	b_1(US$)
A	1	5	200	20	450	540	0.3	0.03	2 000 000	1 000 000
B	1	3	100	10	450	780				

Table 13.3. Optimal solution for Example 13-1

	Module A	Module B
x_i	1	2
r_i	0.9970	0.9965
Cost, US$	61 000	30 000
Total cost, US$	100 000	
R_s	0.9814	

with development cost

$$f(r_{H,A}) = v_A \left[\frac{-t}{\ln r_{H,A}} \right]^{u_A} = 0.040\,302 \left(\frac{-900}{\ln 0.000\,016\,79} \right)^{3.5} = \text{US\$ } 200\,000.$$

Similarly, for module B, $r_{H,B} = 0.98$ and $f(r_{H,B}) = \text{US\$ } 300\,000$. For mission time $t = 900$ CPU hours, the problem becomes $\max_{r_{S,A}, r_{S,B}, x_2} R_s = f(r_{H,A}, r_{H,B}, r_{S,A}, r_{S,B}, x_1, x_2)$ subject to

$$200\,000 x_1 + 300\,000 x_2 \le b_0,$$

$$C_{1A}\theta_{1A} \frac{v_{0A}}{\lambda_{0A}} \ln \frac{\lambda_{0A}}{\lambda_{SA}} + C_{2A}\mu_{1A} \frac{v_{0A}}{\lambda_{0A}} (\lambda_{0A} - \lambda_{SA})$$

$$+ [1 + k(x_2 - 1)] \left[C_{1B}\theta_{1B} \frac{v_{0B}}{\lambda_{0B}} \ln \frac{\lambda_{0B}}{\lambda_{SB}} + C_{2B}\mu_{1B} \frac{v_{0B}}{\lambda_{0B}} (\lambda_{0B} - \lambda_{SB}) \right] \le b_1,$$

where subscripts A and B indicate the parameters associated with modules A and B, respectively, and k is the development cost of redundancy when the original unit development cost is assumed to be 1.

Note that

$$f(r_{H,A}, r_{H,B}, r_{S,A}, r_{S,B}, x_1 = 1, x_2 = 1) = r_{H,A} r_{S,A} r_{H,B} r_{S,B},$$

and

$$f(r_{H,A}, r_{H,B}, r_{S,A}, r_{S,B}, x_1 = 1, x_2 > 1)$$
$$= r_{H,A} r_{S,A} \left(1 - [1 - r_{H,B}(r_{S,B})^{1-\delta}]^{x_2} \right) (r_{S,B})^{\delta},$$

where δ is a fraction of the failures that are a common cause for software utilized in module B. Table 13.2 gives data for software parameters.

This problem can be solved by the randomized Hooke–Jeeves method and the branch-and-bound method. The resulting optimal solution is shown in Table 13.3.

13.6   Discussion

In this chapter, a system approach has been provided for improving software reliability. The software reliability-related cost function and the reliability function of software redundancy with the common-cause failure model have been discussed. During the concurrent coding and testing phase, system and component reliabilities are examined. Since more information about developing software, such as failure intensity or failure rate, is readily available at this time, more accurate system and component reliabilities can be reevaluated with updated data.

If the system and module reliabilities do not meet minimum requirements, resources can be reallocated and the system optimization problem can be solved again with the updated data. A set of solutions along with determined decision variables can be obtained. The management chooses a solution from among the new multiple solutions obtained. The decision is whether to improve module reliabilities or to increase the number of redundancies.

EXERCISES

13.1 What are the typical applications of software reliability measurements in systems engineering?

13.2 The same data have been fitted with both the basic and logarithmic Poisson models proposed by Musa et al. [241]. The parameters obtained are:

Basic	Logarithmic Poisson
$\lambda_0 = 20$ failures/CPU hour	$\lambda_0 = 50$ failures/CPU hour
$\nu_0 = 120$ failures	$\theta = 0.025$/failure

Note that the logarithmic Poisson usually has a higher initial failure intensity: recall that this falls off more rapidly than the failure intensity for the basic model, but finally it falls off more slowly. First, determine the additional failures and additional execution time required to reach a failure intensity objective of ten failures/CPU hour, using both models. Then, repeat this for an objective of one failure/CPU hour. Assume in both cases that you start from the initial failure intensity.

13.3 The figure below represents the event diagram for a time-sharing system running on a machine with two processors. The time-sharing system runs under a general-purpose operating system, and these run on both processors. The reliabilities of the hardware for a 10-hour prime shift are labeled. The operating and time-sharing systems have failure intensities measured in execution times of 0.05 and 0.025 failures/CPU hour, respectively. One runs about 20 percent of the time, and another 60 percent of the processor capacity is used by application programs. Find the prime shift reliability of the entire extended machine (reliability of application programs is not considered).

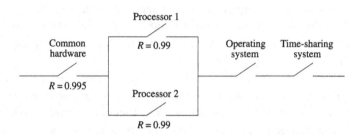

13.4 A system with two partially independent software components (A and B) in parallel can be transformed into a series system with two independent components in parallel and a common-cause component. Set up

(a) the Markov model of this system with common-cause failure, where λ_1 is the failure rate of each independent component and λ_2 is the failure rate of common-cause component;

(b) the differential equation of the Markov process.

13.5 A program has an initial failure intensity of ten failures/CPU hour. We want to test and debug this program until a failure intensity of one failure/ten CPU hours is achieved. Assume the following resource usage parameters.

Resource usage	Per hours	Per failure (hour)
Failure-identification effort	3[a]	2[a]
Failure-correction effort	0	6[a]
Computer time	1.5	1

[a] Measurements are in person hours.

(a) What resources must be expended to achieve the reliability improvement required? Use the basic execution time model. Assume a failure intensity decay parameter of 0.05.

(b) If the failure intensity objective is cut in half, are the resources required doubled?

13.6 The following problem uses the basic execution time model. We are conducting a system analysis for the proposed software system COMPROD to determine the range of cost choices available as a function of the failure intensity objective. We wish to investigate the costs for failure intensity objectives ranging from 1 to 5 failures per 100 CPU hours. The following system test costs (with ranges of variation) have been determined.

Failure intensity objective (per 100 CPU hours)	System test cost (US$K)	Range of system test cost (US$K)
1	1585	1230–1840
2	1490	1160–1710
3	1440	1100–1650
4	1400	1070–1600
5	1360	1040–1560

Compute the overall component of system cost that is dependent on reliability (failure intensity objective). Assume that the system will operate for four years, 250 days/year, ten CPU hours/day. There will be five installations of the system. The total cost impact of a failure is estimated to be US$ 100. What is the optimum value for the failure intensity objective? What range of values does this objective take on as a result of inaccuracies in the knowledge of total failures and initial failure intensity?

14 Case study on an optimal scheduled-maintenance policy

More than one independent (conflicting) criterion is usually important in determining the replacement age of a critical item for a maintained system. In this study, mathematical models have been developed for three such criteria: (1) minimum replacement cost-rate, (2) maximum availability, and (3) lower bound on mission reliability. Four methods are used to obtain solutions for multiple criteria decision making. These are: (1) strictest selection, (2) the lexicographic method, (3) the Waltz lexicographic method, and (4) the sequential multiple objective problem-solving technique (SEMOPS). Using an aircraft engine as an example, the optimal replacement age has been found by the four different methods. The results and the implications of the methods are discussed in this case study, which is largely based on Hwang et al. [137].

14.1 Introduction

For every piece of military or commercial equipment, there exist critical items whose failure could result in a shut-down or in hazardous conditions for individuals using the items. It is often reasonable to use multiple independent (conflicting) criteria to determine the replacement age of a critical item. The three criteria most often used are:

1. minimum replacement cost-rate,
2. maximum availability, and
3. lower bound on mission reliability.

The expected cost of mission failure is linearly related to mission reliability, and thus need not be considered explicitly. In determining the maintenance policy of critical items of aircraft, Ladany and Aharoni [182] selected the most conservative (youngest) replacement age by considering: (1) the minimum replacement cost-rate, (2) the lower bound on mission reliability, and (3) the in-flight failure cost. The following deficiencies exist in their problem statement and decision process:

1. The availability, that is, the principal measure of the effectiveness of a maintained system, is not considered.

2. Their equation for replacement cost-rate does not consider down times for preventive and failure replacements.

3. The strictest selection method is unduly conservative and removes any flexibility from the decision maker.

In this study, four multiple criteria decision-making methods are used to obtain an optimal scheduled-maintenance policy with respect to the above three criteria. For other multiple criteria decision-making methods, such as the surrogate-worth tradeoff method [278], see [48], [134], [135], [138], [183], and [331].

Notation

$A(t_p)$	availability of the critical item
C_{pr}	cost of a preventive replacement
C_{fr}	cost of a failure replacement
$C(t_p)$	total expected replacement cost-rate
E_f	expected cost of mission failure
$F(x, h)$	probability of failure between x and $x + h$, given that the unit has not failed until x
$f(x), F(x)$	probability density and cumulative disribution functions of lifetime of the critical item, respectively
H	mission duration time
$L(t_p)$	length of a cycle
$M(t_p)$	mean-life during cycle $\left[= \int_0^{t_p} R(t)\, dt\right]$
$R(x)$	survival function $[= 1 - F(x)]$
$R(x, h)$	mission reliability $[= 1 - F(x, h)]$
t_{dpr}	mean down time for preventive replacement
t_{dfr}	mean down time for failure replacement
t_p	preventive replacement age

14.2   Criteria functions

The optimal scheduled-maintenance policy will be determined for preventive replacement of critical items. The repair or replacement will return the item to the equivalent new condition. Assuming that the planning horizon is very long, the three criteria previously mentioned are evaluated as follows.

1. Minimum replacement cost-rate:

$$L(t_p) = M(t_p) + t_{dpr} R(t_p) + t_{dfr} F(t_p),$$

$$C(t_p) = \frac{\text{total expected replacement cost per cycle}}{\text{expected cycle length}}$$

$$= \frac{C_{pr} R(t_p) + C_{fr} F(t_p)}{L(t_p)}. \tag{14.1}$$

The t_{p1}^*, optimal t_p, is found by minimizing $C(t_p)$ by the usual calculus method.

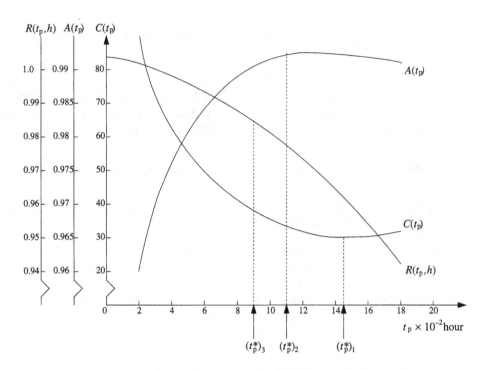

Figure 14.1. The optimal replacement age by the strictest selection method

2. Maximum availability:

$$A(t_p) = \frac{\text{expected operating time in the cycle}}{\text{expected cycle}}$$

$$= \frac{M(t_p)}{L(t_p)}. \tag{14.2}$$

The t_{p2}^*, optimal t_p, is found by maximizing $A(t_p)$ by the usual calculus method.

3. Mission reliability, that is the probability that an item finishes a mission of length h which begins at age x, is:

$$R(x, h) = \frac{R(x + h)}{R(x)}. \tag{14.3}$$

The mission reliability will be minimum at the preventive replacement age. An expected cost of mission failure can be translated into the equivalent bound on mission reliability t_{p3}^*.

Example 14-1

A numerical example from Bell et al. [27] is used here. The lifetimes of aircraft engines are closely represented by a Weibull distribution with a shape parameter $\beta = 3.0$ and a characteristic life α of 1390 hours.

Table 14.1. The optimal preventive replacement ages

i	Criterion	t_{pi}^* (hour)
1	Replacement cost-rate (min)	1455 optimum
2	Availability (max)	1129 optimum
3	Mission reliability (lower bound)	913 absolute max

It is reasonable to assume the following costs (in kilo dollars), times, and lower bound on mission reliability:

$C_{pr} = 25,$

$C_{fr} = 37.5,$

$t_{dpr} = 8$ hours,

$t_{dfr} = 16$ hours,

$R(x, h)_{min} = 0.985.$

It is straight-forward to substitute the Weibull equations and above values into (14.1)–(14.3). For the above data, the values of $C(t_p)$, $A(t_p)$, and $R(x, h)$ are given in Figure 14.1.

14.3   The strictest selection

For each multiple independent (conflicting) criterion, the item will have a different optimal preventive replacement age. The strictest selection method uses the youngest of the optimal preventive replacement ages for each criterion.

The optimal preventive replacement ages are given in Table 14.1. The youngest replacement age, 913 hours, is then selected (see Figure 14.2).

The strictest selection method generally assures that the minimal (or maximal) values of the goals are attained for most criteria. In some cases, where there is a mixture of minimal limits for some criteria and maximal limits for other criteria, the strictest selection method will not work, since there will be logical contradictions. Once the decision maker decides the minimal (or maximal) values for all the criteria, the solution is essentially determined. Then the only way the decision maker can change the solution is by changing the limits (minimal or maximal) for the criteria.

14.4   The lexicographic method

This method requires that the objectives be ranked in order of importance by the decision maker. The preferred solution obtained by this method is one which optimizes

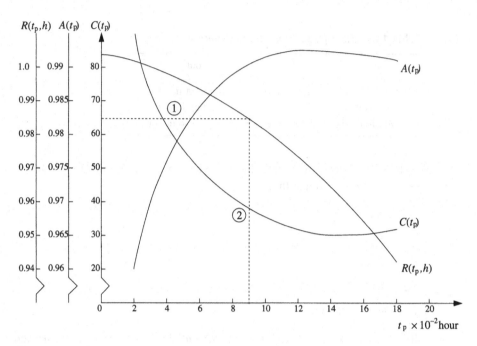

Figure 14.2. Finding the replacement age by the lexicographic method

the criteria starting with the most important and proceeds according to the order of importance of the criteria [135].

Assume that the decision maker ranks the importance of the three criteria in this order: (1) satisfy lower bound mission reliability, (2) minimize the replacement cost-rate, and (3) maximize the availability. Then:

Problem 14.1

Instead of $\text{Max}_{t_p} R(t_p, h)$, find t_p subject to

$$R(t_p, h) \geq R(t_p, h)_{\min}, \qquad t_p \geq 0,$$

where $R(t_p, h)_{\min} = 0.985$. The solution is $t_p \leq 913$ hours (see Figure 14.2).

Since this is not a unique solution, that is, any $t_p \leq 913$ hours satisfies $R(t_p, h) \geq R(t_p, h)_{\min}$:

Problem 14.2

Minimize $C(t_p)$ subject to

$$R(t_p, h) \geq 0.985, \qquad t_p \geq 0.$$

The solution to Problem 14.2 is $t_p = 913$ hours. Since this is a unique solution, the procedure is terminated.

Table 14.2. The results of different rankings of criterion importance from the lexicographic method

	Priority 1	Priority 2	Priority 3	t_p (hour)
Case 1	$C(t_p)$	a	a	1455
Case 2	$A(t_p)$	a	a	1129
Case 3	$R(t_p, h)$	$A(t_p)$	a	913
Case 4	$R(t_p, h)$	$C(t_p)$	a	913

a These criteria do not affect the solution.

Now, $t_p = 913$ hours is the solution to the total problem. The less important criterion, maximize availability, is ignored by this method.

The optimal replacement age of this specific example obtained by the lexicographic method is illustrated in Figure 14.2. Since there are three criteria in the problem, the decision maker could rank their importance in $3! = 6$ different ways. The results in Table 14.2 indicate that, for this example, the solution is dominated by the first priority, or the first and second priorities. The solution will be different if the priorities of the three criteria are changed. Since the solution is very sensitive to the ranking of the criteria determined by the decision maker, the analyst should exercise caution in applying this method when some of the criteria are of nearly equal importance.

14.5   The Waltz lexicographic method

The Waltz lexicographic method [319] is useful for reducing the sensitivity of the solution to the decision maker's priority of criteria. After the first criterion is optimized, the second criterion is optimized subject to keeping the first criterion within a certain percentage of its optimum. The third criterion is then optimized keeping the first two within a certain percentage of the optimum values found in the previous step, and so on.

For the sake of demonstration, suppose the decision maker decides that the importance of the three criteria are ranked as (1) minimize the replacement cost-rate, (2) maximize the availability, and (3) lower bound mission-reliability. Then:

Problem 14.3

Minimize $C(t_p)$, subject to

$$t_p \geq 0.$$

Its solution is $t_p^* = 1455$ hours, and $C(t_p^*) = 28.92$ US\$/hour (see Figure 14.3).

However, if the decision maker decides a replacement cost-rate ≤ 30.5/hour is satisfactory, then:

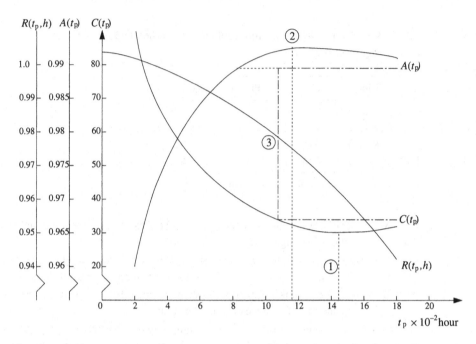

Figure 14.3. Finding the replacement age by the Waltz lexicographic method

Problem 14.4

Maximize $A(t_p)$ subject to

$$C(t_p) \leq 30.5, \qquad t_p \geq 0.$$

The solution to Problem 14.4 is $t_p^* = 1129$ hours, and $A(t_p^*) = 0.9888$ (see Figure 14.3).

If the decision maker thinks that the availability ≥ 0.9875 is acceptable, then:

Problem 14.5

Maximize $R(t_p, h)$ subject to

$$C(t_p) \leq 30.5, \qquad A(t_p) \geq 0.9875, \qquad t_p \geq 0.$$

The solution to Problem 14.5 is $t_p^* = 1057$ hours, $R(t_p^*, h) = 0.98$ (see Figure 14.3).

Finally, the solution to the total problem is $t_p^* = 1057$ hours, for which $C(t_p^*) = 30.5$, $A(t_p^*) = 0.988\,77$, and $R(t_p^*) = 0.98$. The sensitivity of the solution to the priority ranking has been reduced. The optimal replacement age of this example using the Waltz lexicographic method is illustrated in Figure 14.3.

From the decision maker's point of view, the Waltz lexicographic method is very different from other methods. There is information feedback to the decision maker which can be used in determining the amount each criterion is to be relaxed. Obtaining a preferred solution instead of the optimal solution for each criterion is then pursued, and the sensitivity of the solution to the ranking is reduced.

14.6   SEMOPS: an interactive method

SEMOPS was proposed by Monarchi et al. [236]. Here, the decision maker is allowed to tradeoff one objective with another, interactively. Some implicit tradeoff information is generated by the method so that the decision maker is able to determine if the current solution is satisfactory.

The algorithm generates information under the guidance of the decision maker. Information concerning the interrelationships between objectives is in terms of how achievement or nonachievement of one objective affects the aspiration levels (ALs) of other objectives (see [135] for details of this method).

Consider the example where the problem includes three goals and one nonnegative decision variable:

- Criterion functions and goals:

$$f_1 = C(t_p) \leq \text{AL}_1, \qquad f_2 = A(t_p) \geq \text{AL}_2, \qquad f_3 = R(t_p, H) \geq \text{AL}_3.$$

- Goal levels (GL, the initial aspiration levels):

$$\text{GL}_1 = 30.5 \text{ US\$/hour}, \qquad \text{GL}_2 = 0.9885, \qquad \text{GL}_3 = 0.99.$$

- Relevant range of f_i:

$$(f_{1L}, f_{1U}) = (0, 80), \qquad (f_{2L}, f_{2U}) = (0.9, 1), \qquad (f_{3L}, f_{3U}) = (0.9, 1).$$

The criterion functions are transformed into

$$Y_i = \frac{f_i - f_{iL}}{f_{iU} - f_{iL}},$$

that is,

$$Y_1 = \frac{f_1}{80}, \quad Y_2 = \frac{f_2 - 0.9}{1 - 0.9}, \quad \text{and} \quad Y_3 = \frac{f_3 - 0.9}{1 - 0.9}.$$

Initially, the aspiration levels are assumed to be equal to the goal levels, that is,

$$\text{AL}_i = \text{GL}_i, \qquad \text{for } i = 1, 2, 3.$$

Define

$$A_i = \frac{\text{AL}_i - f_{iL}}{f_{iU} - f_{iL}}, \qquad \text{for } i = 1, 2, 3.$$

Then the values of A_is are

$$A_1 = 0.381\,25, \qquad A_2 = 0.885, \qquad \text{and} \quad A_3 = 0.9.$$

Define

$$d_1 = \frac{Y_1}{A_1}, \quad d_2 = \frac{A_2}{Y_2}, \quad \text{and} \quad d_3 = \frac{A_3}{Y_3}.$$

Note that d_2 and d_3 are defined as reciprocals of Y_2/A_2 and Y_3/A_3, respectively, since the second and third objectives involve maximization. Let

$$\mathbf{d} = (d_1, d_2, d_3) \quad \text{and} \quad \mathbf{f} = (f_1, f_2, f_3).$$

Table 14.3. Results of the first cycle (SEMOPS), $AL = GL = (30.5, 0.9885, 0.99)$

s_1	t_p	d	f
$s_1 = 3.9696$	613	$d_1 = (1.4010, 1.0316, 0.9662)$	$f_1 = (42.73, 0.9858, 0.9992)$
$s_{1.1} = 3.7903$	1058	$d_{1.1} = (1.0000, 0.9969, 1.1252)$	$f_{1.1} = (30.49, 0.9888, 0.9800)$
$s_{1.2} = 3.4051$	925	$d_{1.2} = (1.0588, 1.0000, 1.0638)$	$f_{1.2} = (32.29, 0.9885, 0.9846)$
$s_{1.3} = 3.0010$	635	$d_{1.3} = (1.3615, 1.0275, 0.9713)$	$f_{1.3} = (41.53, 0.9861, 0.9927)$

First cycle

The principal problem to be solved for the first cycle is

minimize $s_1 = d_1 + d_2 + d_3$, subject to $t_p > 0$.

We also construct three auxiliary problems in which we attempt to satisfy each of the goals in turn. If inequality $f_1 \leq 30.5$ is entered as a constraint, d_1 is deleted from the surrogate objective function giving:

minimize $s_{1.1} = d_2 + d_3$, subject to $f_1 \leq 30.5$, $t_p > 0$.

Similarly, the second and third auxiliary problems are:

minimize $s_{1.2} = d_1 + d_3$, subject to $f_2 \geq 0.9885$, $t_p > 0$,

minimize $s_{1.3} = d_1 + d_2$, subject to $f_3 \geq 0.9900$, $t_p > 0$.

The optimum results for the four problems are given in Table 14.3. The change in availability is relatively independent of the attainment or nonattainment of the other goals. It seems reasonable to choose an aspiration level for this goal and enter it as a constraint. Assume that the decision maker sets a new aspiration level of 0.987 for goal 2.

Second cycle

The aspiration levels are now $AL_1 = (30.5, 0.987, 0.99)$, and attainment of goal 2 has been entered as a constraint. The principal and the two auxiliary problems to be solved for this cycle are:

minimize $s_{2.2} = d_1 + d_3$, subject to $f_2 \geq 0.987$, $t_p > 0$;

minimize $s_{2.21} = d_3$, subject to $f_2 \geq 0.987$, $f_1 \leq 30.5$, $t_p > 0$;

minimize $s_{2.23} = d_1$, subject to $f_2 \geq 0.987$, $f_3 \geq 0.99$, $t_p > 0$.

The optimum results are presented in Table 14.4. Inspection of Table 14.4 reveals that if $f_1 \leq 38.33$ is satisfactory, then the decision maker has reached a satisfactory solution. However, the decision maker thinks the replacement cost-rate is too high. From Table 14.4, the reduction of the replacement cost can be achieved by reducing the reliability. Therefore, the decision maker has chosen an aspiration level of 0.985 for goal 3.

Table 14.4. Results of the second cycle (SEMOPS), $AL_1 = (30.5, 0.987, 0.99)$

s_2	t_p	d	f
$s_{2.2} = 2.9975$	705	$d_{2.2} = (1.2567, 1.0000, 0.9892)$	$f_{2.2} = (38.33, 0.9870, 0.9910)$
$s_{2.21} = 2.7934$	1058	$d_{2.21} = (1.0000, 0.9800, 1.1252)$	$f_{2.21} = (30.49, 0.9888, 0.9800)$
$s_{2.23} = 2.0081$	705	$d_{2.23} = (1.2567, 1.0000, 0.9892)$	$f_{2.23} = (38.33, 0.9870, 0.9910)$

Table 14.5. Results of the third cycle (SEMOPS), $AL_2 = (30.5, 0.987, 0.985)$

s_3	t_p	d	f
$s_{3.23} = 1.0662$	913	$d_{3.23} = (1.0662, 0.9836, 1.0000)$	$f_{3.23} = (32.52, 0.9885, 0.9850)$

Third cycle and termination

The principal problem that is solved for this cycle is

minimize $s_{3.23} = d_1$, subject to $f_2 \geq 0.987$, $f_3 \geq 0.985$, $t_p > 0$;

and there are no more auxiliary problems. The results are given in Table 14.5. From this table, the decision maker is satisfied with the result where $t_p = 913$ hours, $f_1 = C(t_p) = 32.53$ US\$/hour, $f_2 = A(t_p) = 0.9885$, and $f_3 = R_3(t_p, H) = 0.985$.

SEMOPS generates information so that the decision maker can detect the inconsistencies of the constraint set and select an acceptable alternative. The key feature of SEMOPS is its interactive nature. This avoids the problem of specifying a preference structure by allowing the decision maker to keep secret the transformations made to convert numbers into value judgments. The decision maker can develop a ranking of goals as the information is received concerning the feasible alternatives. The decision maker may also revise the preferences during the course of the interactions. As shown by the numerical example, SEMOPS indicates the modifications necessary to the conflicting aspiration levels, so that an acceptable solution can be determined. SEMOPS accomplishes this by revealing the extent to which the decision maker's aspirations will have to be modified to achieve a feasible alternative.

14.7 Discussion

As to which method the decision maker should use, there is no clear-cut answer. However, the following approach is suggested.

Solve the problem by the strictest selection method first. If the decision maker is satisfied with the solution obtained, the problem is solved. Otherwise, the decision maker must revise or supply more information, such as the order of importance of the criteria, and solve the problem again by the lexicographic method. If the decision maker is still not satisfied, then the goals should be relaxed somewhat for each criterion. If after this, the solution is still not satisfactory, then the interactive method, that is, SEMOPS, should be used. This approach will generate immediate information for the decision maker so that the inconsistent constraint sets can be evaluated and modified so that an acceptable solution can quickly be determined. It becomes obvious that if the decision maker is satisfied with the solutions of the simpler methods, the process should be terminated. Otherwise, the more complex methods should be used to solve the problem.

15 Case studies on reliability optimization

15.1 Case study on maintenance for mission effectiveness

A military system is required to perform a sequence of flight missions with fixed breaks between. This system is composed of four subsystems in series. In order to enhance the system reliability, each subsystem is composed of a number of computer-controlled equipment (components) in parallel. Let r_{ij} be the reliability of the jth component in the ith system. The components located in the same subsystem do not have to be identical because they may be supplied by different vendors. This is particularly true in terms of the software systems. At the end of a given mission, certain components will need to be maintained. Let t_{ij} be the maintenance effort required to fix the jth component in the ith subsystem at the end of the mission. The effort t_{ij} is usually expressed in number of days. However, not all components will need maintenance at the end of a mission. Let S_{ij} denote the maintenance status of the jth component in the ith subsystem as described in Figure 15.1.

Note that S_{ij} is 1 if maintenance is not needed or if the maintenance requested is granted and 0 otherwise. Table 15.1 lists reliability, maintenance data, and the status of the mission requirement.

The maintenance crews can allocate a total of 14 days between missions. Therefore, not all requested maintenance can be accommodated.

Problem 15.1

In order to maximize the system reliability before accomplishing the next mission, we seek to maximize

$$R_s = R_1 R_2 R_3 R_4,$$

subject to

$$\sum_M t_{ij} S_{ij} \leq 14,$$

343

Table 15.1. Data on reliability and maintenance of components

Subsystem i	Component j	Reliability r_{ij}	Maintenance effort t_{ij} (in days)	Maintenance status requested	S_{ij}
1	1	0.85	4	No	1
	2	0.80	3	Yes	1 or 0
2	1	0.90	2	Yes	1 or 0
	2	0.90	2	Yes	1 or 0
	3	0.90	2	Yes	1 or 0
	4	0.95	3	Yes	1 or 0
3	1	0.75	4	Yes	1 or 0
	2	0.80	5	No	1
	3	0.80	5	Yes	1 or 0
4	1	0.98	2	Yes	1 or 0
	2	0.98	2	Yes	1 or 0

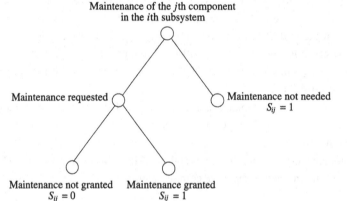

Figure 15.1. Decision tree for determination

where M is the set of all components for which maintenance is requested, $S_{11} = S_{32} = 1$, other S_{ij}s are either 0 or 1, and

$$R_1 = 1 - (1 - 0.85S_{11})(1 - 0.80S_{12}),$$

$$R_2 = 1 - (1 - 0.95S_{24}) \prod_{j=1}^{3}(1 - 0.90S_{2j}),$$

$$R_3 = 1 - (1 - 0.75S_{31}) \prod_{j=2}^{3}(1 - 0.80S_{3j}),$$

$$R_4 = 1 - \prod_{j=1}^{2}(1 - 0.98S_{4j}).$$

Table 15.2. Report on maintenance status requested for the components of ten consecutive missions

Subsystem	Component	Maintenance status requested, k									
i	j	1	2	3	4	5	6	7	8	9	10
1	1	No	No	No	No	Yes	No	No	No	No	Yes
	2	Yes	Yes	Yes	Yes	No	Yes	Yes	Yes	Yes	No
2	1	Yes	Yes	Yes	No	No	No	Yes	Yes	Yes	No
	2	Yes	Yes	No	No	No	Yes	Yes	Yes	No	No
	3	Yes	No	No	No	Yes	Yes	Yes	No	No	Yes
	4	Yes	Yes	Yes	No	No	No	Yes	Yes	Yes	No
3	1	Yes	No	Yes	Yes	No	Yes	No	Yes	No	Yes
	2	No	Yes	No	Yes	No	Yes	No	Yes	No	Yes
	3	Yes	No	Yes	No	Yes	No	Yes	No	Yes	No
4	1	Yes	Yes	Yes	No	No	No	Yes	Yes	Yes	No
	2	Yes	No	No	No	Yes	Yes	Yes	No	No	No

Table 15.3. Optimal maintenance schedule for ten missions

Subsystem	Component	Maintenance status requested, k									
i	j	1	2	3	4	5	6	7	8	9	10
1	1	1^a	1^a	1^a	1^a	1	1^a	1^a	1^a	1^a	1
	2	1	1	1	1	1^a	1	1	1	1	1^a
2	1	1	1	0	1^a	1^a	1^a	1	1	0	1^a
	2	1	1	1^a	1^a	1^a	1	0	0	1^a	1^a
	3	0	1^a	1^a	1^a	1	1	0	1^a	1^a	0
	4	0	0	1	1^a	1^a	1^a	0	0	1	1^a
3	1	0	1^a	1	1	1^a	0	1^a	1	1^a	1
	2	1^a	1	1^a	1	1^a	1	1^a	1	1^a	1
	3	1	1^a	1	1^a	1	1^a	1	1^a	1	1^a
4	1	1	1	0	1^a	1^a	1^a	1	0	1	1^a
	2	0	1^a	1^a	1^a	1	1	1	1^a	1^a	1^a

^{a} Indicates that no maintenance is needed.

Solving the above problem for a set of maintenance requests S_{ij} by dynamic programming, we obtain the optimal solution

$$S_{12} = S_{21} = S_{22} = S_{33} = S_{41} = 1,$$

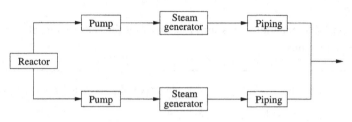

(a) PWR coolant system with two loops

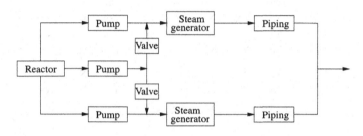

(b) PWR coolant system with two loops, a voting
pump, and two switching valves

Figure 15.2. PWR coolant systems

for which the system reliability is $R_s = 0.903\,45$. Since the maintenance status request will have to be modified from time to time, the optimal solution will change accordingly. The overall mission effectiveness can be determined by considering the optimal maintenance for a series of S_{ij}, such as presented in Table 15.2. Table 15.3 gives the optimal maintenance schedule for the requests in Table 15.2.

15.2   Case study on PWR coolant system

A pressurized water reactor (PWR) typically has two coolant systems: primary and secondary. The major components of a PWR primary coolant system are the coolant pumps, the steam generators, a pressurizer, and the piping equipment. In the primary coolant system, the coolant pump circulates water in order to transport the heat generated in the reactor to the steam generator through piping equipment. The heat transfered to the steam generator is ultimately converted into electricity by a turbine generator located in the secondary coolant system. As the primary system has to perform the above critical function, it is required to have very high reliability.

Figure 15.2(a) gives a flow diagram of a PWR primary coolant system in which the reactor is connected to two parallel loops, each consisting of a pump, a steam generator, and piping. If either of the loops fails, the other one will enable the system

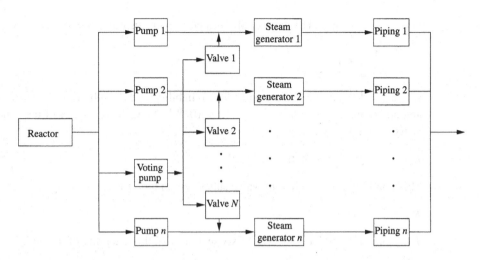

Figure 15.3. PWR coolant systems with n loops

to operate. To increase the system reliability, a voting pump can be connected to each steam generator through a switching valve, as shown in Figure 15.2(b). However, the configuration with two parallel loops, as shown in Figure 15.2(b), may not ensure the required system reliability. In order to increase the system reliability to the desired level, more parallel loops can be provided in addition to parallel redundancy at the component level. However, such an increase leads to higher system cost. Anand and Chidambaram [11] have considered the problem of selecting an optimal number of parallel loops and an optimal redundancy level for each component such that the system cost is subject to a minimum requirement on system reliability. They have formulated the problem as a nonlinear integer programming problem and applied the random search method proposed by Mohan and Shanker [235] and the multi-stage Monte Carlo simulation method of Conely and Kossik [65]. This problem is described below.

Problem description
As shown in Figure 15.3, suppose the system consists of n parallel loops and each one of them is connected to a voting pump through a switching valve. Let:

x_{gi} number of steam generators in parallel in loop i
x_o number of voting pumps in parallel
x_{pi} number of pumps in parallel in loop i
x_{vi} number of switching valves in parallel connecting the voting pump
       subsystem and loop i

The reliability of each pump is 0.7372, and that of the pump subsystem in loop i is

$$R_{pi} = 1 - (0.2628)^{x_{pi}}.$$

Similarly, the reliability of the voting pump subsystem is

$$R_o = 1 - (0.2628)^{x_o}.$$

Let $X_g = \sum_{i=1}^{n} x_{gi}$. The number of tubes per steam generator, connected in series is

$$t = \left\lfloor \frac{2.38}{X_g} \right\rfloor + 1 \tag{15.1}$$

where $\lfloor \cdot \rfloor$ denotes the integer part. The reliability of each tube is $1 - 8.76 \times 10^{-6}$, and that of each generator is $(1 - 8.76 \times 10^{-6})^t$. Now, the reliability of the generator subsystem in loop i is

$$R_{gi} = 1 - [1 - (1 - 8.76 \times 10^{-6})^t]^{x_{gi}}. \tag{15.2}$$

The reliability of the switching valve subsystem in loop i is

$$R_{vi} = 1 - (0.0365)^{x_{vi}}.$$

The piping in each loop consists of six sections and the reliability of the piping in loop i is

$$R_1 = (1 - 8.76 \times 10^{-7})^6.$$

Therefore, the system reliability can be written as

$$R_s = 1 - R_o \prod_{i=1}^{n} (1 - R_{pi} R_{gi} R_1 - R_{vi} R_{gi} R_1 + R_{pi} R_{gi} R_{vi} R_1)$$

$$- (1 - R_o) \prod_{i=1}^{n} (1 - R_{pi} R_{gi} R_1). \tag{15.3}$$

For conditions of:

maximum system operating pressure P_{max} = 1411 kg/cm^2,
coolant pipe diameter D = 2.56 cm, and
secondary steam pressure P_s = 677 kg/cm^2,

the costs (in dollars) as given by Henley and Kumamoto [124] are as follows.

1. The cost per pump is

$$455\,000 + 49.0 P_{max} = 524\,000.$$

The cost of a voting pump is also the same as above.

2. The cost per steam generator is

$$(582\,000 + 83.0 P_{max} + 304.0 P_s)\left[\frac{1}{4}(X_g)^{-0.69}\right] = 220\,000(X_g)^{-0.69}.$$

3. The cost of the main coolant piping per loop is

$$(150\,000 + 37.0 P_{max})(D/2.29)^2 = 253\,000.$$

4. The cost per switching valve is 45 000.

5. The cost per pressurizer is

$$105.0(P_{max}/n) + 26\,000 = 148\,000/n + 26\,000.$$

6. The cost of the reactor core is

$$236\,P_{max} + 30\,000n = 333\,000 + 30\,000n.$$

Now, the total cost of the system is

$$C_s = 524\,000(X_p + x_o) + 220\,000(X_g)^{0.31} + 45\,000X_v + 309\,000n + 481\,000,$$

(15.4)

where $X_p = \sum_{i=1}^n x_{pi}$, $x_g = \sum_{i=1}^n x_{gi}$, and $x_v = \sum i = 1^n x_{vi}$.

The redundancy allocation problem studied by Anand and Chidambaram [11] is as follows.

Problem 15.2

Minimize C_s, subject to

$R_s \geq R_s^0$,

$2 \leq n \leq 7$,

$1 \leq x_o \leq 5$,

$1 \leq x_{gi} \leq 5$,

$1 \leq x_{pi} \leq 5$,

$1 \leq x_{vi} \leq 5$,

for $i = 1, \ldots, n$, where R_s^0 is the minimum system reliability required.

Note that the problem is different from the typical nonlinear integer programming formulations of redundancy allocation in reliability systems. The number of decision variables depends on the value chosen for n. It can be seen from (15.1) and (15.2) that the reliability of a steam generator depends on the total number of steam generators in the system.

Anand and Chidambaram [11] have adopted the multi-stage Monte Carlo simulation method of Conely and Kossik [65] to solve the above problem and compared its performance with the random search technique of Mohan and Shanker [235]. The random search technique involves generation, by a random process, of a large set of solutions which yield system reliability higher than the specified lower limit R_s^0. The efficiency of this method is poor when the minimum system reliability R_s^0 is high. In the multi-stage Monte Carlo simulation method, random solutions are generated iteratively. A solution is generated in each iteration, and the sample space for each variable in an iteration depends on the solution of the previous iteration. The sample space progressively decreases over iterations for each variable.

Heuristic method

Prasad and Kuo [269] have adopted a simple heuristic method to solve the above redundancy allocation problem. In this method, for each value of n in the set $\{2, \ldots, 7\}$, the integer restriction on the other variables is relaxed and the resulting nonlinear programming problem is solved. The optimal solution is rounded off by a simple procedure to derive a solution of the problem under consideration. Finally, the best among the six solutions is taken as the required solution. The optimal solution of each nonlinear programming problem is rounded off in the following fashion.

The rounding-off procedure

If the fractional part of a variable (in the optimal solution) is at least 0.75, round up the value of that variable to the nearest integer. Round down all the variables of which the fractional part is less than 0.75. Let I denote the set of all variables which are rounded-down. If the current integer solution is feasible, stop the rounding off procedure. Otherwise, for each variable in I, compute the increase ΔC in C_s and increase ΔR in R_s due to an increment of 1 in the variable. Select the variable in I that gives a minimum ratio $\Delta C/\Delta R$, and increase that variable by 1. If the resulting solution is feasible, stop the procedure. Otherwise, delete the selected variable from I, and repeat the procedure until a feasible solution is obtained.

The heuristic method is now illustrated for $R_s^0 = 0.9999$. For each value of n, the relaxed problem is solved by a penalty method. In the penalty method, a large penalty is imposed for violation of constraints and the unconstrained problem is solved by the Hooke–Jeeves search method. The solutions are denoted by vector

$$(x_0, x_{g1}, x_{p1}, x_{v1}, x_{g2}, x_{p2}, x_{v2}, \ldots, x_{gn}, x_{pn}, x_{vn}).$$

The optimal solutions and the corresponding rounded-off solutions are presented in Table 15.4. The best among the six integer solutions is

$n = 2,$

$x_0 = 1,$

$(x_{g1}, x_{p1}, x_{v1}) = (1, 3, 1),$

$(x_{g2}, x_{p2}, x_{v2}) = (1, 3, 1),$

and the corresponding cost and reliability of the system are

$$C_s^* = 0.513 \times 10^7 \quad \text{and} \quad R_s^* = 0.999\,912\,87.$$

Anand and Chidambaram [11] have reported that for $R_s^0 = 0.9999$, the multi-stage Monte Carlo method and the random search method have yielded 0.547×10^7 and 0.575×10^7 as the optimal costs, respectively. The optimal solution given by the heuristic method is significantly better than those given by the two probabilistic methods. In terms of computational time, the problem is solved (including rounding off) for each value of n on a PC 486 with 80 MHz within 0.2 seconds, and the entire

Table 15.4. Solutions for different values of n

n	Optimal solution	Cost/10^7	R_s	Rounded-off solution	Cost/10^7	R_s
2	(1, 1, 3.41, 1, 1, 2.50, 1)	0.5083	0.999 902	(1, 1, 3, 1, 1, 3, 1)	0.5130	0.999 913
3	(1.91, 1, 2, 1, 1, 1.50, 1, 1, 1.50, 1)	0.5474	0.999 902	(2, 1, 2, 1, 1, 2, 1, 1, 1, 1)	0.5520	0.999 913
4	(1, 1, 1.41, 1, 1, 1.50, 1, 1, 1.50, 1, 1, 1.50, 1)	0.5856	0.999 903	(1, 1, 2, 1, 1, 2, 1, 1, 1, 1, 1, 1)	0.5903	0.999 913
5	(1, 1, 1, 1, 1, 1, 1, 1, 1, 1, 1.41, 1, 1, 1.50, 1)	0.6235	0.999 903	(1, 1, 1, 1, 1, 1, 1, 1, 1, 1, 2, 1, 1, 1, 1)	0.6281	0.999 913
6	(1, 1, 1, 1, 1, 1, 1, 1, 1, 1, 1, 1, 1, 1, 1, 1, 1, 1)	0.6656	0.999 913	(1, 1, 1, 1, 1, 1, 1, 1, 1, 1, 1, 1, 1, 1, 1, 1, 1, 1)	0.6656	0.999 913
7	(1, 1, 1, 1, 1, 1, 1, 1, 1, 1, 1, 1, 1, 1, 1, 1, 1, 1, 1, 1, 1)	0.7553	0.999 977	(1, 1, 1, 1, 1, 1, 1, 1, 1, 1, 1, 1, 1, 1, 1, 1, 1, 1, 1, 1, 1)	0.7553	0.999 977

problem is solved within 1.2 seconds. These execution times are very short, while those reported by Anand and Chidambaram [11] are at least 120 seconds on a PC/AT computer.

The objective values obtained by the three methods for different values of R_s^0 are given in Table 15.5. It can be seen that the heuristic method of Prasad and Kuo [269] is the best of the three methods with respect to both the objective value and computational time. The solutions yielded by the Prasad and Kuo [269] method for the seven values of R_s^0 are given in Table 15.6. It is interesting to note that n takes the value 2 in all of these cases.

15.3   Case study on design of a gas pipe line

A gas pipe line company needs to install gas pumps for a new facility that extends over a distance of 1200 miles. Each pump can efficiently pump gas for 35 miles. Traditional design suggests that redundancy be added to each station to ensure reliable gas supply to the customers. Incorporating a safety margin, gas pumps are usually installed 30 miles, instead of 35 miles, apart. Two different pump manufacturers make their products (of different reliabilities) available for the gas pipe line company to select from.

Manufacturer A guarantees that its pumps will operate for an MTTF of at least 240 000 hours and that they will require only minimum maintenance. Each pump

Table 15.5. Comparison of three methods for several values of minimum reliability

R_s^0	Prasad and Kuo method		Multi-stage Monte Carlo method		Mohan and Shanker method	
	Cost/10^7	Reliability	Cost/10^7	Reliability	Cost/10^7	Reliability
1 0.9	0.303	0.981 779 56	0.507	0.999 670 55	0.431	0.998 746 39
2 0.99	0.356	0.995 210 39	0.507	0.999 670 55	0.431	0.998 746 39
3 0.999	0.461	0.999 664 02	0.507	0.999 670 55	0.470	0.999 670 49
4 0.9999	0.513	0.999 912 87	0.547	0.999 913 36	0.575	0.998 746 39
5 0.999 99	0.618	0.999 993 55	0.629	0.999 993 95	[a]	[a]
6 0.999 999	0.727	0.999 999 58	0.738	0.999 999 58	[a]	[a]
7 0.999 9999	0.832	0.999 999 97	0.906	0.999 999 97	[a]	[a]

[a] The method fails to yield a solution.

Table 15.6. Optimal solutions for the PWR coolant system

	R_s^0	Optimal n	Optimal solution $(x_0, x_{g1}, x_{p1}, x_{v1}, \ldots, x_{gn}, x_{pn}, x_{vn})$
1	0.9	2	(1, 1, 1, 1, 1, 1, 1)
2	0.99	2	(1, 1, 1, 1, 1, 2, 1)
3	0.999	2	(2, 1, 3, 1, 1, 1, 1)
4	0.9999	2	(1, 1, 3, 1, 1, 3, 1)
5	0.99999	2	(3, 1, 3, 1, 1, 3, 1)
6	0.999999	2	(4, 1, 3, 2, 1, 4, 1)
7	0.9999999	2	(5, 1, 4, 2, 1, 4, 1)

costs US\$ 0.4 million. On the other hand, Manufacturer B guarantees that its pumps will operate for an MTTF of 300 000 hours, and these pumps also require minimum maintenance. The unit cost quoted by Manufacturer B is US\$ 0.45 million. To simplify potential operational difficulties, the pipe line company would like to purchase pumps from either Manufacturer A or Manufacturer B, but not a mix of both. What would be an optimal system configuration for the gas pipe line company to design in order to warrant a system down time of at most 96 hours each year?

Obviously the gas pipe line company intends to seek a minimum purchase cost, yet meet the minimum failure-free operation requirement. The current design for installing pumps at 30 miles apart is equivalent to building $1200/30 = 40$ pumping stations.

The total purchase cost for selecting the pumps from Manufacturer A is

$$C_A = u_A \sum_{i=1}^{40} x_i,$$

where $u_A = $ US\$ 400 000 is the unit cost of pumps produced by Manufacturer A, and x_i is the number of pumps to be installed at station i, $i = 1, 2, \ldots, 40$.

The system reliability for pumps installed by Manufacturer A is

$$R_{sA} = \prod_{i=1}^{40} [1 - (1 - r_A)^{x_i}],$$

where r_A is the reliability of a pump produced by Manufacturer A. The reliability of the pipe line for one year, that is, for $t = 8760$ hours, is given by

$$r_A = \exp(-\lambda t) = \exp(-8760/240\,000) = 0.964\,158.$$

Assume that pumps operate beyond the infant-mortality period but fail before the wear-out periods.

Problem 15.3

Because the down time of 96 hours per year is equivalent to a minimum reliability requirement of

$$R_s^0 = \frac{8760 - 96}{8760} = 0.989\,041,$$

and all of the pumps have to be purchased from the same company, minimize

$$C_A = 400\,000 \left(\sum_{i=1}^{40} x_i \right),$$

subject to

$$R_{sA} = \prod_{i=1}^{40} [1 - (0.035\,842)^{x_i}] \geq 0.989\,041,$$

x_i being a positive integer.

If the integer restriction on x_i is relaxed, the above problem becomes a convex programming problem for which Kuhn–Tucker conditions (described in Chapter 6) give a unique optimal solution. For the relaxed problem, Kuhn–Tucker conditions give the optimal solution

$$x_i = \frac{\ln(1 - 0.989\,041^{1/40})}{\ln(0.035\,842)} = 2.4626.$$

A heuristic solution to Problem 15.3 with integer variables can be obtained by installing two or three pumps at each station. Let m denote the number of pump stations that have two pumps and $(40 - m)$ the number of all other stations (with three pumps). Then the total cost of the pumps is

$$C_A = 400\,000[2m + 3(40 - m)] = 400\,000(120 - m).$$

Now, the problem of finding an optimal m is to maximize m subject to

$$R_{sA} = [1 - (0.035\,842)^2]^m [1 - (0.035\,842)^3]^{40-m} \geq 0.989\,041,$$

m being an integer between 0 and 40.

The optimal solution of this problem is $m = 7$, that is, two pumps must be installed at seven stations and three pumps must be installed at the other 33 stations. It means 113 pumps must be installed in the pipe line as the optimal solution. The selection of the seven stations can be arbitrary. The least cost to meet the minimum reliability requirement using the pumps manufactured by company A is $C_A = \text{US\$ } 45.2$ million.

A similar analysis for Manufacturer B determines that the total purchasing cost is minimized subject to the reliability constraint if two pumps are installed at 12 stations and three pumps are installed at the other 28 stations. The corresponding purchasing cost is $C_B = $ US\$ 43.2 million. Since C_B is lower than C_A, the optimal policy is to install two pumps at 12 stations and three pumps at 28 stations, purchasing all of the required pumps from Manufacturer B.

Other design options

Although each pump is designed to supply gas for 35 miles of distance, we can also install pumps every 15 miles so that the failure of one pump would not necessarily stop the whole gas line. Such a design can best be described as a consecutive 2-out-of-n: F system in which only two consecutive pump failures can crash the system. For Manufacturer A, system reliability with x_0 pumps installed at each station, and each station having the same number of pumps, is given by

$$R_{sA} = \sum_{j=0}^{\lfloor (n+1)/2 \rfloor} \binom{n-j+1}{j} [(1-r_A)^{x_0}]^j [1-(1-r_A)^{x_0}]^{n-j}, \tag{15.5}$$

where r_A is the reliability of each pump supplied by Manufacturer A. Given that there are $n = 1200/15 = 80$ stations and $r_A = 0.964\,158$, again we want to minimize

$$C_A = u_A(80x_0),$$

subject to

$$R_{sA} = \sum_{j=0}^{\lfloor (n+1)/2 \rfloor} \binom{n-j+1}{j} (1-r_A)^{j(x_0)} [1-(1-r_A)^{x_0}]^{n-j} \geq 0.989\,041.$$

When $x_0 = 1$, we have $R_{sA} = 0.906\,471$, which is less than the desired value of $R_s^0 = 0.989\,041$. If we take $x_0 = 2$, that is, add one additional pump at each station, then R_{sA} increases to $0.999\,870$, which far exceeds the desired $R_{s,\min}$. Therefore, we may only add redundancy at *some* stations.

However, when the number of pumps installed at each station is different, eq. (15.5) is no longer valid. In fact, in order to maximize the system reliability, redundancy can be added to the stations according to the following sequence of importance (Zuo and Kuo [337]) until the system reliability reaches $R_{sA} = 0.989\,041$:

stations 2, 79, 4, 77, 6, 75, ...

When selecting pumps from Manufacturer B, the reliability of each pump is

$$r_B = \exp(-8760/300\,000) = 0.971\,222,$$

and the corresponding reliability of the consecutive 2-out-of-80: F system is $0.938\,293$. Notice that this is less attractive than the system configurations of the regular redundant system using either Manufacturer A or B, although 80 pumps are needed in every case. (With two pumps installed at each of the 40 stations, R_{sA} and R_{sB} are $0.949\,881$ and $0.967\,403$, respectively.)

Table 15.7. Summary for the case of the gas pipe line design

	Manufacturer A			Manufacturer B		
	v^a	Reliability	Cost (US$ million)	v^a	Reliability	Cost (US$ million)
40 stations in series with redundancy at each station	113	0.989 537	45.2	108	0.989 447	43.2
2-out-of-80: F with redundancy at important stations	117	0.990 451	46.8	114	0.989 572	51.3

^{a} v is the number of components.

With Manufacturer A, for adding redundancy according to the importance sequence, we need to assign two components in the following stations and one component elsewhere:

2, 4, 6, 8, 10, 12, 14, 16, 18, 20, 22, 24, 26, 28, 30, 32, 34, 36, 38,
79, 77, 75, 73, 71, 69, 67, 65, 63, 61, 59, 57, 55, 53, 51, 49, 47, 45.

Now, the system is a linear consecutive k-out-of-n: F system with nonidentical components and with $k = 2$, for which the system reliability is

$$R_s(n; k) = 1 - Q(n; k),$$ (15.6)

where

$$Q(n; k) = Q(n - 1; k) + [1 - Q(n - k - 1; k)]p_{n-k} \prod_{j=n-k+1}^{n} (1 - p_j),$$

$$Q(j; k) = 0, \quad j < k,$$

$$p_0 \equiv 1.$$

Then a total of $80 + 37 = 117$ pumps is needed for a system reliability of 0.990 45 using eq. (15.6) at a C_A of US$ 46.8 million. If we remove the last component which is added in station 38, then we have a system reliability of 0.988 081, which is less than acceptable.

Similarly, with Manufacturer B, to add redundancy according to the importance sequence, we need to assign two components in the following stations and one component elsewhere: from 2 to 34 at every even numbered station and from 47 to 79 at every odd numbered station.

Then a total of $80 + 34 = 114$ pumps is needed for a system reliability of 0.989 572 using eq. (15.6) at a C_B of US$ 51.3 million. If we remove the last component, which is added in station 47, then we have a system reliability of 0.988 024; which is also less than acceptable. The above results are summarized in Table 15.7.

EXERCISES

15.1 In a given system, some components can be more important than others. Birnbaum [36] defined the component reliability importance (B-importance) as

$$I_i = \frac{\partial R_s}{\partial p_i}(n; k) = R_s(p_1, \ldots, p_{i-1}, 1, p_{i+1}, \ldots, p_n; k)$$
$$- R_s(p_1, \ldots, p_{i-1}, 0, p_{i+1}, \ldots, p_n; k).$$

I_i measures the change of system reliability with respect to the change of component-i reliability.

Given reliabilities of n components, prove that the necessary conditions for the optimal design of a linear consecutive k-out-of-n system are:

(a) to arrange components from position 1 to position $\min\{k, n - k + 1\}$ in nondecreasing order of component reliability;

(b) to arrange components from position n to position $\max\{k, n - k + 1\}$ in nondecreasing order of component reliability;

(c) to arrange the $2k - n$ best (most reliable) components from position $n - k + 1$ to position k in any order if $n < 2k$.

15.2 A railway station has 11 lines (numbered from 1 to 11) that receive and send trains. The *usage* of a line is equivalent to the probability that the line is in use. For oversize trains, the line to receive the train and its two neighboring lines must not be in use; that is, an oversize train requires at least three consecutive lines in order to enter the station. Obviously line 1 or line 11 cannot be used to receive the oversize train since lines 1 and 2 only have one neighboring line. What is the probability that the oversize train can enter the station without delay? Taking into account the restrictions imposed on the station, we can regard the station as a linear consecutive 3-out-of-11: G system. Assume that $u_i = \Pr\{\text{line is in use}\}$ is 0.40 for all i.

A1 Outline of dynamic programming

Dynamic programming provides a powerful tool for solving multi-stage decision processes which arise in a number of fields. It is based on the so-called "principle of optimality" and employs the techniques of invariant embedding. The essential notion of dynamic programming is that problems are linked to a serial structure. As mentioned, its cornerstone is the principle of optimality developed by Bellman [29] in 1957, which states: "An optimal policy has the property that whatever the initial state and initial decisions are, the remaining decisions must constitute an optimal policy with regard to the state resulting from the first decision."

Consider a multi-stage process for which x_n denotes a state vector, which represents a set of variables from stage n, and θ_n is a decision (or control) vector, which stands for a set of decision (or control) variables at stage n.

The notion of a stage is actually an abstract one and the function of each stage is to transform the state variables from the input state to the output state. This transformation can generally be expressed as

$$x_n = T_n(x_{n+1}; \theta), \quad n = N, N-1, \ldots, 2, 1. \tag{A1.1}$$

Equation (A1.1) is of vector form. If there are s state variables and one decision variable, (A1.1) can be written as

$$x_{i,n} = T_{i,n}(x_{1,n+1}, x_{2,n+1}, \ldots, x_{s,n+1}, \theta). \tag{A1.2}$$

The objective for the optimization of a multi-stage process is to seek a set of admissible values of $\theta_1, \theta_2, \ldots, \theta_N$ so that the desired performance criterion or a return function is maximized (or minimized). An inherent feature of multi-stage decision processes is that there is an interval profit or return associated with each stage. The objective function is expressed as the summation of these interval profits, that is

$$S(x_{N+1}; \theta_N, \ldots, \theta_2, \theta_1) = \sum_{n=1}^{N} g_n(x_{n+1}; \theta_n). \tag{A1.3}$$

The value of the objective function depends upon the initial state and the sequence of the decisions, $\theta_N, \ldots, \theta_2, \theta_1$. If we represent the maximum return or objective function by $f_N(x_{N+1})$, then

$$\begin{aligned}
f_N(x_{N+1}) &= f_N(x_{1,N+1}, x_{2,N+1}, \ldots, x_{s,N+1}) \\
&= \max S(x_{N+1}; \theta_N, \ldots, \theta_1) \\
&= \max_{\theta_n} \sum_{n=1}^{N} g_n(x_{n+1}; \theta_n).
\end{aligned} \tag{A1.4}$$

Thus, in general, $f_n(x_{n+1})$ is the maximum return obtainable from the operation of an n-stage process if an optimal policy is followed starting with the initial state, x_{n+1}.

If there is one decision variable at each stage, (A1.4) becomes an N-dimensional optimization problem because it must be optimized with respect to all N decision variables. The dynamic programming technique treats this problem as an N one-dimensional problem. For a one-stage process, eq. (A1.4) becomes

$$f_1(x_2) = \max_{\theta_1} \{g_1(x_2; \theta_1)\}, \tag{A1.5}$$

which is the simplest optimization problem among the sequence of problems for $n = 1, 2, \ldots, N$. The other members of this sequence can be obtained by writing (A1.4) in the form,

$$f_n(x_{n+1}) = \max_{\theta_n} \max_{\theta_{n-1}} \cdots \max_{\theta_1} \{g_n(x_{n+1}; \theta_n) + \cdots + g_1(x_2; \theta_1)\}.$$

Since the inputs to the stages following stage n are all affected by θ_n and the state of stage n is not affected by decisions made at the stages following it, we can rewrite this as

$$f_n(x_{n+1}) = \max_{\theta_n} \left\{ g_n(x_{n+1}; \theta_n) + \max_{\theta_{n-1}} \cdots \max_{\theta_1} \left[g_{n-1}(x_n; \theta_{n-1}) + \cdots + g_1(x_2; \theta_1) \right] \right\}. \tag{A1.6}$$

The expression

$$\max_{\theta_{n-1}} \cdots \max_{\theta_1} [g_{n-1}(x_n; \theta_{n-1}) + \cdots + g_1(x_2; \theta_1)]$$

stands for the maximum return (or objective function) from an $(n-1)$ stage process with an initial state x_n. Hence, we can write

$$f_{n-1}(x_n) = \max_{\theta_{n-1}} \cdots \max_{\theta_1} \left[g_{n-1}(x_n; \theta_{n-1}) + \cdots + g_1(x_2; \theta_1) \right]. \tag{A1.7}$$

Equation (A1.6) can be further simplified to

$$f_n(x_{n+1}) = \max_{\theta_n} \left[g_n(x_{n+1}; \theta_n) + f_{n-1}(x_n) \right],$$

or

$$f_n(x_{n+1}) = \max_{\theta_n} \left\langle g_n(x_{n+1}; \theta_n) + f_{n-1}[T(x_{n+1}; \theta_n)] \right\rangle. \tag{A1.8}$$

This last expression is the so-called functional equation and is a mathematical statement of the principle of optimality. It states the recursive relationship between an n-stage process and an $n - 1$-stage process. The solution yields the value for the maximum return with the corresponding optimal policy as a function of the set $\{\theta_n\}$.

See Bellman [29] and Bellman and Dreyfus [31] for further details dealing with dynamic programming as an optimization tool.

A2 The Hooke–Jeeves (H–J) algorithm

The following algorithm developed by Hooke and Jeeves [130] minimizes a function $f(\mathbf{x})$ over R^n without any constraints.

- Step 0: Select a starting point $\mathbf{x}^{(0)}$ and increments $\Delta_1, \ldots, \Delta_n$. Determine the step reduction factor $\alpha < 1$ and a small value $\epsilon > 0$. Let $k = 0$ and $\mathbf{x}_p^{(k+1)} = \mathbf{x}^{(0)}$.

- Step 1: If $\sum_{j=1}^{n} \Delta_j^2 \leq \epsilon$, go to step 4.

- Step 2: Perform the exploratory move at $\mathbf{x}_p^{(k+1)}$. If the move yields a point $\mathbf{y}$ such that $f(\mathbf{y}) < f[\mathbf{x}^{(k)}]$, then let $\mathbf{x}^{(k+1)} = \mathbf{y}$ and go to step 3. Otherwise, let $\Delta_j = \alpha \Delta_j$ for $j = 1, \ldots, n$, let $\mathbf{x}_p^{k+1} = \mathbf{x}^k$, and go to step 1.

- Step 3: Let $k = k + 1$ and perform a pattern move to obtain a new point

$$\mathbf{x}_p^{(k+1)} = \mathbf{x}^{(k)} + \left[\mathbf{x}^{(k)} - \mathbf{x}^{(k-1)}\right],$$

and go to step 2.

- Step 4: Take $\mathbf{x}^{(k)}$ as the approximate optimal solution and stop.

Exploratory move at point x

- Step 0: Let $j = 1$ and $f^* = f(\mathbf{x})$.

- Step 1: Evaluate

$$f_j^+ = f(x_1, \ldots, x_{j-1}, x_j + \Delta_j, x_{j+1}, \ldots, x_n).$$

If $f_j^+ < f^*$, let $x_j = x_j + \Delta_j$ and go to step 2. Otherwise, evaluate

$$f_j^- = f(x_1, \ldots, x_{j-1}, x_j - \Delta_j, x_{j+1}, \ldots, x_n).$$

If $f_j^- < f^*$, let $x_j = x_j - \Delta_j$.

- Step 2: $j = j + 1$. If $j \leq n$, let $f^* = f(\mathbf{x})$ and go to step 1. Otherwise take $\mathbf{x}$ as the solution yielded by the exploratory move and stop.

The flow diagram for the H–J algorithm is given in Figure A2.1.

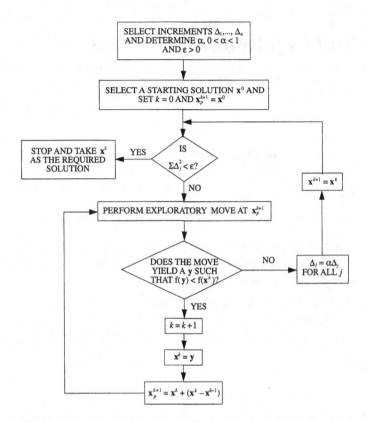

Figure A2.1. Flow diagram for the Hooke–Jeeves method

Derivation of polytope U^{k+1} from U^k

A polytope U^k in R^{m-1} is denoted by the set of its vertices. In order to obtain the polytope

$$U^{k+1} = U^k \cap \{u \in R^{m-1}: \mathbf{h}u > h_0\}$$

from U^k, an incidence matrix W defined for U^k is also used. For the polytope U^k, let

$V = \{1, 2, \ldots, n_v\}$, the set of vertices of U^k,

$P = \{1, 2, \ldots, n_p\}$, the set of supporting hyperplanes of U^k,

$v_j = (v_{j1}, \ldots, v_{j,m-1})^T$, jth vertex, $1 \le j \le n_v$,

$$w_{ij} = \begin{cases} 1, & \text{if the } j\text{th vertex lies on the } i\text{th plane,} \\ 0, & \text{otherwise, and} \end{cases}$$

$W = (w_{ij})_{p \times v}$.

The vertices of U^1 are

$v_1 = (1, 0, 0, \ldots, 0)$,

$v_2 = (0, 1, 0, \ldots, 0)$,

$$\vdots$$

$v_{m-1} = (0, 0, \ldots, 0, 1)$,

$v_m = (0, 0, 0, \ldots, 0)$.

Note that for U^1, $n_p = m$, that is, U^1 has m supporting hyperplanes. The ith supporting hyperplane is $u_i = 0$ for $i = 1, \ldots, m-1$ and the mth one is $\sum_{i=1}^{m-1} u_i = 1$. The corresponding incidence matrix W is given by

$$w_{ij} = \begin{cases} 0, & \text{if } i = j, \\ 1, & \text{otherwise.} \end{cases}$$

The centroid of U^1 is $u^1 = (1/m, \ldots, 1/m)$. The following algorithm can be used to obtain U^{k+1} from U^k.

Algorithm for deriving a new polytope

- Step 1: Obtain the subsets V^-, V^0, and V^+ of vertices U^k as

$$V^- = \{j \in V: \mathbf{h}v_j < h_0\},$$
$$V^0 = \{j \in V: \mathbf{h}v_j = h_0\},$$
$$V^+ = \{j \in V: \mathbf{h}v_j > h_0\}.$$

- Step 2: If $V^+ = \phi$, let $U^{k+1} = \phi$ and stop. Otherwise, go to step 3.
- Step 3: Find the subset P^+ of supporting hyperplanes, which pass through at least one vertex in V^+. Let $\overline{V} = V^0$, $K = V^+$, and $\ell = n_v$.
- Step 4: Select a vertex k from K.
- Step 5: Let $J = V^-$.
- Step 6: Select a vertex j from J and compute

$$t = \sum_{p \in P^+} w_{pj} w_{pk},$$

which is the number of planes in P^+ passing through both j and k. If $t \neq m - 2$, go to step 7. Otherwise, reset $\ell = \ell + 1$, and let

$$v_\ell = v_k + \alpha(v_j - v_k),$$

where $\alpha = (h_0 - hv_k)/(hv_j - hv_k)$. Let $w_{p\ell} = w_{pj}(w_{pk})$ for all $p \in P^+$, and let $\overline{V} = \overline{V} \cup \{\ell\}$.

- Step 7: Let $J = J\backslash\{j\}$. If $J \neq \phi$, go to step 6. Otherwise, go to step 8.
- Step 8: Let $K = K\backslash\{k\}$. If $K \neq \phi$, go to step 4. Otherwise, go to step 9.
- Step 9: Let $P = P^+ \cup \{n_p + 1\}$, $V = \overline{V} \cup V^+$, and, for $j \in V$,

$$w_{(n_p+1)j} = \begin{cases} 1, & \text{if } j \in \overline{V}, \\ 0, & \text{otherwise.} \end{cases}$$

Renumber the vertices in V and the planes in natural order $1, 2, \ldots$, and update W accordingly.

- Step 10: Take P, V, and W as the set of supporting planes, set of vertices, and the incidence matrix of U^{k+1}, respectively, and stop.

A4 Consecutive k-out-of-n systems

Suppose n components are linearly (circularly) connected in such a way that the system malfunctions if, and only if, at least k consecutive components fail. This type of structure was named a linear (or circular) consecutive k-out-of-n: F system. In a case where the system functions if, and only if, at least k consecutive components work, it is called a consecutive k-out-of-n: G system.

Notation

k	minimum number of consecutive good (bad) components required for the system to function (fail)
n	number of components in a system
p	component reliability of a system with independently and identically distributed components
p_i	reliability of component i in the system, $i = 1, 2, \ldots, n$
$Q_C(k, n)$	$1 - R_C(k, n)$
$Q_L(k, n)$	$1 - R_L(k, n)$
q	component unreliability of system with independently and identically distributed components, $q = 1 - p$
q_i	unreliability of component i in the system, $q_i = 1 - p_i$, $i = 1, 2, \ldots, n$
$R_C(k, n)$	reliability of circular consecutive k-out-of-n system
$R_L(k, n)$	reliability of a linear consecutive k-out-of-n system

Chiang and Niu [57] presented the first mathematical formula to compute the exact system reliability of a linear consecutive k-out-of-n: F system with independently and identically distributed components:

$$R_L(k, n) = \sum_{r=1}^{n-k+1} \sum_{m=r+1}^{r+k-1} R_L(k, n - m) p^r q^{m-r} + p^{n-k+1}, \tag{A4.1}$$

where r denotes the first failed component in the sequence and m denotes the first functioning component after position r. This formula is recursive. With proper programming efforts, its complexity is $O(kn)$. There is also a closed formula for the reliability of a consecutive 2-out-of-n: F system:

$$R_L(2, n) = \sum_{j=0}^{[(n+1)/2]} \binom{n - j + 1}{j} q^j p^{n-j}. \tag{A4.2}$$

Circular consecutive k-out-of-n: F systems with identical components

In such a system, the n components are placed in a circle so that the first and nth components become adjacent (consecutive). Derman et al. [79] introduced the concept of circular consecutive k-out-of-n: F systems with independently and identically distributed components as

$$R_C(k, n; p) = p^2 \sum_{i=0}^{k-1} (i+1) q^i R_L(k, n - i - 2; p). \tag{A4.3}$$

Systems with nonidentical components

The reliability evaluation for consecutive k-out-of-n: F systems with independent but not necessarily identical component reliabilities was first reported by Hwang [139] and Shanthikumar [289]. Their approaches are based on disjoint event analysis.

Let E_i be the event that component i is the last functioning one. Since E_is are disjoint events and exactly one of the events $E_{n-k+1}, \ldots, E_n$ must occur for the system to function, then

$$R_L(k, n) = \sum_{i=n-k+1}^{n} \Pr(E_i) R_L(k, i - 1) \tag{A4.4}$$

$$= \sum_{i=n-k+1}^{n} p_i \left(\prod_{j=i+1}^{n} q_j \right) R_L(k, i - 1), \tag{A4.5}$$

with the boundary condition $R_L(v, u) = 1$ for $u < v$.

Alternatively, let F_i be the event that the system first fails at component i. The subscript i is the smallest such that the k consecutive components $\{i - k + 1, i - k + 2, \ldots, i - 1, i\}$ fail. In particular, if $i > k$, then F_i implies that component $(i - k)$ is functioning. Since the F_is are disjoint events and one of them must occur for the first functioning component to fail, we have

$$Q_L(k, n) = \sum_{i=k}^{n} [1 - Q_L(k, i - k - 1)] p_{i-k} \prod_{j=i-k+1}^{i} q_j, \tag{A4.6}$$

with boundary condition: $Q(v, u) = 0$ for $u < v$, and $p_0 \equiv 1$. Therefore,

$$Q_L(k, n) = Q_L(k, n - 1) + R_L(k, n - k - 1) p_{n-k} \prod_{j=i-k+1}^{n} q_j. \tag{A4.7}$$

Shanthikumar [289] also reported the algorithm, given below, for computing the reliability of a linear consecutive k-out-of-n: F system with independent components:

$$R_L(k, n) = R_L(k, n - 1) - R_L(k, n - k - 1) p_{n-k} \prod_{j=n-k+1}^{n} q_j, \tag{A4.8}$$

$$Q_L(k, n) = Q_L(k, n - 1) + R_L(k, n - k - 1) p_{n-k} \prod_{j=n-k+1}^{n} q_j, \tag{A4.9}$$

where $p_0 \equiv 1$. When the components are independently and identically distributed, the following formulae result:

$$R_L(k, n) = R_L(k, n - 1) - pq^k R_L(k, n - k - 1), \tag{A4.10}$$

$$Q_L(k, n) = Q_L(k, n - 1) + pq^k R_L(k, n - k - 1). \tag{A4.11}$$

For circular consecutive k-out-of-n: F systems consisting of statistically independent but not necessarily identical components, the exact reliability can be obtained by using the recursive methods developed by Hwang [139]. The circular system can be expressed as a linear combination of several linear subsystems.

The consecutive k-out-of-n: G system

The concept of the consecutive k-out-of-n: G system is well-explained in Kuo et al. [181]. They establish the relationship between consecutive k-out-of-n: F and consecutive k-out-of-n: G systems, and report similar results on system-reliability evaluation, reliability bound evaluation, and system design for the consecutive k-out-of-n: G systems.

The relationship between the consecutive k-out-of-n: F system and the consecutive k-out-of-n: G system is described below.

Lemma A4.1

If the reliability of component i, p_i, in one type of consecutive k-out-of-n system (say, F system) is equal to the unreliability of component i, q_i, in the other type of consecutive k-out-of-n system (G system) for $i = 1, 2, \ldots , n$, given that both types of systems have the values of n and k, then the reliability of one type of system is equal to the unreliability of the other type of system.

Due to the duality between the consecutive k-out-of-n: F system and the consecutive k-out-of-n: G system described above, and the results available on reliability and reliability bound evaluation for the consecutive k-out-of-n: F systems, the formulae are identified easily, e.g., in Kuo et al. [181].

References

1 J. Abadie. Une méthode arborescente pour les programmes partiellment discrets. *R.I.R.O*, **3**: 24–50, 1969.

2 J. Abadie. Une méthode de résolution des programmes non-linéaires partiellment discrets sans hypothese de convexité. *R.I.R.O*, **1**: 23–38, 1971.

3 J. Abadie, H. Dayan, and J. Akoka. Quelques experiences numeriques sur la programmation non-linéaires en nombres entiers. *R.A.I.R.O Recherche opérationelle*, **10**(10): 65–70, 1976.

4 Aeronautical Radio Inc. (ARINC) *Reliability Engineering*. Prentice-Hall, Englewood Cliffs, NJ, 1964.

5 K. K. Aggarwal. Redundancy optimization in general systems. *IEEE Transactions on Reliability*, **R-25**(25): 330–332, 1976.

6 K. K. Aggarwal and J. S. Gupta. On minimizing the cost of reliable systems. *IEEE Transactions on Reliability*, **R-24**(24): 205, 1975.

7 K. K. Aggarwal, J. S. Gupta, and K. B. Misra. A new heuristic criterion for solving a redundancy optimization problem. *IEEE Transactions on Reliability*, **R-24**(24): 86–87, 1975.

8 K. K. Aggarwal, K. B. Misra, and J. S. Gupta. Reliability evaluation – a comparative study of different techniques. *Microelectronics and Reliability*, **14**(1): 49–56, 1975.

9 A. Albert. *A Measure of the Effort Required to Increase Reliability*. Technical Report 43, Applied Mathematics and Statistics Laboratory; Stanford University, Stanford, 1958.

10 O. G. Alekseev and I. F. Volodos. Combined use of dynamic programming and branch and bound methods in discrete programming problems. *Automation Remote Control*, **37**: 557–565, 1967.

11 S. Anand and M. Chidambaram. Optimal redundancy allocation of a PWR cooling loop using a multi-stage Monte Carlo method. *Microelectronics and Reliability*, **34**(4): 741–745, 1994.

12 K. Andrzejczak. Structure analysis of multistate coherent systems. *Optimization*, **25**(2–3): 301–316, 1992.

13 I. Antonopoulou and S. Papastavridis. Fast recursive algorithm to evaluate the reliability of a circular consecutive k-out-of-n: F system. *IEEE Transactions on Reliability*, **R-36**(36): 83–87, 1987.

14 E. H. L. Arts and J. Korst. *Simulated Annealing and Boltzmann Machines*. Wiley, New York, 1989.

15 N. Ashrafi and O. Berman. Optimization models for selection of programs, considering cost and reliability. *IEEE Transactions on Reliability*, **R-41**(2): 281–287, 1992.

16 N. Ashrafi, O. Berman, and M. Cutler. Optimal design of large software-systems using

N-version programming. *IEEE Transactions on Reliability*, **R-43**(2): 344–350, 1994.

17 M. M. Atiqullah and S. S. Rao. Reliability optimization of communication networks using simulated annealing. *Microelectronics and Reliability*, **33**(9): 1303–1319, 1993.

18 R. J. Aust.   A dynamic programming branch and bound algorithm for pure integer programming. *Computers and Operations Research*, **5**: 27–38, 1976.

19 D. S. Bai, W. Y. Yun, and S. W. Chung. Redundancy optimization of k-out-of-n: G systems with common-cause failures. *IEEE Transactions on Reliability*, **R-40**(1): 56–59, 1991.

20 S. K. Banerjee and K. Rajamani. Optimization of system reliability using a parametric approach. *IEEE Transactions on Reliability*, **R-22**(22): 35–39, 1973.

21 S. K. Banerjee and K. Rajamani. Parametric representation of probability in two dimensions – a new approach in system reliability evaluation.   *IEEE Transactions on Reliability*, **R-21**(21): 56–60, 1973.

22 S. K. Banerjee, K. Rajamani, and S. S. Deshpande. Optimal redundancy allocation for non series-parallel networks. *IEEE Transactions on Reliability*, **R-25**(25): 115–117, 1976.

23 R. E. Barlow and F. Proschan. *Statistical Theory of Reliability and Life Testing*. Holt, Rinehart, and Winston, New York, 1975.

24 L. A. Baxter and F. Harche. On the optimal assembly of series-parallel systems. *Operations Research Letters*, **11**(3): 153–157, 1992.

25 T. Böck. *Evolutionary Algorithms in Theory and Practice*. Oxford University Press, New York, 1996.

26 T. Böck, D. B. Fogel, and Z. Michalewicz. *Handbook of Evolutionary Computation*. Oxford University Press, Oxford, 1997.

27 C. F. Bell, M. Kamins, and J. J. McCall. Some elements of planned replacement theory. In L. S. Gephart, editor, *Proceedings of 1966 Annual Symposium on Reliability*, pp. 98–117, San Francisco, CA, 25–27 January 1966.

28 F. Belli and P. Jedrzejowicz. An approach to the reliability optimization of software with redundancy. *IEEE Transactions on Software Engineering*, **SE-17**(3): 310–312, 1991.

29 R. Bellman. *Dynamic Programming*. Princeton University Press, Princeton, NJ, 1957.

30 R. E. Bellman and S. E. Dreyfus. Dynamic programming and reliability of multicomponent devices. *Operations Research*, **6**: 200–206, 1958.

31 R. Bellman and S. E. Dreyfus. *Applied Dynamic Programming*. Princeton University Press, Princeton, NJ, 1962.

32 R. E. Bellman and L. A. Zadeh. Decision making in a fuzzy environment. *Management Science*, **17**(4) B141–164, 1970.

33 D. Beraha and K. B. Misra. Reliability optimization through random search algorithm. *Microelectronics and Reliability*, **13**(4): 295–297, 1974.

34 O. Berman and N. Ashrafi. Optimization models for reliability of modular software systems. *IEEE Transactions on Software Engineering*, **SE-19**(11): 1119–1123, 1993.

35 D. P. Bertsekas. *Constrained Optimization and Lagrange Multiplier Methods*. Academic Press, New York, 1982.

36 Z. W. Birnbaum. On the importance of different components in a multicomponent system. In P. R. Krishnaiah, editor, *International Symposium on Multivariate Analysis-II*, pp. 581–592. Academic Press, New York, 1969.

37 G. Black and F. Proschan. On optimal redundancy. *Operations Research*, **7**: 581–588, 1959.

38 L. D. Bodin. Optimization procedure for the analysis of coherent structures. *IEEE Transactions on Reliability*, **R-18**(18): 118–126, 1969.

39 P. J. Boland and F. Proschan. Optimal arrangement of systems. *Naval Research Logistics Quarterly*, **31**(3): 399–407, 1984.

40 P. J. Boland, F. Proschan, and Y. L. Tong. Optimal arrangement of components via pairwise rearrangements. *Naval Research Logistics*, **36**(6): 807–815, 1989.

41 D. B. Brown. A computerized algorithm for determining the reliability of redundant configurations. *IEEE Transactions on Reliability*, **R-20**(20): 121–124, 1971.

42 R. L. Bulfin and C. Y. Liu. Optimal allocation of redundant components for large systems. *IEEE Transactions on Reliability*, **R-34**(3): 241–247, 1985.

43 R. M. Burton and G. T. Howard. Optimal system reliability for a mixed series and parallel structure. *Journal of Mathematical Analysis and Applications*, **28**(2): 370–382, 1969.

44 M. F. Cardoso, R. L. Salcedo, and S. F. de Azevedo. Nonequilibrium simulated annealing: a faster approach to combinatorial minimization. *Industrial and Engineering Chemistry Research*, **33**(8): 1908–1918, 1994.

45 L. Chambers. *Practical Handbook of Genetic Algorithms*. Vols. 1 and 2, CRC Press, New York, 1995.

46 V. Chankong and Y. Y. Haimes. *Multiobjective Decision Making: Theory and Methodology*. North-Holland, New York, 1983.

47 R. S. Chen, D. J. Chen, and Y. S. Yeh. A new heuristic approach for reliability optimization of distributed computing systems subject to capacity constraints. *Computers and Mathematics with Applications*, **29**(3): 37–47, 1995.

48 S. J. Chen and C. L. Hwang. *Fuzzy Multiple Attribute Decision Making – Methods and Applications*. Springer-Verlag, New York, 1992.

49 M. S. Chern. On the computational complexity of reliability redundancy allocation in a series system. *Operations Research Letters*, **11**: 309–315, 1992.

50 M. S. Chern and R. H. Jan. Parametric programming applied to reliability optimization problems. *IEEE Transactions on Reliability*, **R-34**(34): 165–170, 1985.

51 M. S. Chern and R. H. Jan. Reliability optimization problems with multiple constraints. *IEEE Transactions on Reliability*, **R-35**(35): 431–436, 1986.

52 M. S. Chern, R. H. Jan, and R. J. Chern. Parametric nonlinear integer programming: the right-hand side case. *European Journal of Operational Research*, **54**(2): 237–255, 1991.

53 D. Chi and Way Kuo. Burn-in optimization under reliability and capacity restrictions. *IEEE Transactions on Reliability*, **R-38**(2): 193–198, 1989.

54 D. H. Chi. Optimization of software system reliability by redundancy and software quality management. Ph.D. dissertation, Iowa State University, Ames, 1989.

55 D. H. Chi and Way Kuo. Optimal design for software reliability and development cost. *IEEE Journal on Selected Areas in Communications*, **8**(2): 276–281, 1990.

56 D. H. Chi, H. H. Lin, and Way Kuo. Software reliability optimization and redundancy. In H. D. Rue, editor, *Proceedings of 1989 Annual Reliability Maintainability Symposium*, pp. 41–45, Atlanta, GA, 24–28 January, 1989.

57 D. T. Chiang and S. Niu. Reliability of a consecutive k-out-of-n: F system. *IEEE Transactions on Reliability*, **R-30**(30): 87–89, 1981.

58 D. W. Coit and A. Smith. Penalty guided genetic search for reliability design optimization.

Computers and Industrial Engineering, **30**(4): 895–904, 1996.

59 D. W. Coit and A. E. Smith. Optimization approaches to the redundancy allocation problem for series-parallel systems. In B. W. Schmeiser and R. Uzsoy, editors, *Proceedings of the Fourth Industrial Engineering Research Conference*, pp. 342–349, Nashville, TN, May 1995.

60 D. W. Coit and A. E. Smith. Reliability optimization of series-parallel systems using a genetic algorithm. *IEEE Transactions on Reliability*, **R-45**(2): 254–260, 1996.

61 D. W. Coit and A. E. Smith. Solving the redundacy allocation problem using a combined neural network/genetic algorithm approach. *Computers and Operations Research*, **23**(6): 515–526, 1996.

62 D. W. Coit and A. E. Smith. Considering risk profiles in design optimization for series-parallel systems. In N. J. McAfee, editor, *Proceedings of the 1997 Annual Reliability and Maintainability Symposium*, pp. 271–277, Philadephia, PA, 13–16 January 1997.

63 D. W. Coit and A. E. Smith. Redundancy allocation to maximize a lower percentile of the system time-to-failure distribution. *IEEE Transactions on Reliability*, **R-47**(1): 79–87, 1998.

64 D. W. Coit, A. E. Smith, and D. M. Tate. Adaptive penalty methods for genetic optimization of constrained combinatorial problems. *INFORMS Journal on Computing*, **8**(2): 173–182, 1996.

65 W. C. Conely and R. Kossik. A 100 000 variable nonlinear problem. *International Journal of Mathematical Education in Science and Technology*, **14**(1): 117–125, 1983.

66 M. W. Cooper. The use of dynamic programming methodology for the solution of a class of nonlinear programming problems. *Naval Research Logistics Quarterly*, **27**: 89–95, 1980.

67 M. W. Cooper. A survey of methods for pure nonlinear integer programming. *Management Science*, **27**: 353–361, 1981.

68 C. J. Dale and A. Winterbottom. Optimal allocation of effort to improve system reliability. *IEEE Transactions on Reliability*, **R-35**(35): 188–191, 1986.

69 L. Davis, editor. *Handbook of Genetic Algorithms*. Van Nostrand Reinhold, New York, 1991.

70 D. L. Deeter and A. E. Smith. Heuristic optimization of network design considering all-terminal reliability. In N. J. McAfee, editor, *Proceedings of the 1997 Annual Reliability and Maintainability Symposium*, pp. 194–199, Philadelphia, PA, 13–16 January 1997.

71 D. L. Deeter and A. E. Smith. Economic design of reliable networks. *IIE Transactions*, **30**(12): 1161–1174, 1998.

72 E. V. Denardo and B. L. Fox. Shortest route methods: 2. Group knapsacks, expanded networks, and branch-and-bound. *Operations Research*, **27**: 548–566, 1979.

73 B. Dengiz, F. Altiparmak, and A. E. Smith. A genetic algorithm approach to optimal topological design of all terminal networks. In C. H. Dagli, M. Akay, C. L. P. Chen, B. R. Férnandez, and J. Ghosh, editors, *Intelligent Engineering Systems Through Artifical Neural Networks*, Vol. 5, pp. 405–410, American Society of Mechanical Engineers, New York, 1995.

74 B. Dengiz, F. Altiparmak, and A. E. Smith. Efficient optimization of all-terminal reliable networks using an evolutionary approach. *IEEE Transactions on Reliability*, **R-46**: 18–26, 1997.

75 B. Dengiz, F. Altiparmak, and A. E. Smith. Local search genetic algorithm for optimal

design of reliable networks. *IEEE Transactions on Evolutionary Computation*, **EC-1**(3): 179–188, 1997.

76   J. E. Dennis and R. B. Schnabel. *Numerical Methods for Unconstrained Optimization and Nonlinear Equations*. Prentice-Hall, New York, 1983.

77   C. Derman, G. J. Lieberman, and S. M. Ross. On optimal assembly of systems. *Naval Research Logistics Quarterly*, **19**: 569–574, 1972.

78   C. Derman, G. J. Lieberman, and S. M. Ross. Assembly of systems having maximum reliability. *Naval Research Logistics Quarterly*, **21**(1): 1–12, 1974.

79   C. Derman, G. J. Lieberman, and S. M. Ross. On the consecutive k-out-of-n: F system. *IEEE Transactions on Reliability*, **R-31**(31): 57–63, 1982.

80   A. K. Dhingra. Optimal apportionment of reliability and redundancy in series systems under multiple objectives. *IEEE Transactions on Reliability*, **R-41**(4): 576–582, 1992.

81   S. E. Dreyfus. *The Art and Theory of Dynamic Programming*. Academic Press, New York, 1977.

82   D. Z. Du and F. K. Hwang. Optimal consecutive 2-out-of-n systems. *Mathematics of Operations Research*, **11**(1): 187–191, 1986.

83   D. Z. Du and F. K. Hwang. Optimal assembly of an s-stage k-out-of-n system. *SIAM Journal of Discrete Mathematics*, **3**: 349–354, 1990.

84   D. E. Echhardt, Jr and L. D. Lee. A theoretical basis for the analysis of multiversion software subject to coincident errors. *IEEE Transactions on Software Engineering*, **SE-11**(12): 1511–1517, 1985.

85   E. El-Neweihi. A relationship between partial derivatives of the reliability function of a coherent system and its minimal path (cut) sets. *Mathematics of Operations Research*, **5**(4): 553–555, 1980.

86   E. El-Neweihi, F. Proschan, and J. Sethuraman. Optimal allocation of components in parallel-series and series-parallel systems. *Journal of Applied Probability*, **23**: 770–777, 1986.

87   E. El-Neweihi, F. Proschan, and J. Sethuraman. Optimal assembly of systems using Schur functions and majorization. *Naval Research Logistics Quarterly*, **34**(5): 705–712, 1987.

88   G. W. Evans. An overview of techniques for solving multiobjective mathematical programs. *Management Science*, **30**(11): 1268–1282, 1984.

89   H. Everett III. Generalized Lagrange multiplier method for solving problems of optimal allocation of resources. *Operations Research*, **11**(3): 399–417, 1963.

90   L. T. Fan, C. S. Wang, F. A. Tillman, and C. L. Hwang. Optimization of systems reliability. *IEEE Transactions on Reliability*, **R-16**(16): 81–86, 1967.

91   A. J. Federowicz and M. Mazumdar. Use of geometric programming to maximize reliability achieved by redundancy. *Operations Research*, **16**(5): 948–954, 1968.

92   A. V. Fiacco and G. P. McCormick. Extension of SUMT for nonlinear programming: equality constraints and extrapolation. *Management Science*, **12**(11): 816–828, 1966.

93   A. V. Fiacco and G. P. McCormick. *Nonlinear Programming: Sequential Unconstrained Minimization Techniques*. Wiley, New York, 1968.

94   C. Floudas. *Nonlinear and Mixed-Integer Optimization: Fundamentals and Applications*. Oxford University Press, New York, 1995.

95   T. Fogarty, editor. *Evolutionary Computing*. Springer-Verlag, Berlin, 1994.

96   D. E. Fyffe, W. W. Hines, and N. K. Lee. System reliability allocation and a computational algorithm. *IEEE Transactions on Reliability*, **R-17**(17): 64–69, 1968.

97   M. R. Garey and D. S. Johnson. *Computers and Intractability: a Guide to the Theory of NP-Completeness*. Freeman, San Francisco, CA, 1979.

98   M. Gen and R. Cheng. Optimal design of system reliability using interval programming and genetic algorithms. *Computers and Industrial Engineering*, **31**(1–2): 237–240, 1996.

99   M. Gen and R. Cheng. *Genetic Algorithms and Engineering Design*. Wiley, New York, 1997.

100 M. Gen, K. Ida, and J. U. Lee. A computational algorithm for solving 0–1 goal programming with GUB structures and its application for optimization problems in system reliability. *Electronics and Communications in Japan*, **73**(12): 88–96, March 1990.

101 M. Gen, K. Ida, M. Sasaki, and J. U. Lee. Algorithm for solving large-scale 0–1 goal programming and its application to reliability optimization problem. *Computers and Industrial Engineering*, **17**(1–4): 525–530, 1989.

102 M. Gen, K. Ida, and T. Taguchi. *Reliability Optimization Problems: a Novel Genetic Algorithm Approach*. Technical Report ISE93-5, Ashikaga Institute of Technology, Ashikaga, Japan, 1993.

103 M. Gen, K. Ida, Y. Tsujimura, and C. E. Kim. Large-scale 0–1 fuzzy goal programming and its application to reliability optimization problem. *Computers and Industrial Engineering*, **24**(4): 539–549, 1993.

104 M. Gen, H. Okuno, and S. Shinofuji. An optimizing method in system reliability with failure-modes by implicit enumeration algorithm. *Journal of Operations Research, Japan*, **19**: 99–116, 1976.

105 Y. G. Genis and I. A. Ushakov. Optimization of the reliability of multipurpose systems. *Engineering Cybernetics*, **21**(3): 54–61, 1983.

106 A. M. Geoffrion. An improved implicit enumeration approach for integer programming. *Operations Research*, **17**: 437–454, 1969.

107 P. M. Ghare and R. E. Taylor. Optimal redundancy for reliability in series system. *Operations Research*, **17**: 838–847, 1969.

108 A. Glankwahmdee, J. S. Liebman, and G. L. Hogg. Unconstrained discrete nonlinear programming. *Engineering Optimization*, **4**: 95–108, 1979.

109 R. J. Glauber. Time dependent statistics of the Ising model. *Journal of Mathematical Physics*, **4**: 294–307, 1963.

110 F. Glover and M. Laguna. *Tabu Search*. Kluwer Academic Publishers, Boston, MA, 1997.

111 D. E. Goldberg. *Genetic Algorithms in Search, Optimization, and Machine Learning*. Addison-Wesley, Reading, MA, 1989.

112 K. Gopal, K. K. Aggarwal, and J. S. Gupta. An improved algorithm for reliability optimization. *IEEE Transactions on Reliability*, **R-27**(27): 325–328, 1978.

113 K. Gopal, K. K. Aggarwal, and J. S. Gupta. A new method for solving reliability optimization problem. *IEEE Transactions on Reliability*, **R-29**(29): 36–38, 1980.

114 R. Gordon. Optimum component redundancy for maximum system reliability. *Operations Research*, **5**: 229–243, 1957.

115 K. K. Govil and R. A. Agarwala. Lagrange multiplier method for optimal reliability allocation in a series system. *Reliability Engineering*, **6**(3): 181–190, 1983.

116 D. L. Grosh. *A Primer of Reliability Theory*. Wiley, New York, 1989.

117 O. K. Gupta and A. Ravindran. Nonlinear integer programming algorithms: a survey. *Opsearch*, **20**(4): 189–206, 1983.

118 O. K. Gupta and A. Ravindran. Branch-and-bound experiments in convex nonlinear programming. *Management Sciences*, **31**(12): 1533–1546, 1985.

119 S. D. Gupta and M. J. Al-Musawi. Reliability optimization in cable system design using a fuzzy uniform-cost algorithm. *IEEE Transactions on Reliability*, **R-37**(37): 75–80, 1988.

120 G. Hadley. *Nonlinear and Dynamic Programming*. Addison-Wesley, Reading, MA, 1964.

121 P. Hansen. Methods of nonlinear 0–1 programming. *Annals of Discrete Mathematics*, **5**: 53–70, 1979.

122 P. Hansen, B. Jaumard, and V. Mathon. Constrained nonlinear 0–1 programming. *ORSA Journal on Computing*, **5**(2): 97–119, 1993.

123 Ir. R. N. von Hees and Ir. H. W. von den Meerendonk. Optimal reliability of parallel multi-component systems. *Operations Research Quarterly*, **12**(1): 16–26, 1961.

124 E. J. Henley and H. Kumamoto. *Designing for Reliability and Safety Control*. Prentice-Hall, Englewood Cliffs, NJ, 1985.

125 M. R. Hestenes. Multiplier and gradient method. *Journal of Optimization Theory and Applications*, **4**(5): 303–320, 1969.

126 M. Hikita, Y. Nakagawa, K. Nakashima, and H. Narihisa. Reliability optimization of systems by a surrogate-constraints algorithm. *IEEE Transactions on Reliability*, **R-41**(3): 473–480, 1992.

127 D. S. Hochbaum. Lower and upper bounds for the allocation problem and other nonlinear optimization problems. *Mathematics of Operations Research*, **19**(2): 390–409, 1994.

128 D. P. Holcomb and J. C. North. An infant mortality and long-term failure rate model for electronic equipment. *AT&T Technical Journal*, **64**(1): 15–31, 1985.

129 J. H. Holland. *Adaptation in Natural and Artificial Systems*. University of Michigan Press, Ann Arbor, MI, 1975.

130 R. Hooke and T. A. Jeeves. Direct search solutions of numerical and statistical problems. *Journal of the Association of Computing Machinery*, **8**: 212–229, 1961.

131 Y. C. Hsieh, T. C. Chen, and D. L. Bricker. *Genetic Algorithms for Reliability Design Problems*. Technical Report, Department of Industrial Engineering, University of Iowa, 1997.

132 C. L. Hwang, L. T. Fan, F. A. Tillman, and S. Kumar. Optimization of life support system reliability by an integer programming method. *AIIE Transactions*, **3**(3): 229–238, 1971.

133 C. L. Hwang, K. C. Lai, F. A. Tillman, and L. T. Fan. Optimization of system reliability by the sequential unconstrained minimization technique. *IEEE Transactions on Reliability*, **R-24**(24): 133–135, 1975.

134 C. L. Hwang and M. J. Lin. *Group Decision Making under Multiple Criteria – Methods and Applications*. Springer-Verlag, New York, 1987.

135 C. L. Hwang and A. S. M. Masud. *Multiple Objective Decision Making–Methods and Applications, a State-of-the Art Survey*. (in collaboration with S. R. Paidy and K. Yoon) Springer-Verlag, New York, 1979.

136 C. L. Hwang, F. A. Tillman, and Way Kuo. Reliability optimization by generalized Lagrangian-function and reduced-gradient methods. *IEEE Transactions on Reliability*,

R-28(28): 316–319, 1979.

137 C. L. Hwang, F. A. Tillman, W. K. Wei, and C. H. Lie. Optimal scheduled maintenance policy based on multiple-criteria decision-making. *IEEE Transactions on Reliability*, R-28(28): 394–397, 1979.

138 C. L. Hwang and K. Yoon. *Multiple Attribute Decision Making – Methods and Application.* Springer-Verlag, New York, 1981.

139 F. K. Hwang. Fast solutions for consecutive k-out-of-n: F systems. *IEEE Transactions on Reliability*, R-31(5): 447–448, 1982.

140 F. K. Hwang. Optimal assignment of components to a two-stage k-out-of-n system. *Mathematics of Operations Research*, 14(2): 376–382, 1989.

141 F. K. Hwang and Dinghua Shi. Redundant consecutive-k systems. *Operations Research Letters*, 6(6): 293–296, 1987.

142 F. K. Hwang and U. G. Rothblum. Optimality of monotone assemblies for coherent systems composed of series modules. *Operations Research*, 42(4): 709–720, 1994.

143 K. N. Hyun. Reliability optimization by 0–1 programming for a system with several failure modes. *IEEE Transactions on Reliability*, R-24(24): 206–210, 1975.

144 K. Ida, M. Gen, and T. Yokota. System reliability optimization with several failure modes by genetic algorithm. In M. Gen and T. Kobayashi, editors, *Proceedings of the 16th International Conference on Computers and Industrial Engineering*, pp. 349–352, Ashikaga, Japan, March 1994.

145 T. Inagaki, K. Inoue, and H. Akashi. Interactive optimization of system reliability under multiple objectives. *IEEE Transactions on Reliability*, R-27(27): 264–267, 1978.

146 K. Inoue, S. L. Gandhi, and E. J. Henley. Optimal reliability design of process systems. *IEEE Transactions on Reliability*, R-23(23): 29–33, 1974.

147 R.-H. Jan, F.-J. Hwang, and S.-T. Cheng. Topological optimization of a communication network subject to a reliability constraint. *IEEE Transactions on Reliability*, R-42(1): 63–70, 1993.

148 P. A. Jensen. Optimization of series-parallel-series networks. *Operations Research*, 18: 471–482, 1970.

149 Filus Jerzy. A problem in reliability optimization. *Journal of the Operational Research Society*, 37(4): 407–412, 1986.

150 L. Jianping. A bound heuristic algorithm for solving reliability redundancy optimization. *Microelectronics and Reliability*, 36(3): 335–339, 1996.

151 K. C. Kapur and L. R. Lamberson. *Reliability in Engineering Design.* Wiley, New York, 1977.

152 J. D. Kettelle, Jr Least-cost allocation of reliability investment. *Operations Research*, 10: 249–265, 1962.

153 J. H. Kim and B. J. Yum. A heuristic method for solving redundancy optimization problems in complex systems. *IEEE Transactions on Reliability*, R-42(4): 572–578, 1993.

154 J. Y. Kim and L. C. Frair. Optimal reliability design for complex systems. *IEEE Transactions on Reliability*, R-30(30): 300–302, 1981.

155 Y. H. Kim, K. E. Case, and P. M. Ghare. A method for computing complex system reliability. *IEEE Transactions on Reliability*, R-21(21): 215–219, 1972.

156 K. E. Kinnear, editor. *Advances in Genetic Programming.* MIT Press, Cambridge, MA, 1994.

157 S. Kirkpatrick. Optimization by simulated annealing: quantitative studies. *Journal of Statistical Physics*, **34**(5–6): 975–986, 1984.

158 S. Kirkpatrick, C. D. Gelatt Jr, and M. P. Vecchi. Optimization by simulated annealing. *Science*, **220**: 671–680, 1983.

159 S. Kirkpatrick, C. D. Gelatt Jr, and M. P. Vecchi. Optimization by simulated annealing. *Science*, **220**(4598): 671–680, 1983.

160 Sun Wah Kiu and D. F. McAllister. Reliability optimization of computer-communication networks. *IEEE Transactions on Reliability*, **R-37**(5): 475–483, 1988.

161 S. G. Kneale. Reliability of parallel systems with repair and switching. In W. T. Sumerlin, editor, *Proceedings of the Seventh National Symposium on Reliability and Quality Control*, pp. 129–133, Philadelphia, PA, 9–11 January 1961.

162 J. C. Knight and N. G. Leveson. An experimental evaluation of the assumption of independence in multiversion programming. *IEEE Transactions on Software Engineering*, **SE-12**(1): 96–109, 1986.

163 T. Kohda and K. Inoue. A reliability optimization method for complex systems with the criterion of local optimality. *IEEE Transactions on Reliability*, **R-31**(31): 109–111, 1982.

164 P. J. Kolesar. Linear programming and the reliability of multicomponent systems. *Naval Research Logistics Quarterly*, **14**(3): 317–327, 1967.

165 J. M. Kontoleon. Optimum link allocation of fixed topology networks. *IEEE Transactions on Reliability*, **R-28**(28): 145–147, 1979.

166 John R. Koza. *Genetic Programming: on the Programming of Computers by Means of Natural Selection*. MIT Press, Cambridge, MA, 1992.

167 John R. Koza. *Genetic Programming II: Automatic Discovery of Reusable Programs*. MIT Press, Cambridge, MA, 1994.

168 D. K. Kulshrestha and M. C. Gupta. Use of dynamic programming for reliability engineers. *IEEE Transactions on Reliability*, **R-22**(22): 240–241, 1973.

169 A. Kumar, R. M. Pathak, and Y. P. Gupta. Genetic-algorithm-based reliability optimization for computer network expansion. *IEEE Transactions on Reliability*, **R-44**(1): 63–72, 1995.

170 A. Kumar, R. M. Pathak, Y. P. Gupta, and H. R. Parsaei. A genetic algorithm for distributed system topology design. *Computers and Industrial Engineering*, **28**(3): 659–670, 1995.

171 Way Kuo. Optimization techniques for systems reliability with redundancy. M.S. Thesis, Kansas State University, Manhattan, 1977.

172 Way Kuo. Software reliability estimation: a realization of competing risk. *Microelectronics and Reliability*, **23**(2): 249–260, 1983.

173 Way Kuo. Reliability enhancement through optimal burn-in. *IEEE Transactions on Reliability*, **R-33**(33): 145–156, 1984.

174 Way Kuo, K. Chien, and T. Kim. *Reliability, Yield, and Stress Burn-in: a Unified Approach for Microelectronics Systems Manufacturing and Software Development*. Kluwer Academics, Boston, MA, 1998.

175 Way Kuo, C. L. Hwang, and F. A. Tillman. A note on heuristic methods in optimal system reliability. *IEEE Transactions on Reliability*, **R-27**(27): 320–324, 1978.

176 Way Kuo and R. Kim. An overview of manufacturing yield and reliability modeling for semiconductor products. *Proceedings of the IEEE*, **87**(11): 1329–1344, 1999.

177 Way Kuo and Y. Kuo. Facing the headaches of early failures: a state-of-the-art review of

burn-in decisions. *Proceedings of the IEEE*, **71**(11): 1257–1266, 1983.

178 Way Kuo, H. Lin, Z. Xu, and W. Zhang. Reliability optimization with the Lagrange multiplier and branch-and-bound technique. *IEEE Transactions on Reliability*, **R-36**(36): 624–630, 1987.

179 Way Kuo and L. Oh. Design for reliability. *IEEE Transactions on Reliability*, **R-44**(2): 170–171, 1995.

180 Way Kuo and V. R. Prasad. An annotated overview of system reliability optimization. *IEEE Transactions on Reliability*, **R-49**(2): 176–191, 2000.

181 Way Kuo, W. Zhang, and M. J. Zuo. Consecutive k-out-of-n: G: the mirror image of a consecutive k-out-of-n: F system. *IEEE Transactions on Reliability*, **R-39**(2): 244–253, 1990.

182 S. P. Ladany and M. Aharoni. Maintenance policy of aircraft according to multiple criteria. *International Journal of Systems Science*, **6**(11): 1093–1101, 1975.

183 Y. J. Lai and C. L. Hwang. *Fuzzy Multiple Objective Decision Making – Methods and Applications*. Springer-Verlag, New York, 1994.

184 B. K. Lambert, A. G. Walvekar, and J. P. Hirmas. Optimal redundancy and availability allocation in multistage systems. *IEEE Transactions on Reliability*, **R-20**(20): 182–185, 1971.

185 A. H. Land and A. G. Doig. An automatic method of solving discrete programming problems. *Econometrica*, **28**: 497–520, 1960.

186 E. L. Lawler and M. D. Bell. A method for solving discrete optimization problems. *Operations Research*, **14**: 1098–1112, 1966.

187 J. P. Lawrence, III and K. Steiglitz. Randomized pattern search. *IEEE Transactions on Computers*, **C-21**(4): 382–385, 1972.

188 K. W. Lee, J. J. Higgins, and F. A. Tillman. Stochastic modeling of human-peformance reliability. *IEEE Transactions on Reliability*, **R-37**(5): 501–504, 1988.

189 K. W. Lee, J. J. Higgins, and F. A. Tillman. Stochastic models for mission effectiveness. *IEEE Transactions on Reliability*, **R-39**(3): 321–324, 1990.

190 K. W. Lee, F. A. Tillman, and J. J. Higgins. A literature survey of the human reliability component in a man–machine system. *IEEE Transactions on Reliability*, **R-37**(1): 24–34, 1988.

191 K. W. Lee, F. A. Tillman, and J. J. Higgins. System effectiveness model with degrees of failure. *IEEE Transactions on Reliability*, **R-37**: 24–34, 1988.

192 G. Leipins and M. Hilliard. Genetic algorithm: foundations and applications. *Annals of Operations Research*, **21**: 31–58, 1989.

193 C. E. Lemke and K. Spielberg. Direct search algorithms for zero–one and mixed integer programming. *Operations Research*, **15**: 892–914, 1967.

194 D. Li. Iterative parametric dynamic programming and its applications in reliability optimization. *Journal of Mathematical Analysis and Applications*, **191**(3): 589–607, 1995.

195 D. Li and Y. Y. Haimes. A decomposition method for optimization of large-system reliability. *IEEE Transactions on Reliability*, **R-41**(2): 183–188, 1992.

196 C. H. Lie, Way Kuo, F. A. Tillman, and C. L. Hwang. Mission effectiveness model for a system with several mission types. *IEEE Transactions on Reliability*, **R-33**(4): 346–352, 1984.

197 F. H. Lin and Way Kuo. *Reliability Importance and Invariant Optimal Allocation.* Technical Report, Texas A&M University, College Station, TX, 1996.

198 F. H. Lin, Way Kuo, and F. Hwang. Structure importance of consecutive-k-out-of-n systems. *Operations Research Letters*, **25**: 101–107, 1999.

199 H. H. Lin and Way Kuo. A comparison of heuristic reliability optimization methods. *Proceedings of 1987 International Industrial Engineering Conference.* pp. 583–589, Institute of Industrial Engineers, Washington, D.C., 1987.

200 H. H. Lin and Way Kuo. Reliability related software life cycle cost model. In J. S. Sindt, editor, *Proceedings of 1987 Annual Reliability Maintainability Symposium*, pp. 364–368, Philadephia, PA, 27–29 January, 1987.

201 J. M. Littschwager. Dynamic programming in the solution of a multistage reliability problem. *Journal of Industrial Engineering*, **15**: 168–175, 1964.

202 D. K. Lloyd and M. Lipow. *Reliability: Management, Methods and Mathematics.* Prentice-Hall, Englewood Cliffs, NJ, 1962.

203 D. G. Luenberger. Quasi-convex programming. *SIAM Journal of Applied Mathematics*, **16**: 1090–1095, 1968.

204 R. Luus. Optimization of system reliability by a new nonlinear integer programming procedure. *IEEE Transactions on Reliability*, **R-24**(24): 14–16, 1975.

205 S. R. V. Majety and J. Rajagopal. *Dynamic Penalty Function for Evolutionary Algorithms with an Application to Reliability Allocation.* Technical Report, Department of Industrial Engineering, University of Pittsburgh, Pittsburgh, PA, 1997.

206 D. M. Malon. Optimal consecutive 2-out-of-n: F component sequencing. *IEEE Transactions on Reliability*, **R-33**(33): 414–418, 1984.

207 D. M. Malon. Optimal consecutive k-out-of-n: F component sequencing. *IEEE Transactions on Reliability*, **R-34**(1): 46–49, 1985.

208 D. M. Malon. When is greedy module assembly optimal? *Naval Research Logistics Quarterly*, **37**: 847–854, 1990.

209 A. W. Marshall and I. Olkin. *Inequalities: Theory of Majorization and Its Applications.* Academic Press, New York, 1979.

210 R. E. Marsten and T. L. Morin. A hybrid approach to discrete mathematical programming. *Mathematical Programming*, **14**(1): 21–40, 1978.

211 G. P. McCormick. *Nonlinear Programming: Theory, Algorithms and Applications.* Wiley, New York, 1983.

212 D. W. McLeavey. On an algorithm of Ghare and Taylor. *Operations Research*, **21**: 1315–1318, 1973.

213 D. W. McLeavey. Numerical investigation of optimal parallel redundancy in series systems. *Operations Research*, **22**: 1110–1117, 1974.

214 D. W. McLeavey and J. A. McLeavey. Optimization of system reliability by branch-and-bound. *IEEE Transactions on Reliability*, **R-25**(25): 327–329, 1976.

215 M. Messinger and M. L. Shooman. Techniques for optimum spares allocation: a tutorial review. *IEEE Transactions on Reliability*, **R-19**(19): 156–166, 1970.

216 N. Metropolis, A. W. Rosenbluth, M. N. Rosenbluth, A. Teller, and E. Teller. Equation of state calculations by fast computing machines. *Journal of Chemical Physics*, **21**: 1087–1092, 1953.

217 Z. Michalewicz. *Genetic algorithms + data structure = evolution programs*. Springer-Verlag, New York, 1996.

218 MIL-STD-721B. Definitions of effectiveness terms for reliability, maintainability, human factors, and safety. *Microelectronics and Reliability*, **11**(5): 429–433, 1972.

219 H. Mine. Reliability of physical system. *IRE Transactions on Circuit Theory*, **CT-6**: 138–151, 1959. (special supplement).

220 K. Misra and V. Misra. A procedure for solving general integer programming problems. *Microelectronics and Reliability*, **34**(1): 157–163, 1994.

221 K. B. Misra. A method of solving redundancy optimization problems. *IEEE Transactions on Reliability*, **R-20**(20): 117–120, 1971.

222 K. B. Misra. Reliability optimization of a series-parallel system. Part I: Lagrange multiplier approach; Part II: maximum principle approach. *IEEE Transactions on Reliability*, **R-21**(21): 230–238, 1972.

223 K. B. Misra. A simple approach for constrained redundancy optimization problems. *IEEE Transactions on Reliability*, **R-21**(21): 30–34, 1972.

224 K. B. Misra. A method for redundancy allocation. *Microelectronics and Reliability*, **12**(4): 389–393, 1973.

225 K. B. Misra. Reliability optimization through sequential simplex search. *International Journal of Control*, **18**: 173–183, 1973.

226 K. B. Misra. Optimum reliability design of a system containing mixed redundancies. *IEEE Transactions on Power Apparatus Systems*, **PAS-94**(3): 983–993, 1975.

227 K. B. Misra. On optimal reliability design: a review. *System Science*, **12**(4): 5–30, 1986.

228 K. B. Misra. An algorithm to solve integer programming problems: an efficient tool for reliability design. *Microelectronics and Reliability*, **31**(2–3): 285–294, 1991.

229 K. B. Misra and M. D. Ljubojevic. Optimal reliability design of a system: a new look. *IEEE Transactions on Reliability*, **R-22**(22): 255–258, 1973.

230 K. B. Misra and U. Sharma. An efficient algorithm to solve integer programming problems arising in system reliability design. *IEEE Transactions on Reliability*, **R-40**(1): 81–91, 1991.

231 K. B. Misra and U. Sharma. An efficient approach for multiple criteria redundancy optimization problems. *Microelectronics and Reliability*, **31**(2–3): 303–321, 1991.

232 K. B. Misra and U. Sharma. Multicriteria optimization for combined reliability and redundancy allocation in systems employing mixed redundancies. *Microelectronics and Reliability*, **31**(2–3): 323–335, 1991.

233 M. Mitchell. *An Introduction to Genetic Algorithms*. MIT Press, Cambridge, MA, 1996.

234 K. Mizukami. Optimum redundancy for maximum system reliability by the method of convex and integer programming. *Operations Research*, **16**: 392–406, 1968.

235 C. Mohan and K. Shanker. Reliability optimization of complex systems using random search technique. *Microelectronics and Reliability*, **28**(4): 513–518, 1988.

236 D. E. Monarchi, C. C. Kisiel, and L. Duckstein. Interactive multi-objective programming in water resources: a case study. *Water Resources Research*, **9**(4): 837–850, 1973.

237 T. L. Morin and R. E. Marsten. An algorithm for nonlinear knapsack problems. *Management Science*, **22**(10): 1147–1158, 1976.

238 T. L. Morin and R. E. Marsten. Branch-and-bound strategies for dynamic programming. *Operations Research*, **24**: 611–627, 1976.

239 D. F. Morrison. The optimum allocation of spare components in system. *Technometrics*, **3**(3): 399–406, 1961.

240 F. Moskowitz and J. B. McLean.   Some reliability aspects of system design.   *IRE Transactions on Reliability and Quality Control*, **PGRQC-8**: 7–35, 1965.

241 J. D. Musa, A. Iannino, and K. Okumoto. *Software Reliability: Measurement, Prediction, Application*. McGraw-Hill, New York, 1987.

242 B. L. Myers and N. L. Enrich. Algorithmic optimization of system reliability. *22nd Annual Technical Conference Transactions, American Society for Quality Control*, pp. 455–460, 1968.

243 Y. Nakagawa. Studies on optimal design of high reliability system: single and multiple objective nonlinear integer programming. Ph.D. thesis, Kyoto University, 1978.

244 Y. Nakagawa, M. Hikita, and H. Kamada. Surrogate constraints algorithm for reliability optimization problem with multiple constraints. *IEEE Transactions on Reliability*, **R-33**(33): 301–305, 1984.

245 Y. Nakagawa and S. Miyazaki. An experimental comparison of the heuristic methods for solving reliability optimization problems. *IEEE Transactions on Reliability*, **R-30**(30): 181–184, 1981.

246 Y. Nakagawa and S. Miyazaki. Surrogate constraints algorithm for reliability optimization problem with two constraints. *IEEE Transactions on Reliability*, **R-30**(30): 175–180, 1981.

247 Y. Nakagawa and K. Nakashima. A heuristic method for determining optimal reliability allocation. *IEEE Transactions on Reliability*, **R-26**(26): 156–161, 1977.

248 Y. Nakagawa, K. Nakashima, and Y. Hattori. Optimal reliability allocation by branch-and-bound techniques. *IEEE Transactions on Reliability*, **R-27**(27): 31–38, 1978.

249 A. D. Narasimhalu and H. Sivaramakrishnan. A rapid algorithm for reliability optimization of parallel redundant systems. *IEEE Transactions on Reliability*, **R-27**(27): 263–268, 1978.

250 D. S. Necsulescu and M. Krieger. Reliability optimization – a case study. *IEEE Transactions on Reliability*, **R-31**(31): 101–104, 1982.

251 J. A. Nelder and R. Mead. A simplex method for function minimization. *The Computer Journal*, **7**(4): 308–313, 1965.

252 A. C. Nelson, Jr, I. R. Batts, and R. L. Beadles. A computer program for approximating system reliability. *IEEE Transactions on Reliability*, **R-19**(19): 61–65, 1970.

253 G. L. Nemhauser and L. A. Wolsey. *Integer and Combinatorial Optimization*. Wiley, New York, 1988.

254 G. E. Neuner and R. N. Miller.   Resource allocation for maximum reliability.   In L. S. Gephart, editor, *Proceedings of 1966 Annual Symposium on Reliability*, pp. 332–346, San Francisco, CA, 25–27 January 1966.

255 C. A. Ntuen, E. H. Park, and W. Byrd. A heuristic program for reliability and maintainability allocation in complex hierarchical systems. *Computers and Industrial Engineering*, **25**(1–4): 345–348, 1993.

256 P. D. O'Connor. *Practical Reliability Engineering*. 3rd edition, Wiley, Chichester, NY, 1995.

257 L. Painton and J. Campbell. Identification of components to optimize improvements in system reliability. In G. E. Apostrolakis and J. S. Wu, editors, *Proceedings of the SRA PSAM-II Conference on System-based Methods for the Design and Operation of Technological Systems and Processes*, pp. 10.15–10.20, San Diego, CA, March 1994.

258 L. Painton and J. Campbell. Genetic algorithms in optimization of system reliability. *IEEE Transactions on Reliability*, **R-44**(2): 172–178, 1995.

259 S. G. Papastavridis and M. Sfakianakis. Optimal-arrangement and importance of the components in a consecutive-k-out-of-r-from-n: F system. *IEEE Transactions on Reliability*, **R-40**(3): 277–279, 1991.

260 K. S. Park. Fuzzy apportionment of system reliability. *IEEE Transactions on Reliability*, **R-36**(36): 129–132, 1987.

261 D. Petrovic. Decision support for improving systems reliability by redundancy. *European Journal of Operational Research*, **55**(3): 357–367, 1991.

262 H. Pham. Optimal design of parallel-series systems with competing failure modes. *IEEE Transactions on Reliability*, **R-41**(4): 583–587, 1992.

263 H. Pham. On the optimal design of n-version software systems subject to constraints. *Journal of Systems and Software*, **27**(1): 55–61, 1994.

264 M. J. D. Powell. An efficient method for finding the minimum of a function of several variables without calculating derivatives. *The Computer Journal*, **7**(2): 155–162, 1964.

265 M. J. D. Powell. A method for nonlinear constraints in minimization problems. In R. Fletcher, editor, *Symposium of the Institute of Mathematics and its Applications, University of Keele, England, 1968*, pp. 283–298. Academic Press, London, 1969.

266 V. R. Prasad, Y. P. Aneja, and K. P. K. Nair. Optimization of bicriterion quasi-concave function subject to linear constraints. *Opsearch*, **27**(2): 73–92, 1990.

267 V. R. Prasad, Y. P. Aneja, and K. P. K. Nair. A heuristic approach to optimal assignment of components to a parallel-series network. *IEEE Transactions on Reliability*, **R-40**(5): 555–558, 1991.

268 V. R. Prasad and Way Kuo. *An Evolutionary Algorithm for Reliability–Redundancy Allocation Problem.* Technical Report, Department of Industrial Engineering, Texas A&M University, College Station, TX, 1997.

269 V. R. Prasad and Way Kuo. *Redundancy Allocation in a Power Cooling System.* Technical Report, Department of Industrial Engineering, Texas A&M University, College Station, TX, 1998.

270 V. R. Prasad and Way Kuo. Reliability optimization of coherent systems. *IEEE Transactions on Reliability*, **R-49**(3): 2000.

271 V. R. Prasad, Way Kuo, and K. M. Oh Kim. Optimal allocation of s-identical, multifunctional spares in a series system. *IEEE Transactions on Reliability*, **R-48**(2): 118–126, 1999.

272 V. R. Prasad, K. P. K. Nair, and Y. P. Aneja. Optimal assignment of components to parallel-series and series-parallel systems. *Operations Research*, **39**: 407–414, 1991.

273 V. R. Prasad and M. Raghavachari. Optimal allocation of interchangeable components in a series-parallel system. *IEEE Transactions on Reliability*, **R-47**(3): 255–260, 1998.

274 F. Proschan and T. A. Bray. Optimum redundancy under multiple constraints. *Operations Research*, **13**: 800–814, 1965.

275 V. Ravi, B. Murty, and P. Reddy. Nonequilibrium simulated-annealing algorithm applied reliability optimization of complex systems. *IEEE Transactions on Reliability*, **R-46**(2): 233–239, 1997.

276 G. V. Reklaitis, A. Ravindran, and K. M. Ragsdell. *Engineering Optimization: Methods and Applications.* Wiley, New York, 1983.

277 D. F. Rudd. Reliability theory in chemical system design. *Industrial and Engineering Chemistry Fundamental*, **1**(2): 138–143, 1962.

278 M. Sakawa. Multiobjective optimization by the surrogate worth trade-off method. *IEEE Transactions on Reliability*, **R-27**(27): 311–314, 1978.

279 M. Sakawa. Multiobjective reliability and redundancy optimization of a series-parallel system by the surrogate worth trade-off method. *Microelectronics and Reliability*, **17**(4): 465–467, 1978.

280 M. Sakawa. Decomposition approaches to large-scale multi-objective reliability design. *Journal of Information and Optimization Science*, **1**(2): 103–120, 1980.

281 M. Sakawa. Multiobjective optimization for a standby system by the surrogate worth trade-off method. *Journal of Operational Research Society*, **31**: 153–158, 1980.

282 M. Sakawa. Optimal reliability-design of a series-parallel system by a large-scale multiobjective optimization method. *IEEE Transactions on Reliability*, **R-30**(30): 173–174, 1981.

283 M. Sakawa. Interactive multiobjective optimization by sequential proxy optimization technique (SPOT). *IEEE Transactions on Reliability*, **R-31**(31): 461–464, 1982.

284 H. Sarper. No special schemes are needed for solving software reliability optimization models. *IEEE Transactions on Software Engineering*, **SE-21**(8): 701–702, 1995.

285 M. Sasaki. A simplified method of obtaining highest system reliability. In M. P. Smith, editor, *Proceedings of the Eighth National Symposium on Reliability and Quality Control*, pp. 489–502, Washington, D.C., 9–11 January, 1962.

286 M. Sasaki. An easy allotment method achieving maximum system reliability. In R. E. Kuehn, editor, *Proceedings of the Ninth National Symposium on Reliability and Quality Control*, pp. 109–124, San Francisco, CA, 22–24 January, 1963.

287 M. Sasaki, M. Gen, and M. Ida. A method for solving reliability optimization problem by fuzzy multiobjective 0–1 linear programming. *Electronics and Communications in Japan Part II: Fundamental Electronic Science*, **74**(12): 106–116, 1992.

288 V. Selman and N. T. Grisamore. Optimum system analysis by linear programming. *1966 Processing Annual Symposium on Reliability, American Society for Quality Control*, pp. 696–703, 1966.

289 J. G. Shanthikumar. Reliability of systems with consecutive minimal cutsets. *IEEE Transactions on Reliability*, **R-36**(36): 546–549, 1987.

290 J. Sharma and K. V. Venkateswaran. A direct method for maximizing the system reliability. *IEEE Transactions on Reliability*, **R-20**(20): 256–259, 1971.

291 U. Sharma and K. B. Misra. An efficient algorithm to solve integer-programming problems in reliability optimization. *International Journal of Quality and Reliability Management*, **7**(5): 44–56, 1990.

292 U. Sharma, K. B. Misra, and A. K. Bhattacharjee. Application of an efficient search technique for optimal design of a computer communication network. *Microelectronics and Reliability*, **31**(2–3): 337–341, 1991.

293 J. Shen and M. Zuo. Optimal design of series consecutive k-out-of-n: G systems. *Reliability Engineering and System Safety*, **45**(3): 277–283, 1994.

294 A. C. Shershin. Mathematical optimization techniques for the simultaneous apportionments of reliability and maintainability. *Operations Research*, **18**(1): 95–106, 1970.

295 H. V. K. Shetty and D. P. Sengupta. Reliability optimization using SLUMT. *IEEE Transactions on Reliability*, **R-24**(24): 80–82, 1975.

296 D. H. Shi. A new heuristic algorithm for constrained redundancy-optimization in complex systems. *IEEE Transactions on Reliability*, **R-36**(36): 621–623, 1987.

297 A. E. Shura-Bura. The method of sequential optimization for solving problems of optimal multilevel redundancy. *Engineering Cybernetics*, **20**(2): 66–71, 1982.

298 D. Simmons, N. Ellis, H. Fujihara, and W. Kuo. *Software Measurement: a Visualization Toolkit for Project Control and Process Improvement*. Prentice-Hall, NJ, 1998.

299 G. Syswerda. Uniform crossover in genetic algorithms. In J. Schaffer, editor, *Proceedings of the Third International Conference on Genetic Algorithms*, pp. 2–9, San Mateo, CA, May 1989, Morgan Kaufmann, 1989.

300 T. Taguchi, K. Ida, and M. Gen. Method for solving nonlinear goal programming with interval coefficients using genetic algorithm. *Computers and Industrial Engineering*, **33**(3–4): 597–600, 1997.

301 F. A. Tillman. Optimization by integer programming of constrained reliability problems with several modes of failure. *IEEE Transactions on Reliability*, **R-18**(18): 47–53, 1969.

302 F. A. Tillman, C. L. Hwang, L. T. Fan, and S. A. Balbale. System reliability subject to multiple nonlinear constraints. *IEEE Transactions on Reliability*, **R-17**(17): 153–157, 1968.

303 F. A. Tillman, C. L. Hwang, L. T. Fan, and K. C. Lai. Optimal reliability of a complex system. *IEEE Transactions on Reliability*, **R-19**(19): 95–100, 1970.

304 F. A. Tillman, C. L. Hwang, and Way Kuo. Determining component reliability and redundancy for optimum system reliability. *IEEE Transactions on Reliability*, **R-26**(26): 162–165, 1977.

305 F. A. Tillman, C. L. Hwang, and Way Kuo. Optimization techniques for system reliability with redundancy – a review. *IEEE Transactions on Reliability*, **R-26**(26): 148–155, 1977.

306 F. A. Tillman, C. L. Hwang, and Way Kuo. *Optimization of Systems Reliability*. Marcel Dekker, New York, 1980.

307 F. A. Tillman, C. L. Hwang, and Way Kuo. System effectiveness models: an annotated bibliography. *IEEE Transactions on Reliability*, **R-29**(29): 295–304, 1980.

308 F. A. Tillman, Way Kuo, and C. L. Hwang. A numerical simulation of the system effectiveness-a renewal theory approach. In H. J. Kennedy, editor, *Proceedings of the 1982 Annual Reliability and Maintainability Symposium*, pp. 252–261, Los Angeles, CA, 26–28 January, 1982.

309 F. A. Tillman, C. H. Lie, and C. L. Hwang. Analysis of pseudo-reliability of a combat tank system and its optimal design. *IEEE Transactions on Reliability*, **R-25**(25): 239–242, 1976.

310 F. A. Tillman, C. H. Lie, and C. L. Hwang. Simulation model of mission effectiveness for military systems. *IEEE Transactions on Reliability*, **R-27**(27): 191–194, 1978.

311 F. A. Tillman and J. M. Littschwager. Integer programming formulation of constrained reliability problems. *Management Science*, **13**(11): 887–899, 1967.

312 F. A. Tillman, Way Kuo, R. F. Nassar, and C. L. Hwang. Numerical evaluation of instantaneous availability. *IEEE Transactions on Reliability*, **R-32**(32): 119–123, 1983.

313 S. G. Tzafestas. Optimization of system reliability: a survey of problems and techniques. *International Journal of Systems Science*, **11**(4): 455–486, 1980.

314 I. A. Ushakov. A heuristic method of optimization of the redundancy of multifunction systems. *Engineering Cybernetics*, **10**(4): 612–613, 1972.

315 P. J. M. van Laarhoven and E. H. L. Arts. *Simulated Annealing: Theory and Applications*.

D. Reidel Publishing Company, Dordrecht, Holland, 1987.

316 V. L. Volkovich and V. A. Zaslavskii. An algorithm for solving the problem of the optimization of the reliability of a complex system using reserve elements of different types in the subsystems. *Kibernetika (Kiev)*, **134**(5): 54–61, 1986.

317 A. F. Voloshin, V. A. Zaslavskii, and S. V. Kudryavtsev. Combined algorithms for discrete programming in solving problems of the optimization of the reliability of complex systems. *Veslnik Kievskogo Universiteta Modelirovaniei Optimizatsiya Slozhnykh Sistem*, **115**(6): 53–56, 1987.

318 G. A. Walters and D. K. Smith. Evolutionary design algorithm for optimal layout of tree networks. *Engineering Optimization*, **24**: 261–281, 1995.

319 F. M. Waltz. An engineering approach: Hierarchical optimization criteria. *IEEE Transactions on Automatic Control*, **AC-12**(2): 179–180, 1967.

320 L. R. Webster. Choosing optimum system configurations. In E. F. Jahr, editor, *Proceedings of the Tenth National Symposium on Reliability and Quality Control*, pp. 345–359, Washington, D.C., 7–9 January, 1964.

321 L. R. Webster. Optimum system reliability and cost effectiveness. In S. Zwerling, editor, *Proceedings of 1967 Annual Symposium on Reliability*, pp. 489–500, Washington, D.C., 10–12 January, 1967.

322 V. K. Wei, F. K. Hwang, and V. T. Sos. Optimal sequencing of items in a consecutive 2-out-of-n: system. *IEEE Transactions on Reliability*, **R-32**(32): 30–33, 1983.

323 P. Wolfe. Methods of nonlinear programming. In R. Graves and P. Wolfe, editors, *Recent Advances in Mathematical Programming*, pp. 67–86, McGraw-Hill, New York, 1963.

324 C. F. Woodhouse. Optimal redundancy allocation by dynamic programming. *IEEE Transactions on Reliability*, **R-21**(21): 60–62, 1972.

325 Weapon Systems Effectiveness Industry Advisory Committee (WSEIAC). *Final Report of Task Group I Requirements Methodology*. Technical Report AFSC-TR-65-1, System Effectiveness Division, System Policy Directorate; Headquarters, US Air Force Systems Command, Andrews, AFB, MD, 1965.

326 Z. Xu, Way Kuo, and H. Lin. Optimization limits in improving system reliability. *IEEE Transactions on Reliability*, **R-39**(1): 51–60, 1990.

327 S. Yamada and S. Osaki. Optimal software release policies with simultaneous cost and reliability requirements. *European Journal of Operational Research*, **31**(1): 46–51, 1987.

328 T. Yokota, M. Gen, and K. Ida. System reliability of optimization problems with several failure modes by genetic algorithm. *Japanese Journal of Fuzzy Theory and Systems*, **7**(1): 117–135, 1995.

329 T. Yokota, M. Gen, K. Ida, and T. Taguchi. Optimal design of system reliability by an approved genetic algorithm. *Transactions of Institute of Electronics, Information and Communication Engineers*, **J78A**(6): 702–709, 1995.

330 T. Yokota, M. Gen, and Y. X. Li. Genetic algorithm for non-linear mixed integer programming problems and its applications. *Computers and Industrial Engineering*, **30**(4): 905–917, 1996.

331 K. Yoon and C. L. Hwang. Multiple attribute decision making: an introduction. In M. S. Lewis-Beck, editor, *Quantitative Applications in the Social Science Series*. SAGE, Thousand Oaks, CA, 1995.

332 D. K. Yun and K. S. Park. Redundancy optimization by linear knapsack approach.

International Journal of Systems Science, **13**(8): 839–848, 1982.

333 V. A. Zaslavskiy.   Optimization of the reliability of nonseries systems when there are constraints. *Soviet Automatic Control*, **15**(5): 50–59, 1983.

334 F. Zettwitz. An enumeration algorithm for combinational problems of the reliability analysis of binary coherent systems. *Optimization*, **19**(6): 875–888, 1988.

335 W. Zhang, C. Miller, and Way Kuo. Application and analysis for consecutive k-out-of-n: G structure. *Reliability Engineering and System Safety*, **33**(2): 189–197, 1991.

336 H. Zimmermann.   Fuzzy programming and linear programming with several objective functions. *Fuzzy Sets and Systems*, **1**: 45–55, 1978.

337 M. J. Zuo and Way Kuo.   Design and performance analysis of consecutive k-out-of-n structure. *Naval Research Logistics Quarterly*, **37**(2): 203–230, 1990.

Index